Hazardous Waste Risk Assessment

D. Kofi Asante-Duah
Environmental Consultant
Newport Beach, California

CRC Press is an imprint of the
Taylor & Francis Group, an **informa** business

First published 1993 by Lewis Publishers, Inc.

Published 2019 by CRC Press
Taylor & Francis Group
6000 Broken Sound Parkway NW, Suite 300
Boca Raton, FL 33487-2742

First issued in paperback 2019

ISBN-13: 978-0-367-44997-1 (pbk)
ISBN-13: 978-0-87371-570-6 (hbk)

Visit the Taylor & Francis Web site at
http://www.taylorandfrancis.com

and the CRC Press Web site at
http://www.crcpress.com

Library of Congress Cataloging-in-Publication Data

Asante-Duah, D. Kofi.
Hazardous waste risk assessment / D. Kofi Asante-Duah.
p. cm.
Includes bibliographical references and index.
ISBN 0-87371-570-5
1. Hazardous wastes—Risk assessment. I. Title.
TD1050.R57A82 1993
363.72′87—dc20 92-35415
CIP

Publisher's Note
The publisher has gone to great lengths to ensure the quality of this reprint but points out that some imperfections in the original may be apparent.

To my Mother, Alice Adwoa Twumwaa
To my Father, George Kwabena Duah
To All the Duah Brothers and Sisters
To All the Good Friends

D. Kofi Asante-Duah, Ph.D., C.E., is currently an environmental consultant in Newport Beach, CA; the technical director of Novelty Engineering and Environmental Services, Accra; and a member of the Institute for Risk Research (IRR) at the University of Waterloo, Ontario. Dr. Asante-Duah has served internationally in various capacities on several projects, including Senior Project Engineer, Project Manager, Principal Investigator, and Consultant. He has more than 10 years of diversified experience with consulting firms and research institutions. His fields of expertise span several topics in the areas of hazardous waste risk assessment and risk management; development of cleanup criteria for site remediation; probabilistic risk assessment; dam safety evaluation; stochastic simulation model applications for solving water management problems; decision analysis approaches to environmental assessment; and statistical evaluation for environmental impact assessment. Dr. Asante-Duah has previously been on visiting appointments to the University of Pittsburgh, Civil Engineering Department (as Visiting Scholar/Scientist in 1985/86) and the University of Waterloo, Civil Engineering Department (as Visiting Research Assistant Professor in 1990/91). He has previously been a research assistant at the Utah Water Research Laboratory (UWRL) at Utah State University (1986 to 1988). Dr. Asante-Duah is a past UNESCO fellow (1984) and also a World Bank's McNamara fellow (1990 to 1991). He has affiliations with several professional bodies internationally and is the author or co-author of several technical papers and presentations relating to water resources, risk assessment, and hazardous waste management.

Preface

Hazardous waste management is of great concern in view of the risks associated with waste disposal activities. Hazardous waste sites and facilities may pose significant risks to the public because of the potential health and environmental effects, and to the potentially responsible parties because of the potential financial liabilities resulting from their effects. The effective management of hazardous wastes has therefore become an important environmental priority and will be a growing social challenge for years to come. However, it has also become evident that the proper management of hazardous wastes poses great challenges. Risk assessment seems to be one of the fastest evolving tools for developing appropriate management strategies that will aid hazardous waste management decisions.

This book provides a comprehensive review of currently available methods and applications of risk assessment in hazardous waste management. It contains an elaboration of pertinent concepts and techniques/methodologies in risk assessment that may be applied to the management of hazardous waste problems. These include methods for developing and performing health and environmental risk assessment tasks; the application of risk assessment procedures to evaluate corrective action programs; the use of risk assessment results to make risk management decisions; and methods for the safety evaluation of hazardous waste facilities. Both qualitative and quantitative methods for the analyses of hazardous conditions and consequences are presented. The main emphasis in this book, however, is on the use of health risk assessment to support management decisions on hazardous waste disposal and the remediation of hazardous waste sites. The elaboration specifically includes a literature review of risk assessment methodologies applicable to hazardous waste management. It also includes specific numerical evaluations and case studies to demonstrate typical applications of risk assessment used to aid hazardous waste management decisions.

This book should serve as a comprehensive and practical reference resource for many a professional encountering risk assessment. Most of the literature available on hazardous waste risk assessment, often written for professionals with some prior knowledge of the subject, do little to cautiously and systematically lead the beginner who wishes to apply risk assessment techniques and tools to resolve hazardous waste management problems. This book is an attempt to give a simplified presentation of risk assessment methods and applications. This will facilitate the understanding and justification for the application of risk assessment by consultants, regulators, policy makers, and indeed, practitioners of environmental management in general. This book can also serve as a resource and text for graduate students taking programs in hazardous waste management and risk assessment studies. The book covers risk assessment applications to hazardous waste management in an easy-to-follow manner insofar as possible, and still remains pragmatic. A number of step-by-step numerical case evaluations and example problems are included to demonstrate typical applications of risk assessment to hazardous waste problems.

D. Kofi Asante-Duah, Ph.D., C.E.
Newport Beach, CA
July 1992

Acknowledgments

Financial, administrative, and/or material support was provided by Novelty Engineering and Environmental Services. Sincere thanks are due the Duah family and several friends and colleagues who provided much-needed moral and enthusiastic support throughout preparation of the manuscript for this book. Some of the work discussed in this book resulted from previous assignments performed at the Utah Water Research Laboratory (UWRL), Utah State University (for which Dr. L. R. Anderson, Dr. D. S. Bowles, Dr. L. D. James, and Dr. R. C. Sims provided initial review comments); and the Institute for Risk Research (IRR), University of Waterloo (in which Dr. M. Haight, Dr. F. F. Saccomanno, and Dr. J. H. Shortreed provided some review comments). Section 3.5 and Appendices D.5 and E.2 were contributed by Professor Edward A. McBean of the Department of Land, Air, and Water Resources, University of California, Davis. Finally, review comments and suggestions on an earlier draft of the manuscript for this book were provided by Dr. R. Nichols Hazelwood (Distinguished Technical Associate, International Technology Corporation, Irvine, CA), Dr. Ed A. McBean (Professor, University of California, Davis, CA), and Dr. L. Douglas James (Professor, UWRL, Utah State University, Logan, UT).

Contents

CHAPTER 3

CHAPTER 4

CHAPTER 5

CHAPTER 1

Introduction

Waste production is an inevitable characteristic of an industrial society. The effective management of hazardous wastes, and the associated treatment, storage, and disposal facilities (TSDFs), is of major concern not only to the industry producing such material, but also to governments and individual citizens alike due to the nature and potential impact of such wastes on the environment and public health. In particular, hazardous waste management is of great concern in view of the uncertainties and risks associated with waste disposal activities. Apart from its immediate and direct health and environmental hazards, hazardous waste disposal could lead to the long-term contamination of the ambient air, soils, surface and ground waters, and the food chain if disposal facilities are not properly designed and maintained or if remedial actions are not taken in an effective manner. In fact, this is in a way summarized by Clapham's (Bhatt et al., 1986) view of hazardous waste management facilities as the kidneys of industrial societies. A responsible system for dealing with hazardous wastes is therefore essential to sustain the modern way of life, inasmuch as a well-functioning kidney is necessary to rid the human body of certain toxins.

Hazardous waste sites and facilities, in particular, and potentially contaminated sites, in general, may pose significant risks to the public because of the potential health and environmental effects and to the potentially responsible parties (PRPs) because of the potential financial liabilities resulting from their effects. The effective management of such sites and facilities has therefore become an important environmental priority and will be a growing social challenge for years to come. However, it has also become evident that the proper management of hazardous wastes poses great challenges. Risk assessment, which encompasses varying degrees and types of qualitative and quantitative analyses, is one of the fastest evolving tools for developing appropriate management strategies relating to hazardous waste management decisions. The U.S. Environmental Protection Agency

(EPA) recognizes the use of risk assessment to facilitate decisions on whether or not remedial actions are needed to abate site-related risks, and also in the enforcement of regulatory standards. Risk assessment techniques have been used in various regulatory programs employed by federal, state, and local agencies. For instance, both the feasibility study process under the Comprehensive Environmental Response, Compensation, and Liability Act of 1980 (CERCLA, or "Superfund") and alternate concentration limit (ACL) demonstrations under the Resource Conservation and Recovery Act of 1976 (RCRA) involve the use of risk assessment to establish cleanup standards for contaminated sites.

The primary application of quantitative risk assessment in the U.S. EPA Superfund program is to evaluate the potential risk posed at each National Priorities List (NPL) facility, so that the appropriate remedial alternative can be identified (Paustenbach, 1988); NPL is the list of uncontrolled or abandoned hazardous waste sites identified for possible long-term remedial actions under Superfund. The U.S. EPA uses a risk-based evaluation method, the Hazard Ranking System (HRS), to identify uncontrolled and abandoned hazardous waste sites falling under Superfund programs. The HRS allows the selection or rejection of a site for placement on the U.S. EPA NPL; it is used for prioritizing sites so that those posing the greatest risks receive quicker response. Another application of risk assessment is in the selection of appropriate sites for hazardous waste facilities; sites are ranked for their appropriateness for stipulated purpose(s) according to the levels of risk that each potentially poses under different scenarios. Furthermore, ACLs (which can be considered as surrogate values for the maximum allowable health and environmental exposure levels) can be established when hazardous constituents are identified in groundwater at RCRA facilities, by applying risk assessment procedures in the analytical processes involved. In fact, nearly every process for developing cleanup criteria incorporates some concept that can be classified as a risk assessment. Thus, all decisions on setting cleanup standards for potentially contaminated sites include, implicitly or explicitly, some aspect of risk assessment. In all situations, to ensure public safety, contaminant migration beyond a compliance boundary into the public exposure domain must be below some stipulated health-based standard, or a maximum exposure level (MEL). Indeed, to ensure public health and environmental sustainability, decisions relating to hazardous waste management should be based on a systematic and scientifically valid process, such as is offered by risk assessment.

1.1 THE NATURE OF HAZARDOUS WASTES

Broadly speaking, a hazardous material is one which is capable of producing adverse effects and/or reactions in potential biological receptors; toxic concentrations of substances generally present unreasonable risk of harm to human health and/or the environment. Such substances need to be regulated. Specifically, hazardous waste is that by-product which has the potential of causing detrimental effects on human health and/or the environment if not managed efficiently. Such wastes may belong to one or more of several categories, including

- Toxic organic or inorganic chemicals
- Bioaccumulative materials
- Nondegradable and persistent chemicals
- Radioactive substances

Hazardous waste disposal practices may result in the release of chemicals into air (via volatilization and fugitive dust emissions), surface water (from surface runoff or overland flow and groundwater seepage), groundwater (through leaching/infiltration), soils (due to erosion, including fugitive dust generation/deposition, and tracking), sediments (from surface runoff/overland flow, seepage, and leaching), and biota (due to biological uptake and bioaccumulation). For instance, typical chemicals finding their way into drinking water aquifers have been reported elsewhere to include aliphatic hydrocarbons (e.g., chloroform, vinyl chloride); aromatic hydrocarbons (e.g., benzene, toluene, xylene, DDT, benzo[a]pyrene [BaP] from coal tar); chlorinated solvents; pesticides; polychlorinated biphenyls (PCBs); trace metals and other inorganic compounds (e.g., arsenic, cadmium, chromium, cyanide, lead, mercury); and other organic compounds (e.g., acetone, methyl ethyl ketone). These and several other chemicals may be present in the environment, and can lead to several detrimental effects.

The identification of potentially hazardous waste streams is important in the investigation of potential risks that such wastes may present. The wastes generally originate from any of a number of industries (Table 1.1). These industries generate several waste types such as organic waste sludges and still bottoms (containing chlorinated solvents, metals, oil, etc.); oil and grease (with PCBs, polyaromatic hydrocarbons [PAHs], metals, etc.); heavy metal solutions (of arsenic, cadmium, chromium, lead, mercury, etc.); PCB wastes; pesticide and herbicide wastes; anion complexes (with cadmium, copper, nickel, zinc, etc.); paint and organic residuals; and several miscellaneous chemicals and products. There is always some risk of chemicals from these waste streams escaping into the environment during treatment storage, or final disposal.

1.2 WASTE CLASSIFICATION SYSTEMS

In order to develop effective pollution control strategies, wastes must be appropriately categorized. A typical categorization will comprise putting the wastes into high-, intermediate-, and low-risk classes (Figure 1.1). The high-risk wastes will be those of priority concern, known to contain significant concentrations of constituents that are highly toxic, mobile, persistent, and/or bioaccumulative. Examples of this type of waste are chlorinated solvent wastes from metal degreasing (due to their toxicity, mobility, and to some extent, persistence in the environment, etc.); cyanide wastes (due to their high toxicity, etc.); dioxin-based wastes (due to potential high toxicity and carcinogenicity effects, etc.); PCB wastes (due to persistence and bioaccumulative properties, etc.). Intermediate-risk wastes will include metal hydroxide sludges (excluding categories such as hexavalent chromium, Cr^{+6}, which will be included under the high-risk wastes due to its extreme toxicity) for which the toxic metals are in relatively insoluble physical form with

Table 1.1 Typical Industries Generating Potentially Hazardous Wastes

Aerospace
Automobile
Batteries (storage and primary)
Beverages
Computer manufacture
Electronic components manufacturing
Electroplating and metal finishing
Explosives
Food processing and dairy products
Ink formulation
Inorganic chemicals
Inorganic pigments
Iron and steel
Leather tanning and finishing
Metal smelting and refining
Organic chemicals
Paints and coatings
Perfumes and cosmetics
Pesticides and herbicides
Petroleum refining
Pharmaceuticals
Photographic equipment and supplies
Printing
Pulp and paper mills
Rubber products, plastic materials, and synthetics
Ship building
Soap and detergent manufacturing
Textile mills
Wood preservation and processing

low mobility. Low-risk wastes include primarily high-volume, low-hazard wastes and some putrescible wastes, for which the cutoff between a "hazardous" and "nonhazardous" waste is least clear cut (Batstone et al., 1989). In fact, the degree of hazard posed by wastes may be dependent on several factors, such as

- Physical form, composition, reactivity (fire and explosion), and quantities
- Biological and ecological effects (i.e., toxicity, ecotoxicity)
- Mobility (i.e., transport in various environmental media, leaching potential)
- Persistence (including fate in environment, detoxification potential, etc.)
- Indirect health effects (that may result from pathogens, vectors, etc.)
- Local conditions (e.g., temperature, soil type, groundwater table conditions, humidity, light)

It is important to recognize the fact that there are varying degrees of hazards associated with different waste streams, and there are good economic advantages for ranking wastes according to the level of hazards they present. It is also important to identify the compatibilities and/or incompatibilities of the chemical constituents of waste streams to allow effective management. For example, some wastes (such as aluminium and chemical or alkaline cleaners) will, when mixed, form a more dangerous toxic or hazardous substance. In contrast, waste containing some chemicals can help attenuate the hazards of other chemicals' toxicity. For instance,

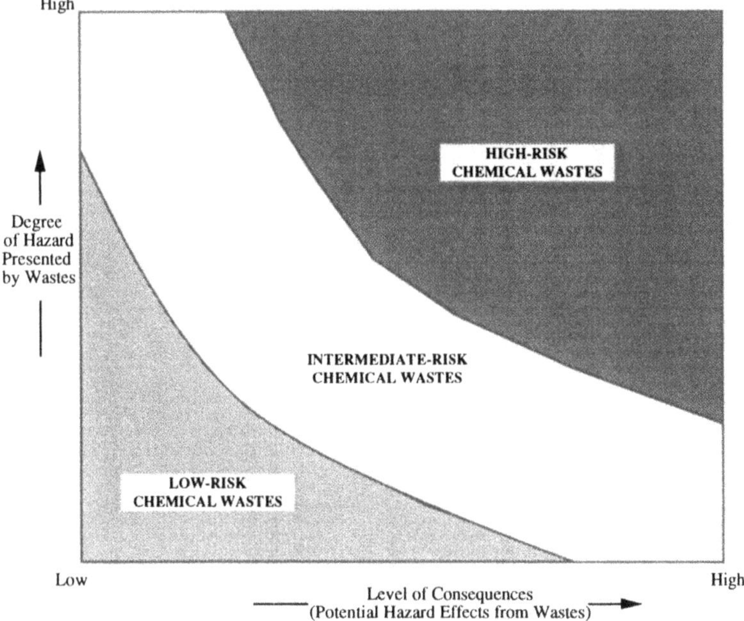

Figure 1.1 Conceptual categorization scheme for hazardous wastes classification (Adapted from Asante-Duah, 1990.)

waste containing both selenium and mercury would be less likely to create toxicity problems than waste containing only mercury. A matrix of waste classes that identifies compatible and incompatible wastes should be developed during the waste categorization process. Potentially incompatible wastes should be identified, together with the potential consequences of such mixtures. In addition, compatible and even "friendly" wastes should be identified. This will all add up to aid the development of effective waste management programs for a region.

1.3 HEALTH AND ENVIRONMENTAL IMPLICATIONS OF WASTE MANAGEMENT PRACTICES

Bitter lessons from the past, such as the Love Canal incident in New York, clearly demonstrate the dangers arising from unsafe disposal and/or management of hazardous wastes. Love Canal was the site for disposal of chemical wastes for a period of about 25 to 30 years (e.g., Gibbs, 1982; Levine; 1982). Subsequent use of the site resulted in residents in a township in the area suffering very serious health impairments, including several children in the neighborhood apparently being born with serious birth defects. Analogous incidents are recorded in other U.S. locations, in Europe, in Japan and other locations in Asia, etc. (Table 1.2). Inadequacies in waste management practices can therefore create potentially hazardous situations and pose significant risks of concern to society. If dumped indiscriminately in any

Table 1.2 Selected Examples of Potential Human Exposures to Hazardous Wastes, Chemicals, and/or Materials Due to Improper Management, Handling, or Disposal Practices

Site Location	Site Conditions	Chemicals of Concern	Exposure Scenario	Comments
Love Canal, Niagara Falls, NY	Industrial waste landfill located in residential setting	Various, including hydrocarbon residues from pesticide manufacture	Potential human exposure routes by direct contact and also various water pathways	Several health impairments, including birth defects to residents living in vicinity of the contaminated site
Bloomington, IN	Industrial waste entering municipal sewage system, used for garden manure/fertilizer	PCBs	Direct human contacts and also via the food chain from human ingestion of contaminated food	PCB-contaminated sewage sludge used as fertilizer, resulting in plant uptakes; also discharges and runoff into rivers lead to potential fish contamination
Triana, AL	Industrial wastes dumped in local stream	DDT and other compounds	Potential for human exposure via food chain — consumption of fish	High DDT metabolite residues detected in fish consumed by community residents
Woburn, MA	Abandoned waste lagoon with several dumps	Arsenic compounds, heavy metals, organic compounds	Potential human receptors and ecosystem exposure via direct contacts and water pathways	Potential for leachate generated contaminating groundwater resources and also surface runoff carrying contamination to surface water bodies
Times Beach, MO	Dioxins (TCDD) in waste oils sprayed on public access areas for dust control	Dioxins (TCDD)	Direct contacts, inhalation, and probable ingestion of contaminated dust and soils	Waste oils contaminated with dioxins (TCDD) was sprayed in several public areas (residential, recreational, and work areas) for dust control

Kamioka Zinc Mine, Japan	Contaminated surface waters	Cadmium	Ingestion of contaminated water and consumption of rice contaminated by crop uptake of contaminated irrigation water	Water containing large amounts of cadmium discharged from the Kamioka Zinc Mine into river used for drinking water and also for irrigating paddy rice; long-term exposures resulted in kidney problems for population
Minamata Bay and Agano River at Niigata, Japan	Effluents from wastewater treatment plants entering coastal waters	Mercury, giving rise to the presence of highly toxic methylmercury	Human consumption of contaminated seafood	Accumulation of methylmercury in fish and shellfish; human consumption of the contaminated seafood resulted in health impairments, particularly severe neurological symptoms

Sources: Long and Schweitzer, 1982; Canter et al., 1988.

environmental media, hazardous wastes may have both short- and long-term effects on both human and ecological systems. In addition, improper treatment, storage, and disposal of hazardous wastes can result in contaminant releases, possible exposures, and potential adverse health and environmental impacts. In general, any chemical can cause severe health impairment or even death if taken by humans in sufficiently large amounts. On the other hand, there are those chemicals of primary concern which, even in small doses, can cause adverse health impacts.

The potential for adverse health effects in populations contacting hazardous wastes may involve any organ system, depending on the specific chemicals contacted, the extent of exposure, the characteristics of the exposed individual (e.g., age, sex, genetic makeup), the metabolism of the chemicals involved, and the presence or absence of confounding variables such as other diseases (Grisham, 1986). In general, the following pertinent factors influence human response to toxic chemicals present in the environment:

- Dosage (because a large dose may mean more immediate effects)
- Age (since the elderly and children are more susceptible to toxins)
- Gender (since each sex has hormonally-controlled hypersensitivities)
- Body weight (which is inversely proportional to toxic responses/effects)
- Psychological status (because stress increases vulnerability)
- Genetics (because different metabolic rates affect receptor responses)
- Immunological status and presence of other diseases (because health status influences general metabolism)
- Weather conditions (since temperature, humidity, barometric pressure, season, etc. potentially affect absorption rates)

Several health effects of primary concern may affect populations exposed to hazardous chemicals present in a region, including (Grisham, 1986)

- Carcinogenesis (i.e., causing cancers)
- Genetic defects, including mutagenesis (i.e., causing alterations in genes which are transmitted from one generation to another, or causing heritable genetic damage)
- Reproductive abnormalities, including teratogenesis (i.e., causing damage to developing fetus)
- Alterations of immunobiological homeostasis
- Central nervous system (CNS) disorders
- Congenital anomalies

Table 1.3 lists typical symptoms, health effects, and other biological responses caused by toxic chemicals in the environment. Invariably, exposures to chemicals escaping into the environment can lead to a reduction of life-expectancy and possibly a period of reduced quality of life (due to anxiety from exposures, diseases, etc.). An uncontrolled waste disposal practice can therefore be perceived as a potential source of several health and environmental problems and a disbenefit to any region or nation.

1.4 THE NEED FOR REGULATIONS AND PRESCRIPTIONS FOR ENVIRONMENTAL PROGRAMS

There is increasing public concern about the numerous problems and potentially dangerous situations associated with hazardous waste management in general and disposal practices in particular. Such concerns, together with the legal provisions of various legislative instruments and regulatory programs, have all compelled industry and governmental authorities to carefully formulate waste management plans. Adequate techniques are therefore needed to provide good initial assessments of potential impacts and abatement measures. Indeed, waste management planning is becoming less conceptual and more quantitative, and that is where risk assessment is appropriate for use in the decision-making process.

Once the diagnostic assessment of potential environmental problems are completed for a hazardous waste facility or site, plans can be made toward effective corrective actions where warranted. The overall goal invariably is to minimize health and environmental impacts with the concurrent optimization of economic, social, and psychological impacts on society. Several determinations are important for the investigation and planning of hazardous waste management facilities, including the following:

- The best practical determination of waste characteristics, including waste types, degree of hazards, waste compatibilities, and the ability to segregate ignitable, reactive, or incompatible wastes. Wastes should be well characterized so as to separate incompatible wastes and to select suitable treatment and disposal techniques.
- Fate and transport characteristics of chemical constituents of wastes and their projected degradation products (including an identification of anticipated reaction or decomposition by-products).
- The critical media of concern (such as air, surface water, groundwater, soils and sediments, and terrestrial and aquatic biota).
- An evaluation of potential exposure pathways of waste constituents and the potential for human and ecosystem exposures.
- Assessment of the environmental and health impacts of the wastes, if such waste should reach critical human and ecological receptors.
- Characterization of disposal sites, including site geology, topography, hydrogeology, and meteorological conditions.
- Determination of extent of service area for proposed waste facility — that is, whether it will handle wastes from local industry only or from regional and/or national generators also.
- Satisfaction that the location proposed for a waste facility is suitable, given environmental, social, and economic concerns — including proximity to populations, ecological systems, water resources, etc.; active seismic areas and floodplains should be excluded from the siting of waste management facilities.
- Indication that a proposed waste facility represents the best available technology (BAT) for handling the particular wastes. In addition, there should be contingency plans and emergency procedures in the design of waste management plans.
- Provision for effective long-term monitoring and surveillance programs, including postclosure maintenance of facilities, are a necessary part of the overall design plans.

Table 1.3 Some Typical Health and Ecological Effects Due to Selected Toxic Chemicals Potentially Present in the Environment

Chemical	Typical Health Effects and Toxic Manifestations (Symptoms/Response)
Aldrin/dieldrin	Convulsions, kidney damage, tremors; bioaccumulates in aquatic organisms
Antimony	Heart disease
Aromatics	Cirrhosis (liver), mild fatty metamorphosis
Arsenic and compounds	Acute hepatocellular injury, anemia, angiosarcoma, cirrhosis, developmental disabilities, embryotoxicity, heart disease, hyperpigmentation, peripheral neuropathies
Asbestos	Asbestosis (scarring of lung tissue), fibrosis (lung and respiratory tract), emphysema, irritations, pneumonia/pneumoconioses
Benzene	Aplastic anemia, CNS depression, embryotoxicity, leukemia and lymphoma, skin irritant
Beryllium	Granuloma (lungs and respiratory tract)
Cadmium	Developmental disabilities, kidney damage, neoplasia (lung and respiratory tract), neonatal death/fetal death, Pulmonary edema; bioaccumulates in aquatic organisms
Carbon tetrachloride	Narcosis, hepatitis, renal damage, liver tumors
Chlorinated aliphatics	Cirrhosis (liver), mild fatty metamorphosis
Chromium and compounds	Asthma, cholestasis (of liver), neoplasia (lung and respiratory tract), skin irritant
Copper	Gastrointestinal irritant, liver damage; toxic to fish
Cyanide	Asthma, asphyxiation, hypersensitivity, pneumonitis, skin irritant; toxicity to fish
Dichlorodiphenyl trichloroethane (DDT)	Ataxic gait, convulsions, human infertility/reproductive effects, kidney damage, neurotoxin, peripheral neuropathies, tremors; bioaccumulation in aquatic organisms
Dimethylacetimide	Birth defects/fetal death
Dioxins/TCDD	Hepatitis, neoplasia, spontaneous abortion/fetal death; bioaccumulative
Formaldehyde	Allergic reactions; gastrointestinal upsets; tissue irritation
Lead and compounds	Anemia, bone marrow depression, CNS symptoms, convulsions, embryotoxicity, neoplasia, neuropathies, kidney damage, seizures; biomagnifies in food chain
Lindane	Convulsions, coma and death, disorientation, headache, nausea and vomiting, neurotoxin, paresthesias
Lithium	Gastroenteritis, hyperpyrexia, nephrogenic diabetes, Parkinson's disease
Manganese	Bronchitis, cirrhosis (liver), influenza (metal-fume fever), pneumonia
Mercury and compounds	Ataxic gait, contact allergen, CNS symptoms; developmental disabilities, neurasthenia, kidney and liver damage, Minamata disease; biomagnification of methylmercury
Methylene chloride	Anesthesia, respiratory distress, death
Naphthalene	Anemia

Nickel and compounds	Asthma, CNS effects, gastrointestinal effects, headache, neoplasia (lung and respiratory tract)
Nitrate	Methemoglobinemia (in infants)
Organochlorine pesticides	Hepatic necrosis, hypertrophy of endoplasmic reticulum, mild fatty metamorphosis
Pentachlorophenol (PCP)	Malignant hyperthermia
Phenol	Asthma, skin irritant
Polychlorinated biphenyls (PCBs)	Embryotoxicity/infertility/fetal death, dermatoses, hepatic necrosis, hepatitis, immune suppression; toxicity to aquatic organisms
Silver	Blindness, skin lesions, pneumonoconiosis
Toluene	Acute renal failure, ataxic gait, CNS depression, memory impairment
Trichloroethylene	CNS depression, deafness, liver damage, paralysis, respiratory and cardiac arrest, visual effects
Vinyl chloride	Leukemia and lymphoma, neoplasia, spontaneous abortion/fetal death, tumors, death
Xylene	CNS depression, memory impairment
Zinc	Corneal ulceration, esophagus damage, pulmonary edema

Sources: Rowland and Cooper, 1983; Blumenthal, 1985; Grisham, 1986; Lave and Upton, 1987.

Figure 1.2 shows a comprehensive process flow for developing a general waste management program appropriate for a hazardous waste facility. Subsequent chapters in this book will provide methods relevant to completing these steps.

1.5 PERTINENT STATUTES AND REGULATIONS

Several items of legislation have been implemented to deal with the regulation of toxic substances present in modern society. Requirements under several of these regulations, such as the Clean Air Act (CAA), Safe Drinking Water Act (SDWA), Clean Water Act (CWA), Resource Conservation and Recovery Act (RCRA), and other federal and state environmental laws are potential applicable or relevant and appropriate requirements (ARARs) for hazardous waste management programs. Some of the relevant statutes and regulations which, directly or indirectly, affect hazardous waste management decisions in the U.S., and therefore hazardous waste risk assessment programs, are briefly referenced below (U.S. EPA, 1974; 1985; 1987; 1988; 1989).

1.5.1 Clean Air Act (CAA)

The objective of the CAA of 1970 is to protect and enhance air quality in order to promote and maintain public health and welfare and the productive capacity of the population. Under Section 109, the CAA requires that National Ambient Air Quality Standards (NAAQS) be set and ultimately met for any air pollutant which, if present in the air, may reasonably be anticipated to endanger public health or welfare and whose presence in the air results from numerous or diverse mobile and/or stationary sources. Two types of NAAQS are provided for: primary standards, designed to protect public health, and secondary standards, designed to protect public welfare (e.g., vegetation, visibility, materials). Under Section 111 of the CAA, EPA sets New Source Performance Standards (NSPS) for new or modified stationary source categories whose emissions cause or significantly contribute to air pollution which may endanger public health or welfare. Section 112 of the CAA also requires the establishment of National Emission Standards for Hazardous Air Pollutants (NESHAP), where a hazardous air pollutant is defined as a pollutant not covered by a NAAQS and exposure to which may reasonably be anticipated to result in an increase in mortality or an increase in serious irreversible, or incapacitating reversible, illness. This covers all pollutants that may cause significant risks.

1.5.2 Safe Drinking Water Act (SDWA)

The SDWA was enacted in 1974 in order to assure that all people served by public water systems would be provided with a supply of high-quality water. The SDWA amendments of 1986 established new procedures and deadlines for setting national primary drinking water standards and established a national monitoring program for unregulated contaminants, among others. The statute covers public water systems, drinking water regulations, and the protection of underground sources of drinking water. It offers regulations with regard to drinking water standards. The SDWA requires the EPA to specify drinking water contaminants that may have adverse

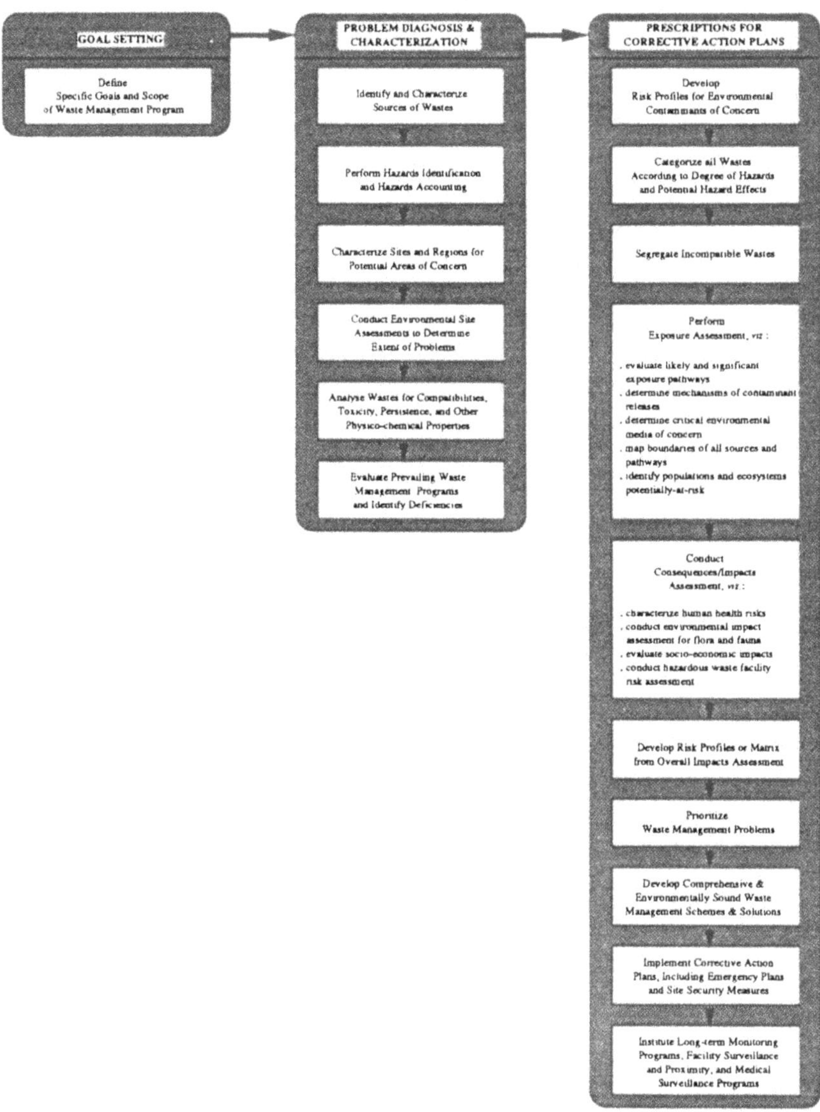

Figure 1.2 Process flow for general waste management planning at a waste facility.

human health effects and are known or anticipated to occur in water. For each contaminant, the EPA must set a maximum contaminant level goal (MCLG) and a maximum contaminant level (MCL). The MCLG (also the recommended maximum contaminant level, RMCL) is a nonenforceable health goal and is set at a level at which no known or anticipated adverse human health effect would occur and which allows an adequate margin of safety. For contaminants classified as human or probable carcinogens, the MCLG is set at zero. For chemicals not classified as probable or human carcinogens, the MCLG is derived from the reference dose

(RfD) or allowable daily intake (ADI). MCLs represent the enforceable drinking water standard and are set as close to the MCLGs as technologically or economically feasible. For substances other than human or probable human carcinogens, the MCL generally equals the MCLG. For human or probable human carcinogens, MCLs will generally fall in the 10^{-7} to 10^{-4} theoretical excess cancer risk range.

1.5.3 Clean Water Act (CWA)

The CWA was enacted in 1977, and an amendment to this introduced the Water Quality Act (WQA) of 1987. The objective of the CWA is to restore and maintain the chemical, physical, and biological integrity of the U.S. waters. This objective is achieved through the control of discharges of pollutants to navigable waters. This control is implemented through the application of federal, state, and local discharge standards. The statute covers the limits on waste discharge to navigable waters; standards for discharge of toxic pollutants; and the prohibition on discharge of oil or hazardous substances into navigable waters. The CWA comprises the Ambient Water Quality Criteria (AWQC) for human health and the AWQC for aquatic organisms. The health criterion is an estimate of the ambient surface water concentration that will not result in adverse health effects in humans. Derivation of numerical water quality criteria for the protection of aquatic organisms uses information from many areas of aquatic toxicology. Water quality criteria developed by the EPA under CWA authority are not by themselves enforceable standards, but can be used in the development of enforceable standards. Another closely associated legislation, the Federal Water Pollution Control Act (FWPCA), deals solely with the regulation of effluent and water quality standards.

1.5.4 Resource Conservation and Recovery Act (RCRA)

The RCRA was enacted in 1976 (as an amendment to the Solid Waste Disposal Act of 1965, later amended in 1970 by the Resource Recovery Act) to regulate the management of hazardous waste, to ensure the safe disposal of wastes, and to provide for resource recovery from the environment by controlling hazardous wastes "from cradle to grave." This is a federal law designed to prevent new uncontrolled hazardous waste sites. It establishes a regulatory system to track hazardous substances from the time of generation to disposal. The law requires safe and secure procedures to be used in treating, transporting, sorting, and disposing of hazardous substances. It covers treatment methods, techniques, or processes; storage and holding periods; disposal methods (such as discharge, injection, spilling, leakages, dumping, etc.); and federal authorization to seek an injunction or bring suit against owners/operators of facilities that endanger public health or the environment. Basically, RCRA regulates hazardous waste generation, storage, transportation, treatment, and disposal. As the major federal statute creating standards for the treatment, storage, and disposal of hazardous waste, RCRA is the most important source of ARARs for actions taken pursuant to CERCLA (c.f.)

response activities. RCRA requirements that may be ARARs with respect to CERCLA response actions are found primarily in RCRA regulations (40 CFR Part 260-271).

The 1984 Hazardous and Solid Waste Amendments (HSWA) to RCRA, Subtitle C, covers a management system that regulates hazardous wastes from the time it is generated until the ultimate disposal — the so-called "cradle-to-grave" system. This, more or less, forms the centerpiece of RCRA. Thus, under RCRA, a hazardous waste management program is based on a "cradle-to-grave" concept, that allows all hazardous wastes to be traced and equitably accounted for.

1.5.5 Comprehensive Environmental Response, Compensation, and Liability Act (CERCLA, or "Superfund")

CERCLA establishes a broad authority to deal with releases or threats of releases of hazardous substances, pollutants, or contaminants from vessels, containments, or facilities. This legislation deals with the remediation of hazardous waste sites, by providing for the cleanup of inactive and abandoned hazardous waste sites. The objective is to provide a mechanism for the federal government to respond to uncontrolled releases of hazardous substances to the environment. The statute covers reporting requirements for past and present owners/operators of hazardous waste facilities and the liability issues for owners/operators for cost of removal or remedial action and damages, in case of release or threat of release of hazardous wastes. Remedial actions are to be cost-effective responses that provide a balance between the need to protect public health and welfare and the environment and the availability of financial resources from the Superfund itself.

The Superfund Amendments and Reauthorization Act (SARA) of 1986 strengthens and expands the cleanup program under CERCLA, focuses on the need for emergency preparedness and community right-to-know, and changes the tax structure for financing the Hazardous Substance Response Trust Fund established under CERCLA to pay for the cleanup of abandoned and uncontrolled hazardous waste sites.

1.5.6 Toxic Substances Control Act (TSCA)

The TSCA of 1976 provides for a wide range of risk management actions to accommodate the variety of risk/benefit situations confronting the EPA. The risk management decisions under TSCA would consider not only the risk factors, such as probability and severity of effects, but also nonrisk factors, such as benefits derived from use of the material and availability of alternative substances. TSCA regulates the manufacture, use, and disposal of chemical substances. It authorizes the EPA to establish regulations pertaining to the testing of chemical substances and mixtures, premanufacture notification for new chemicals or significant new uses of existing substances, control of chemical substances or mixtures that pose an imminent hazard, and record keeping and reporting requirements; the regulations controlling hazardous chemicals are potential ARARs for CERCLA actions.

1.5.7 Federal Insecticide, Fungicide, and Rodenticide Act (FIFRA)

Under FIFRA, EPA has published procedures for the disposal and storage of excess pesticides and pesticide containers in 40 CFR Part 165, Subpart C. EPA has also promulgated tolerance levels for pesticides and pesticide residues in or on raw agricultural commodities under authority of the Federal Food, Drug, and Cosmetic Act (40 CFR Part 180). These tolerance levels are potential ARARs for sites at which agricultural commodities and wildlife are obtained for consumption. Three principal "criteria" are generally used relating to the following:

- Toxicity category
- Reference dose
- Tolerance/action level

FIFRA provides the EPA with broad authorities to regulate all pesticides. This legislation basically deals with pesticide regulations — including the presence of pesticides at facilities that constitute hazardous waste sites.

1.5.8 Endangered Species Act (ESA)

The ESA of 1973 (reauthorized in 1988) provides a means for conserving various species of fish, wildlife, and plants that are threatened with extinction. The ESA considers an endangered species as that which is in danger of extinction in all or a significant portion of its range; a threatened species is that which is likely to become an endangered species in the near future. Also, the ESA provides for the designation of critical habitats (specific areas within the geographical area occupied by the endangered or threatened species) on which are found those physical or biological features essential to the conservation of the species in question. When an endangered or threatened species or critical habitat is present, the ESA is identified as an ARAR.

1.5.9 Other Programs

Several other programs provide platforms for developing standards to protect environmental media potentially affected by hazardous chemicals. Other laws and regulations also work toward preventing or limiting the chances of releases into the environment, for example, the Hazardous Materials Transport Act (HMTA), which is legislation that deals with the regulation of transport of hazardous materials; the Fish and Wildlife Conservation Act of 1980, which requires states to identify significant habitats and develop conservation plans for these areas; the Marine Mammal Protection Act of 1972, which protects all marine mammals — some but not all of which are endangered species; the Migratory Bird Treaty Act of 1972, which implements many treaties involving migratory birds — to protect almost all species of native birds in the U.S.; and the Wild and Scenic Rivers Act of 1972, which preserves select rivers declared as possessing outstanding remarkable scenic, recreational, geologic, fish and wildlife, historic, cultural, or other similar values.

Depending on the type of program being evaluated, one or more of these regulations may dominate the decision-making process. It is noteworthy that the U.S. EPA, acting for the federal government and several state legislators and regulators, are continually refining their risk assessment guidance and specifications for several programs (such as RCRA, CERCLA, TSCA, and FIFRA). Each may be carried out by a different office within the EPA, for different purposes, and under different legal authority and limitations. Consequently, it is imperative that risk assessors familiarize themselves with the appropriate national and regional EPA guidance, as well as the state and local guidance for case-specific projects. It is also important to be aware of the fact that regulations can and do change periodically. Therefore, one needs to be brought up to date by examining the *Federal Register* and other relevant documents issued periodically by state and local regulatory agencies. Furthermore, it is important that the risk assessor has some understanding of the legal criteria specified for case-specific programs, to facilitate the process of choosing and using the applicable guidance for conducting risk assessment.

1.6 ORGANIZATION OF THE BOOK

This text specifically includes a literature review of risk assessment methodologies and concepts applicable to hazardous waste management. It also includes specific numerical evaluations to demonstrate typical applications of risk assessment to aid hazardous waste management decisions. The protocols presented will aid hazardous waste management personnel to formulate programs for managing hazardous waste problems more efficiently. The main objectives are to

- Present concepts and techniques in risk assessment that may be applied to hazardous waste management issues.
- Provide a guidance framework for the formulation of risk assessments in the management of hazardous waste problems.
- Present steps for developing health and environmental risk assessment and site clean-up criteria for potentially contaminated sites.
- Evaluate case-specific risk assessment tasks to demonstrate typical applications of risk assessment to hazardous waste management problems.

Chapter 1, Introduction, presents background information on hazardous waste management issues. This includes identification of hazardous waste generation sources, a discussion of the health and environmental impacts of waste mismanagement, and a review of selected legislation relevant to the conduct of hazardous waste risk assessment.

Chapter 2, Fundamentals of Hazard, Exposure, and Risk Assessment, presents some important definitions that are basic to understanding the nature of hazards and risks associated with hazardous waste facilities and/or sites.

Chapter 3, Concepts in Risk Assessment, provides a discussion of various risk assessment concepts finding applications in hazardous waste management. These include the various measures of risks, risk acceptability criteria, conservatism in risk assessment, uncertainty issues, risk tradeoffs, utility optimization schemes, and the use of risk-time curves.

Chapter 4, The Risk Assessment Process, discusses the generic protocol used for completing a risk assessment, including the levels of detail required for specific risk assessment programs.

Chapter 5, Risk Assessment Techniques and Methods of Approach, elaborates on specific methods employed in evaluating risks associated with hazardous waste management decisions. This includes a review of the state-of-the-art, the risk assessment process and mechanics, risk assessment classification systems, the principles and methods for evaluating health and environmental risks, and the use of probabilistic risk assessment for evaluating aspects of technological risks associated with hazardous waste facilities. It also identifies typical potential applications of the risk assessment methods and techniques.

Chapter 6, Hazardous Waste Management Decisions from Risk Assessment, gives a brief discussion of risk management and risk communication concepts and tools as they affect the ultimate decision-making process in hazardous waste management. It also includes discussions on the general risk management and risk prevention programs.

Chapter 7, Selected Case Studies and Applications, contains a number of example problems and diversified numerical evaluations pertaining to the potential applications of risk assessment in hazardous waste management.

Chapter 8, Epilogue, gives brief recapitulations of the concepts, methods, and applications of risk assessment protocols in the management of hazardous waste problems. It consists of several closing remarks, suggestions, and some contemporary issues pertaining to hazardous waste risk assessment.

A set of appendices are included that contain selected listing of abbreviations, acronyms, glossary of terms and definitions; relevant equations commonly utilized in health risk assessments; carcinogen classification and identification systems of the U.S. EPA and the International Agency for Research on Cancer (IARC); information on selected data bases and models of potential use to risk assessment programs; and some selected units and measures of potential interest to the hazardous waste management professional.

1.7 REFERENCES

Asante-Duah, D. K. "Quantitative Risk Assessment as a Decision Tool for Hazardous Waste Management," in *Proc. 44th Purdue Industrial Waste Conf. (May 1989),* (Chelsea, MI: Lewis Publishers, 1990), pp. 111–123.

Batstone, R., J. E. Smith, Jr., and D. Wilson, Eds. *The Safe Disposal of Hazardous Wastes — The Special Needs and Problems of Developing Countries,* Vols. I, II, and III, A Joint Study Sponsored by the World Bank, the World Health Organization (WHO), and the United Nations Environment Programme (UNEP), World Bank Technical Paper 0253-7494, No. 93 (Washington, DC: The World Bank, 1989).

Bhatt, H. G., R. M. Sykes, and T. L. Sweeney, Eds. *Management of Toxic and Hazardous Wastes* (Chelsea, MI: Lewis Publishers, 1986).

Blumenthal, D. S., Ed. *Introduction to Environmental Health* (New York: Springer-Verlag, 1985).

Canter, L. W., R. C. Knox, and D. M. Fairchild. *Ground Water Quality Protection* (Chelsea, MI: Lewis Publishers, 1988).

Gibbs, L. M. *Love Canal: My Story* (Albany, NY: State University of New York Press, 1982).

Grisham, J. W., Ed. *Health Aspects of the Disposal of Waste Chemicals* (Oxford, England: Pergamon Press, 1986).

Lave, L. B. and A. C. Upton, Eds. *Toxic Chemicals, Health, and the Environment* (Baltimore: Johns Hopkins University Press, 1987).

Levine, A. G. *Love Canal: Science, Politics, and People* (Lexington, MA: Lexington Books, 1982).

Long, F. A. and G. E. Schweitzer. *Risk Assessment at Hazardous Waste Sites* (Washington, DC: American Chemical Society, 1982).

Paustenbach, D. J., Ed. *The Risk Assessment of Environmental Hazards: A Textbook of Case Studies* (New York: John Wiley & Sons, 1988).

Rowland, A. J. and P. Cooper. *Environment and Health* (England: Edward Arnold, 1983).

U.S. EPA. Safe Drinking Water Act, Public Law 93-523 (1974).

U.S. EPA. "National Primary Drinking Water Regulations; Volatile Synthetic Organic Chemicals; Final Rule and Proposed Rule," *Fed. Register.* 50:46830–46901; "National Primary Drinking Water Regulations, Synthetic Organic Chemicals, Inorganic Chemicals and Microorganisms; Proposed Rule," *Fed. Register.* 50:47025–56936 (1985).

U.S. EPA. "The New Superfund: What It Is, How It Works," U.S. Environmental Protection Agency, Washington, DC (1987).

U.S. EPA. "CERCLA Compliance with Other Laws Manual (Inetrim Final)," EPA/540/G-89/006, Office of Solid Waste and Emergency Response, Washington, DC (1988).

U.S. EPA. "CERCLA Compliance with Other Laws Manual: Part II — Clean Air Act and Other Environmental Statutes and State Requirements," EPA/540/G-89/009, OSWER Directive 9234.1-02, Office of Solid Waste and Emergency Response, Washington, DC (1989).

CHAPTER 2

Fundamentals of Hazard, Exposure, and Risk Assessment

Hazard is something with the potential for creating undesired adverse consequences; exposure is the vulnerability to hazards; and risk is the probability or likelihood of an adverse effect due to some hazardous situation. The assessment of potential hazards posed by a substance or an object involves, among several other things, a critical evaluation of available scientific and technical information on the substance or object of concern, the vulnerabilities of potential receptors likely to be exposed, as well as the possible modes of exposure. In addition, potential receptors will have to be exposed to the hazards of concern before any risk could be said to exist. The availability of the best possible information is a prerequisite to sound hazard, exposure, and risk assessments. The integrated assessment of hazards, exposures, and risks are important contributions to decisions aimed at managing hazardous situations.

2.1 THE NATURE OF HAZARD AND RISK

Hazard is defined as the potential for a substance or situation to cause harm or to create adverse impacts on persons, the environment, and/or property. It represents the unassessed loss potential and may comprise a condition, a situation, or a scenario with the potential for creating undesirable consequences. The degree of hazard will normally be determined by the exposure scenario and the potential effects or responses resulting from any exposures.

Risk may be considered as the probability of an adverse effect, or an assessed threat to persons, the environment, and/or property, due to some hazardous situation. It is a measure of the probability and severity of adverse consequences of exposure to potential receptors due to a system failure; it may simply be represented by the measure of the frequency of an event. Risk represents the assessed loss potential, often estimated by the mathematical expectation of the consequences of an adverse event occurring (defined by the product of the two components of the probability of occurrence [p] and the consequence or severity of occurrence [S]), *viz*:

$$\text{Risk} = p \times S$$

In fact, the level of risk is dependent on the degree of hazard as well as on the amount of safeguards or preventative measures against adverse effects; consequently, risk can also be defined by the following conceptual relationships:

$$\text{Risk} = \frac{(\text{Hazard})}{(\text{Preventative Measures})}$$

or

$$\text{Risk} = f(\text{Hazard, Exposure, Safeguards})$$

where "Preventative Measures" or "Safeguards" are considered to be a function of exposure — or rather inversely proportional to the degree of exposure.

2.1.1 Basis for Measuring Risks

Risk measures give an indication of the probability and severity of adverse effects (to health, environment, or property), and generally are established with varying degrees of confidence according to the importance of the decision involved. Risk estimation involves an integration of information on the intensity, frequency, and duration of exposure for all identified exposure routes for the exposed or affected group(s). Measures used in risk assessment assume various forms, depending on the type of problem, degree of resolution appropriate for the situation on hand, and the analysts' preference. It may be expressed in quantitative terms, in which case it takes on values from zero (associated with certainty for no- adverse effects) to unity (associated with certainty for adverse effects to occur); in several other cases, risk is only described qualitatively, by use of descriptors such as "high," "moderate," "low," etc.; or indeed, risk may be described in semiquantitative/ semiqualitative terms.

The typical basis for risk qualification or quantification is normally in reference to several measures, parameters, and/or tools represented by one or more of the following:

- Probability distributions (based on probabilistic analyses)
- Expected values
- Economic losses or damages
- Risk profile diagrams

- Relative risk (defined by the ratio [incidence rate in exposed group]:[incidence rate in non-exposed group])
- Individual lifetime risk (equal to the product of exposure level and severity, e.g. [dose × potency])
- Population risk (defined by the product of the individual lifetime risk and the population exposed)
- Frequency-consequence diagrams (also known as F-N curves for fatalities, to define societal risk)
- Quality of life adjustment
- Loss of life expectancy (given by the product of individual lifetime risk and the average remaining lifetime)

Individual lifetime risk is about the most commonly used measure of risk, presented as the probability that the individual will be subjected to an adverse effect from exposure to identified hazards.

2.1.2 Analyzing Hazards and Risks

Procedures for analyzing hazards and risks may comprise several steps (Figure 2.1), consisting of the following elements:

Hazard identification
- Identify hazards (including nature/identity of hazard, location, etc.)
- Identify initiating events (i.e., causes)
- Identify resolutions for hazard

Vulnerability analysis
- Identify vulnerable zones
- Identify concentration/impact profiles for affected zones
- Determine populations potentially at risk (such as human and ecological populations, and critical facilities)

Analysis of risks/consequences/impacts
- Determine sequence of events and system response
- Determine probability of adverse outcome (including exposures)
- Estimate consequences (including severity, uncertainties, etc.)

Some or all of these elements may have to be analyzed in a systematic and comprehensive manner, depending on the variation and level of detail of risk assessment that is appropriate or is being performed. The variations fall into two broad categories: endangerment assessment (which is contaminant based, such as health and environmental risk assessment), and safety assessment (which is system failure based, such as probabilistic risk assessment).

2.2 WHAT IS RISK ASSESSMENT?

Several definitions of risk assessment have been published in the literature by various authors (e.g., Rowe, 1977; NRC, 1982; NRC, 1983; OTA, 1983; U.S. EPA,

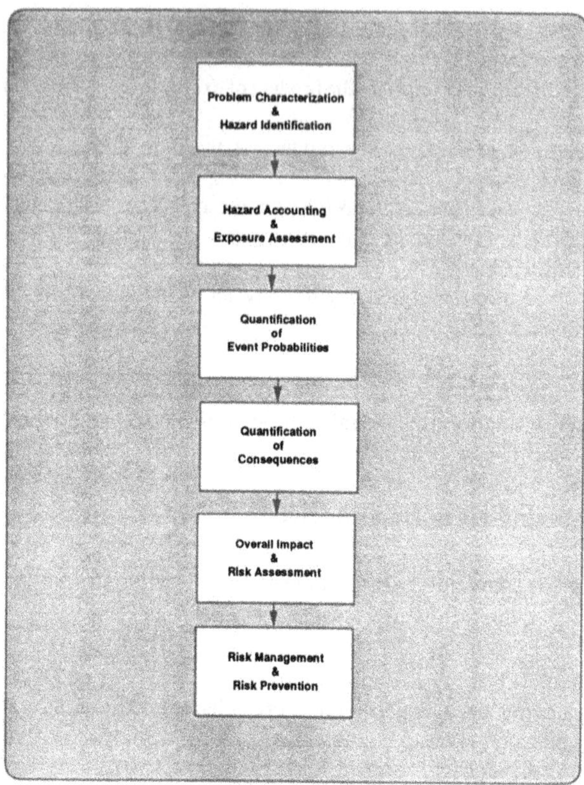

Figure 2.1. Steps in the analyses of hazards and risks.

1984; Hallenbeck and Cunningham, 1988; Bowles et al., 1987; Asante-Duah, 1990) in describing their risk assessment models and/or protocols. In a generic sense, risk assessment is defined as a systematic process for making estimates of all the significant risk factors that prevail over an entire range of failure modes and/or exposure scenarios due to the presence of some type of hazard. It is a qualitative or quantitative evaluation of consequences arising from some initiating hazard(s) that could lead to specific forms of system response(s), outcome(s), exposure(s), and consequence(s).

Risk assessment is a powerful tool for developing insights into the relative importance of the various types of exposure scenarios of potentially hazardous situations. It involves the characterization of potential adverse consequences or impacts to humans and ecological receptors due to exposure to environmental, technological, or other hazards. The risk assessment process is a mechanism that utilizes the best available scientific knowledge to establish case-specific responses that will ensure justifiable and defensible decisions necessary for managing hazardous situations in a cost-efficient manner. The process concerns the assessment of the importance of all identified risk factors to various stakeholders whose interests are embedded in a candidate problem situation (Petak and Atkisson, 1982).

2.3 WHY PERFORM RISK ASSESSMENT?

The overall purpose of risk assessment is to provide, in as far as is feasible, complete information set to risk managers to allow the best decision to be made concerning a potentially hazardous problem. In view of the health, environmental, and socio-economic implications of hazardous waste problems, systematic and valid analytical protocols should be used in the applicable management decisions. In this regard, it is apparent that risk assessment provides one of the best mechanisms known for estimating risks posed by hazardous wastes (Volpp, 1988). Risk assessment methodologies provide a potentially useful analytical tool for studying situations that could result in adverse consequences. This can be used for quantifying and/or qualifying risks associated with different hazards for comparison with common, everyday risks.

The process of quantifying risk does, by its very nature, give an understanding of the degree of importance of the potential hazards being examined. It shows where a given effort can provide the most benefit in modifying a system to improve safety and efficiency. The assessment of risk associated with hazardous waste management decisions allows the identification of those sequences of potential causative events and exposure scenarios contributing most significantly to risks. Information developed in the assessment could help in making decisions about the allocation of resources for safety improvements and hazard/risk prevention by directing attention and efforts to the features and exposure pathways that dominate the risks. The results of the analysis will generally provide decision makers with more justifiable bases for determining risk acceptability and to aid in choices between possible corrective measures aimed at risk reduction and abatement programs. This can then be used to satisfy the principal objective of risk assessment, which is to provide a basis for actions to minimize the impairment of public health and safety.

2.4 RISK ASSESSMENT GOALS, REQUIREMENTS, AND ATTRIBUTES

Whyte and Burton (1980) succintly indicate that a major objective of risk assessment is to help develop risk management decisions that are more systematic, more comprehensive, more accountable and more self-aware of what is involved than has often been the case in the past. The overall goal in a risk assessment is to identify potential accident or failure modes and exposure scenarios that will aid in the development of methods to reduce the probability of failure and the attending human, economic, and environmental consequences of any failure and/or exposure events. Tasks performed during a risk evaluation are intended to help answer the questions "How safe is safe enough?" or "How clean is clean enough?" In general, risk assessment seeks to answer three basic questions:

- What could potentially go wrong, or how can various factors link together to cause things to go wrong?

- What are the chances for this to happen?
- What are the anticipated consequences, if this should happen?

A complete analysis of risks associated with a given situation or activity will generate answers to these questions; subsequently, mitigative measures can be developed to abate the potential adverse effects.

The appropriate type and degree of detail for a risk assessment will depend on the intended use of the information generated. Its purpose will shape the data needs, the protocol, the rigor, and related efforts. Current regulatory requirements at the federal, state, and local levels are particularly important considerations in the risk assessment. The process generally requires a multidisciplinary approach, covering several expertise in most cases. Methods used in estimating risks may be either qualitative or quantitative, depending on the scope of the analysis and the availability of background information. The process may be one of data analysis or modeling, or a combination of the two.

Risk assessment has several specific benefits; some of the major attributes include the following:

- Identifies and ranks all existing and anticipated potential hazards
- Identifies and ranks all potential failure modes
- Explicitly considers all current and possible future exposure scenarios
- Quantifies and/or qualifies risks associated with the full range of loading conditions, system responses, and exposure scenarios, rather than deal only with extreme events
- Identifies factors and exposure routes contributing most to total risk of failure and/or exposure (i.e., the identification of major contributors to risk, critical pathways, and all significant factors involved)
- Facilitates determination of cost-effective risk reduction through remedial alternatives and/or risk management and prevention programs
- Identifies and analyzes sources of uncertainties

All these will eventually play an important role in implementing appropriate risk management programs for case-specific problems. Regarding any limitations, it must be noted that the assessment of risks are inherently imprecise due to the fact that the risk assessor's knowledge of the causative effects and controlling factors usually is limited, and also because the results depend to some extent on the methodology and assumptions adopted. Also, risk assessment can impose potential delays in program implementation; however, the overall gain in program efficiency is likely to more than compensate for any such delays.

2.5 RISK ASSESSMENT FOR HAZARDOUS WASTE SITE PROBLEMS

Hazardous waste site risk assessment is a process used to evaluate the collective demographic, geographic, physical, chemical, biological, and related factors at a potentially contaminated site, so as to determine and characterize possible risks to public health and the environment. Based on such an assessment, it can be determined if corrective actions are needed at the site and/or if compliance with applicable regulatory requirements are violated. The overall objective of this type of assessment is to determine the magnitude and probability of actual or potential harm that a hazardous waste site poses to human health and the environment. Important factors to consider in assessing risks and establishing hazardous waste site cleanup criteria for a potentially contaminated site include the following:

- Degree to which human health, safety, or welfare may be affected by exposure to chemical constituents
- Effect of contamination on the environment
- Individual site characteristics
- Current and future beneficial uses of the affected land and subsurface resources
- Application of appropriate, applicable, or relevant regulatory standards, requirements, and/or guidelines

CERCLA (or "Superfund") establishes a national program for responding to releases of hazardous substances into the environment, with the overall mandate to protect human health and the environment from current and potential threats posed by uncontrolled hazardous substance releases; risk assessment aids decisions on what to do when contaminated sites are identified. The basic approaches to performing risk assessment at both Superfund and non-Superfund sites are fundamentally the same, except for the degree of detail required for the various steps involved.

2.6 LEGISLATIVE-REGULATORY PERSPECTIVES IN RISK ASSESSMENT OF CONTAMINATED SITES

The National Contingency Plan (NCP) (40 CFR 300) requires that ARARs and other standards or criteria be considered when developing remedial actions as part of a remedial investigation/feasibility study (RI/FS) process for a potentially contaminated site. The definitions of ARARs are as follows (U.S. EPA, 1988):

- *Applicable requirements* are those cleanup standards, standards of control, and other substantive environmental protection requirements, criteria, or limitations promulgated under federal or state laws that specifically address a hazardous substance, pollutant, contaminant, remedial action, location, or other circumstances at a CERCLA site.
- *Relevant and appropriate requirements* are those cleanup standards, standards of control, and other substantive environmental protection requirements, criteria, or limitations promulgated under federal or state laws that, while not "applicable" to a hazardous substance, pollutant, contaminant, remedial action, location, or other circumstances at a CERCLA site, address problems or situations sufficiently similar to those encountered at the CERCLA site, that their use is well suited to the particular site.

The requirement can be either one of these categories, but not both. Potential ARARs are identified through a review of current local, state, and federal standards. In making decisions regarding the chemicals of potential concern present in the environment, the concentrations of all the chemicals detected in field samples are compared to potential ARARs or other criteria such as federal, state, and local guidelines. This comparison (with potential ARARs or other guidelines) serves as one type of public health/environmental evaluation for those chemicals for which standards have been promulgated. Also, To-Be-Considered materials (TBCs) may be used as appropriate; TBCs are nonpromulgated advisories or guidance issued by federal or state government that are not legally binding and do not have the status of potential ARARs. However, in many circumstances TBCs will be considered along with ARARs as part of the site risk assessment and may be used in determining the necessary level of cleanup for protection of public health and the environment.

A comprehensive and generic tabulation of the requirements, the administering agencies, and summaries of potential federal and state ARARs, together with other federal, state, and local criteria, advisories, and TBC guidance are necessary to support risk assessment of potentially contaminated sites. The identification of ARARs must be done on a site-specific basis. ARARs (and TBCs to some extent) may be chemical specific, location specific (e.g., based on sensitive habitats or ecosystems, wetlands, floodplains, etc.), or action specific (i.e., based on remedial actions). Protective remedies may be developed using risk assessment, ARARs, and TBCs. Requirements under several regulations, such as the CWA, SDWA, CAA, RCRA, TSCA, and other federal and state environmental laws are potential ARARs for hazardous waste management programs. U.S. EPA (1988; 1989) discusses the details of procedures and criteria for selecting ARARs, TBCs, and other parameters. In general, no universally acceptable standards are promulgated for soil contaminations; site-specific health-based criteria and other guidelines, therefore, become an important basis for comparison with the level of site contamination encountered.

ARARs will define the cleanup goals for specified point of compliance when they give an acceptable level with respect to site-specific factors (including the characteristics of the remedial action, the hazardous chemicals present at the site, or physical setup of the site). For instance, MCLs under the SDWA will ordinarily be acceptable levels for specific contaminants. However, cleanup goals for some chemicals may have to be based on nonpromulgated criteria and advisories (e.g., health advisories such as RfDs) rather than ARARs, either because ARARs do not exist for those chemicals or because an ARAR by itself would not be adequately protective under the prevailing circumstances (e.g., where additive effects from multiple chemicals and/or pathways are involved). Consequently, the cleanup requirements necessary to meet cleanup goals will be based not only on ARARs, but also on health-based criteria, TBCs, state criteria, advisories, and guidance for the state or region in which a site is located. At each potential exposure point, a reasonable maximum exposure scenario should be assumed and/or established and cleanup goals set accordingly to ensure maximum protectiveness.

2.7 REFERENCES

Asante-Duah, D. K., "Quantitative Risk Assessment as a Decision Tool for Hazardous Waste Management," *Proc. 44th Purdue Industrial Waste Conf. (May 1989)*, (Chelsea, MI: Lewis Publishers, 1990), pp. 111–123.

Bowles, D. S., L. R. Anderson, and T. F. Glover. "Design Level Risk Assessment for Dams" in *Proc. Struct. Congr. ASCE* 210–25.

Hallenbeck, W. H. and K. M. Cunningham, *Quantitative Risk Assessment for Environmental and Occupational Health*, 4th Printing (Chelsea, MI: Lewis Publishers, 1988).

NRC (National Research Council) Committee on Risk and Decision-Making. *Risk and Decision-Making: Perspective and Research* (Washington, DC: National Academy Press, 1982).

NRC (National Research Council). *Risk Assessment in the Federal Government: Managing the Process* (Washington, DC: National Academy Press, 1983).

OTA (Office of Technology Assessment). "Technologies and Management Strategies for Hazardous Waste Control," Congress of the U.S., Office of Technology Assessment, Washington, DC (1983).

Petak, W. J. and A. A. Atkisson. *Natural Hazard Risk Assessment and Public Policy: Anticipating the Unexpected* (New York: Springer-Verlag, 1982).

Rowe, W. D. *An Anatomy of Risk* (New York: John Wiley & Sons, 1977).

U.S. EPA. "Risk Assessment and Management: Framework for Decision Making," EPA 600/9-85-002, Washington, DC (1984).

U.S. EPA. "CERCLA Compliance with Other Laws Manual (Inetrim Final)," EPA/540/G-89/006, Office of Solid Waste and Emergency Response, Washington, DC (1988).

U.S. EPA. "CERCLA Compliance with Other Laws Manual: Part II — Clean Air Act and Other Environmental Statutes and State Requirements," EPA/540/G-89/009, OSWER Directive 9234. 1-02, Office of Solid Waste and Energy Response, Washington, DC (1989).

Volpp, C. "'Is It Safe or Isn't It?': An Overview of Risk Assessment," *Water Resour. News*, 4(1) (1988).

Whyte, A. V. and I. Burton, Eds. *Environmental Risk Assessment*, SCOPE Report 15, (New York: John Wiley & Sons, 1980).

2.7 REFERENCES

CHAPTER 3

Concepts in Risk Assessment

It has long been recognized that nothing is either wholly safe or dangerous per se, but that the object involved and the manner and conditions of use determine the degree of hazard or safety. Consequently, it may rightly be concluded that there is no escape from all risk no matter how remote, but that there only are choices among risks (Daniels, 1978). Risk assessment is designed to offer an opportunity to understand a system better by adding an orderliness and completeness to a problem evaluation. It generally embodies the heuristic approach of empirical learning that will provide a "best knowledge" estimate of the relative importance of risks. The full value of risk assessment is realized if it is used properly. In its applications, it becomes important to discern between individual vs. societal risks, and then to use that which the analyst considers most appropriate for the case-specific problem.

Individual risks are considered to be the frequency at which a given individual could potentially sustain a given level of adverse consequence from the realization or occurrence of specified hazards. Societal risk, on the other hand, relates to the frequency and the number of individuals sustaining some given level of adverse consequence due to the occurrence of specified hazards. Individual risk estimates are more appropriate in cases where individuals face relatively high risks. However, when individual risks are not inequitably high, then it becomes important during resources allocation to consider possible society-wide risks which might be relatively higher.

3.1 RISK ACCEPTABILITY CRITERIA: THE DE MINIMUS AND ACCEPTABLE RISK CONCEPTS

Every system has a target risk level that represents tolerable limits to danger that society is prepared to accept in consequence of potential benefits that could accrue;

this may be represented by the *"de minimis"* or "acceptable" risk levels. Risk is *de minimis* if the incremental risk produced by an activity is sufficiently small so that there is no incentive to modify the activity (Whipple, 1987). These are levels assumed to be so insignificant to be of any social concern or to justify use of risk management resources to control them, compared with other beneficial uses for the limited resources available in practice. Simply stated, the *de minimus* principle assumes that extremely low risks are trivial and need not be controlled. A *de minimis* risk level would therefore represent a cutoff, below which a regulatory agency could simply ignore alleged problems or hazards. Thus, in the process of establishing risk levels, it is possible to use *de minimus* levels below which one need not be concerned (Rowe, 1983). There are several approaches to deriving such *de minimus* levels, but that which is selected should be justifiable based on the expected socio-economic, environmental, and health impacts.

An important issue in risk assessment is the risk acceptability level, i.e., what level of risk society can allow for a specified hazard situation, also recognizing that the desirable level is not always attainable. With maintenance of public health and safety being crucial, it should be realized that financial or budgetary constraints may not by themselves be justifiable enough reason for setting acceptable levels on the higher side of the risk spectrum. *De minimis*, a lower bound on the range of acceptable risk for a given activity would generally be preferred. When properly utilized, a *de minimis* risk concept can help prioritize focus of attention of risk management decisions in a socially beneficial way. It may define the threshold for regulatory involvement. A common approach for placing risks in perspective will be to list many risks (which are considered similar in nature) along with some quantitative measures of the degree of risk. Typically, risks below the level of one in one million (i.e., 10^{-6}) risk of premature death will often be considered insignificant or "*de minimis*" by regulatory agencies, since this compares favorably with risk levels from "normal" human activities (e.g., 10^{-3} for smoking a pack/day, or rock climbing; 10^{-4} for heavy drinking, home accidents, driving motor vehicles, or farming; 10^{-5} for truck driving, home fires, skiing, living downstream of a dam, or using contraceptive pills; 10^{-6} for diagnostic X-rays, or fishing; and 10^{-7} for drinking about 10 L of diet soda containing saccharin). In considering a *de minimis* risk level, the possibility of multiple *de minimis* exposures with consequential large aggregate risk should not be overlooked. Whipple suggests the use of a *de minimis* probability idea to help develop a workable *de minimis* policy; the details of its application is discussed in Paustenbach (1988).

The concept of *de minimis* or *acceptable* risk is essentially a threshold concept, in that it postulates a threshold of concern below which there would be indifference to changes in the level of risk. In fact, considerable controversy exists concerning the concept of "acceptable risk" in the risk and decision analysis literature. In practice, acceptable risk is the risk associated with the most acceptable decision — rather than being acceptable in an absolute sense. It has been pointed out (Massmann and Freeze, 1987) that acceptable risk is decided in the political arena and that "acceptable" risk really means "politically acceptable" risk. Current regulatory requirements at the federal, state, and local levels are particularly important considerations in establishing acceptable risk levels. The selection of a *de minimis* risk level is contingent upon the nature of the risks, the stakeholders involved, and a host of other contextual variables, such as other risks being compared against. This means that *de minimis* levels will be fuzzy (in that they can never be precisely specified) and relative (in that they will depend on the special circumstances).

3.2 USE OF CONSERVATIVE ASSUMPTIONS: NOMINAL AND WORST-CASE CONCEPTS

Many of the parameters and assumptions used in hazard, exposure, and risk evaluation studies tend to have high degrees of uncertainties associated with them. Thus, it is common practice for safe design and analysis to model risks such that risk levels estimated for management decisions are overestimated. Such conservative (also worst-case or plausible upper bound) estimates used in risk assessment are based on the premise that pessimism in risk assessment (with resultant high estimates of risks) is more protective of public health and/or the environment.

In performing risk assessment, scenarios are usually developed that will often reflect the worst possible exposure pattern; this notion of "worst-case scenario" in risk assessment refers to the event or series of events resulting in the greatest exposure or potential exposure. On the other hand, gross exaggeration of actual risks could lead to poor decisions being made with respect to limited resources available for general risk mitigation purposes. For instance, simply using upper-bound estimates based on compounded conservative assumptions may lead to the control of insignificant risks, depleting the limited resources. Thus, after establishing a worst-case scenario, it is often desirable to also develop and analyze more realistic or nominal scenarios, so that the level of risk posed by a hazardous situation can be better bounded by selecting "best" or "most likely" sets of assumptions for risk assessment. In deciding on the assumptions to be used in a risk assessment, it is imperative that the analyst chooses parameters that will, at worst, result in erring on the side of safety.

Maxim (Paustenbach, 1988) provides an elaboration on ways and means of making risk assessments more realistic, rather than depending on wholesale compounded conservative assumptions. Among others, there is the need to undertake sensitivity analyses, including the use of multiple assumption sets that reflect a wider spectrum of exposure scenarios. This is important because regulations based on the so-called upper-bound estimate may address risks that are almost nonexistent and infeasible. Indeed, risk assessment using extremely conservative biases do not provide risk managers with the quality information needed to formulate efficient and cost-effective management strategies. Also, using plausible upper-bound risk estimates may lead to spending scarce and limited resources to regulate or control insignificant risks, while more serious risks are probably being ignored.

3.3 UNCERTAINTY ISSUES IN RISK ASSESSMENT

A major difficulty in decision making resides in the uncertainties of system characteristics for the situation at hand. Uncertainty is the absence of the best estimate for the magnitude or probability of occurrence of a given variable. Engineering judgment becomes an important factor in problem solving under uncertainty, and decision analysis provides a means of representing the uncertainties in a manner that allows informed discussion. The presence of uncertainty means, in general, that the best outcome obtainable from an evaluation and/or analysis cannot be guaranteed. Nonetheless, as has been noted by Bean (1988), decisions ought to be made even in an uncertain setting, otherwise action at hazardous waste sites would be completely paralyzed. Specifically, there are inevitable uncertainties associated with risk estimates, but these uncertainties do not invalidate the use of risk estimates in the decision-making process. However, it is important to identify and define the confidence levels associated with the evaluation.

The uncertainties that arise in risk assessments can be of three main types:

- Uncertainties in parameter values (e.g., use of incomplete or biased values)
- Uncertainties in parameter modeling (e.g., issue of model adequacy/inadequacy)
- Uncertainties in the degree of completeness (e.g., representativeness of evaluation scenarios)

Depending on the specific level of detail of a risk assessment being carried out, the type of uncertainty that dominates at each stage of the analysis can be different. Uncertainty analysis can be performed qualitatively or quantitatively; sensitivity analysis is often a useful adjunct to uncertainty analysis. Thus, in addition to presenting the best estimate, the risk model may also provide a range of likely estimates in the form of a sensitivity analysis. Generally, a sensitivity analysis (to help cater for the inevitable uncertainties) should become an integral part of the risk evaluation process. Sensitivity analysis entails the determination of how rapidly the output of an analysis changes with respect to variations in the input. Sensitivity studies do not necessarily incorporate the error range or uncertainty of the input.

Uncertainty can be characterized via a sensitivity analysis and/or probability analysis (e.g., Monte Carlo simulation) techniques (U.S. EPA, 1989). The technique selected depends on the availability of input data statistics. Sensitivity analyses require data on the range of values for each exposure factor in the scenario. Probabilistic analyses require data on the range and probability function (or distribution) of each exposure factor within the scenario. Through sensitivity analyses, uncertainties can be assessed properly, and their effects on given decisions accounted for systematically. In this manner the risk associated with given decision alternatives may be delineated and then appropriate corrective measures taken accordingly. In view of the fact that risk assessment constitutes a very important part of the hazardous waste management decision-making process, it is essential that all apparent sources of uncertainty and error are well documented. This ensures that the limitations of quantitative results are clearly understood.

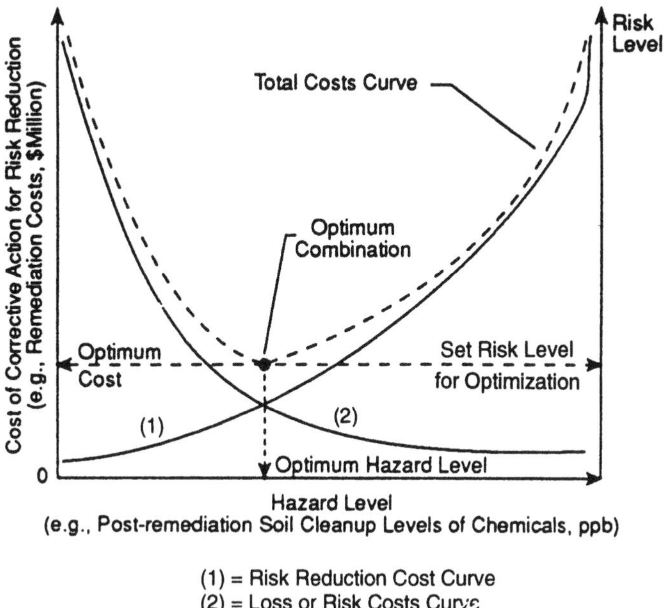

Figure 3.1 A schematic of corrective action costs (e.g., cleanup costs) and risk levels for varying hazard levels (e.g., chemical concentrations in environmental media).

3.4 APPLICATION OF RISK-BASED DECISION ANALYSES TO RISK STUDIES

Hazardous waste management generally involves competing objectives. The prime objective for hazardous waste disposal consists of minimizing both hazards and waste management costs, given appropriate constraints. In general, however, reducing hazards would require increasing costs and cost minimization during hazard abatement will likely leave higher degrees of unmitigated hazards (Figure 3.1). Hazard assessment is performed to evaluate the consequences of failure (i.e., loss of life, economic loss, environmental impact, etc.) for a given hazardous waste management problem, so that adequate corrective action can be taken. Once the minimum acceptable and/or achievable level of protection has been established via hazard assessment, alternative courses of action are developed that weigh the magnitude of adverse consequences against the cost of corrective measures. Generally, a decision is made based on the alternative that accomplishes the desired objectives at the least total cost (total cost here being the sum of hazard cost and remedial cost).

Comparison between risks, benefits, and costs for various corrective action strategies is a necessary part of an overall risk management program. These may include the evaluation of relative risks among available decision alternatives; evaluating cost effectiveness of corrective action plans; or determining and comparing the risks and costs with the benefits of several management options. Subjective and controversial as it might appear to express hazard in terms of cost, especially where public health and/or safety is concerned, it nevertheless has been used to provide an objective way of evaluating the problem. Decisions based strictly on cost-benefit analysis utilize a single criterion to judge alternative management programs. Because the use of monetary value alone as a measure of effectiveness precludes certain vital intangibles from consideration in the analysis, cost-benefit analysis has received a great deal of criticism as an unreliable and inappropriate tool for decision making in environmental problems (e.g., Ashby, 1980). Consequently, alternative methods of analyses are normally employed under such circumstances.

Decision analysis is a management tool that consists of a conceptual and systematic procedure for analyzing complex sets of alternatives in a rational manner so as to improve the overall performance of the decision-making process. Decision theory provides a logical and systematic framework to structure objectives and to evaluate and rank alternative potential solutions to a problem. Multiattribute decision analysis and utility theory have been suggested (e.g., Lifson, 1972; Keeney and Raiffa, 1976) for the evaluation of problems involving multiple conflicting objectives, such as is the case for decisions on hazardous waste management programs. Under such circumstances, the decision maker is faced with the problem of having to trade off the performance of one objective against another. A mathematical structure may be developed around utility theory that presents a deductive philosophy for risk-based decisions (Lifson, 1972; Keeney and Raiffa, 1976; Starr and Whipple, 1980; Keeney, 1984). It should be acknowledged that, although decision analysis presents a systematic and flexible technique that incorporates the decision makers' judgment, it does not provide a complete analysis of the public's perception of risk — an unfortunate shortcoming.

3.4.1 Risk Tradeoffs in Hazardous Waste Management

Choices between alternative disposal methods and/or facilities for risk/hazard reduction and concurrent cost reduction often has to be made by the environmental analyst. Risk tradeoffs may be assessed by the application of multiattribute decision analysis and utility theory in the investigation of tradeoffs between increased expenditure on waste disposal and/or remedial action and the hazard reduction achieved. The tradeoff may be determined by applying weighting factors of preferences in a utility-attribute analysis. The weighting factors are changed to reflect varying tradeoff values associated with alternative decisions. The most cost-effective solution can be selected from this analysis. It should be noted that, generally, risk abatement costs increase dramatically once certain threshold risk levels are attained.

3.4.2 Cost Effectiveness of Risk Reduction

Cost effectiveness analysis involves a comparison of the costs of alternative methods to achieve some set goal(s) of risk reduction, such as the acceptable risk level or cleanup criteria. The process compares the costs associated with different methods of achieving a specific goal. The analysis can be used to allocate limited resources among several programs of risk abatement that will achieve the greatest results per unit cost. It may also be used to project and compare total costs of several corrective action plans. A fixed goal is established and policy options are analyzed on the ability to achieve that goal in a most cost-effective manner. The goal is generally at a specified level of acceptable risk, and the options are compared on the basis of the monetary costs necessary to reach that level of risk. Cost constraints can also be imposed so that the options are assessed on their ability to control the risk most effectively for that set cost. The efficacy of alternatives in the hazard reduction process can then be assessed, and the most cost effective (i.e., one with minimum cost meeting the constraint of acceptable risk/hazard level) can then be chosen for implementation. This would then guarantee the objective of meeting the constraints at the lowest feasible cost.

3.4.3 Optimization in Risk-Cost-Benefit Analyses

Risk-cost-benefit analysis is a generic term for techniques encompassing risk assessment and the inclusive evaluation of risks, costs, and benefits of alternative projects or policies. In performing risk-cost-benefit analysis, one attempts to measure risks, costs, and benefits, to identify uncertainties and potential tradeoffs, and then to present this information coherently to decision makers. This represents yet another available concept in risk assessment, though much less used presently in the field of hazardous waste management. A general form of objective function for use in the risk-cost-benefit analysis that treats the stream of benefits, costs, and risks in a net present value calaculation is given by (Crouch and Wilson, 1982; Massmann and Freeze, 1987)

$$\Phi = \sum_{t=0}^{T} \frac{1}{(1+r)^t} \{B(t) - C(t) - R(t)\} \tag{3.1}$$

where

Φ = objective function ($)
t = time, spanning 0 to T (years)
T = time horizon (years)
r = discount rate
B(t) = benefits in year t ($)
C(t) = costs in year t ($)
R(t) = risks in year t ($)

The risk term is defined by the expected cost associated with the probability of failure, and is a funtion of the costs or consequences of failure in year t, ECD_t. That is,

$$R(t) = f\left(ECD_t\right) \tag{3.2}$$

Assuming a constancy in time of the dollar value, if the optimization is performed on a per annum basis, then Equation 3.1 reduces to

$$\Phi = TB - TC - RC \tag{3.3}$$

where

TB = total annual benefits
TC = total annual costs
RC = annual risk costs

A risk-cost-benefit technique may generally be used to evaluate the extent to which a new policy, on achieving a balance, can improve the overall well-being of those potentially affected by the program. Indeed, the overall expected net benefits of a hazardous waste management policy may be represented by the following simplistic conceptual relationship:

$$ENB = \left\{MOC + \sum_i P_i B_i\right\} - \left\{\sum_j P_j C_j + TOC\right\} \tag{3.4}$$

where

ENB = expected net benefits from program implementation
MOC = monetary and other compensations to region from facility location (e.g., state and local business taxes)
P_i = probability that benefit i (e.g., employment, improved waste management, etc.) will be realized as a result of new policy
B_i = value of benefit i within region due to the given activity
P_j = probability of detrimental impact j (e.g., reduced property values, health/environmental costs, etc.) in region due to the given activity
C_j = costs of adverse impact j in performing the activity due to new waste management policy
TOC = fixed transport, storage, disposal, administrative/regulations, and miscellaneous costs anticipated from the activity and/or for new policy

An important objective with respect to compensation and other benefits is to maximize ENB with the important associated conditions for minimizing P_j and C_j. P_j can be reduced by mitigative measures or protection; other kinds of abatement policies can be used to reduce C_j. Conditional compensations/payments such as insurance may also be used to reduce C_j, while in-kind compensations may increase B_i with respect to nonmonetary benefits. A criterion should be involved, by which

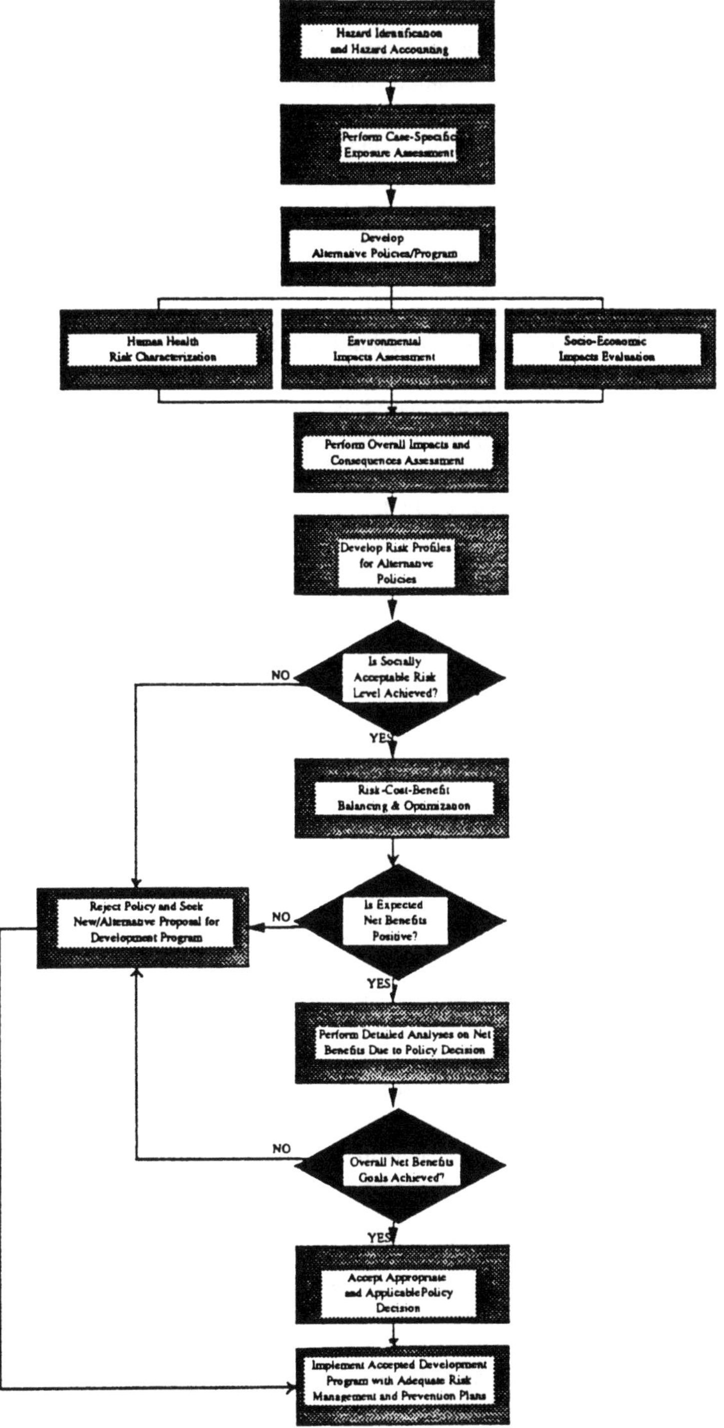

Figure 3.2 A conceptual framework for risk-cost-benefit tradeoffs analyses in a waste management program.

a decision to proceed with program implementation makes some people better off and no one worse off. Beyond that, there should be equitable distribution of net benefits. A representative evaluation framework for performing risk-cost-benefit tradeoffs evaluation is depicted in Figure 3.2. Such a structure will facilitate the process of balancing the risks, costs, and benefits for a given waste management program. Tradeoff decisions made in the process will be directed at improving both short- and long-term benefits of the development program.

3.4.4 Utility Theory Applications in Risk Studies

Multiattribute decision analysis and utility theory may be applied in the investigation of hazardous waste management programs. In using expected utility maximization, the preferred alternative is the one that maximizes the expected utility — or equivalently, the one that minimizes the loss of expected utility. In a way, this is a nonlinear generalization of cost-benefit or risk-benefit analysis. Unfortunately, there have been limited applications of the expected utility approach in risk assessment studies in the past even though this may have great potentials to aid management decisions. While utility theory offers a rational procedure, it may, however, transfer the burden of decision to the assessment of utility functions. Also, several subjective assumptions are used in the application of utility functions that are a subject of debate; the details of the paradoxes surrounding conclusions from expected utility applications are beyond the scope of this elaboration and are not discussed here.

Utility Attributes

Attributes measure how well the objectives are being achieved. Through the use of multiple attributes scaled in the form of utilities and weighted according to their relative importance, a decision analyst can describe an expanded set of consequences associated with hazardous waste management. Adopting utility as the criterion of choice among alternatives allows a multifaceted representation of each possible consequence. Hence, in its application to decisions involving hazardous waste disposal, both hazards and costs can be converted to utility values, as measured by the relative importance that the decisionmaker attaches to either attribute. In assigning utility value to hazard, it would be better to use various social and environmental goals which can help determine the threats the hazard poses to health, the environment, etc., rather than using the direct concept of hazard. These utility values can then be used as the basis for selection among alternatives. The utility function need not be linear since the utility is not necessarily proportional to the attribute; thus, curves of the forms shown in Figure 3.3 can be generated for the utility function. An arbitrary value (e.g., 0 or 1 or 100) of 1 can be assigned to the best situation, i.e., no hazard, no cost, while the worst scenarios, i.e., high hazard, high cost, are assigned a corresponding relative value of 0. The shape of the curves is determined by the relative value given each attribute. The range in utilities is the same for each attribute, and attributes should, strictly speaking, be expressed as specific functions of system characteristics.

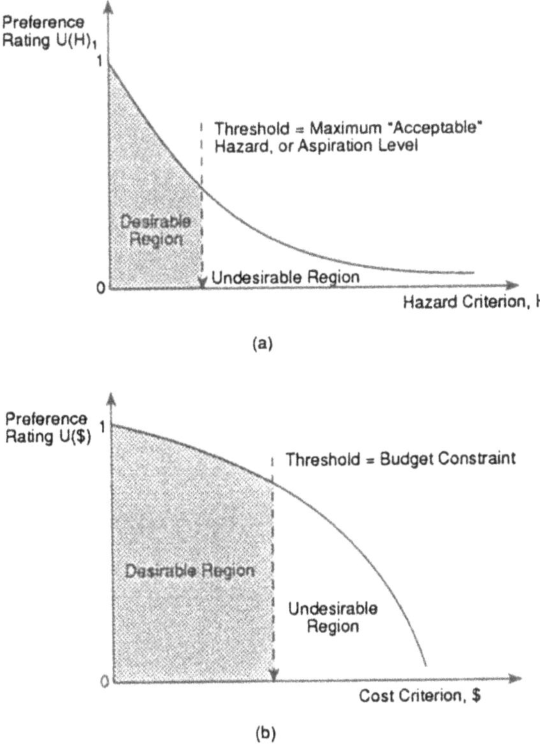

Figure 3.3 Utility functions giving the relative values of hazards and costs in similar (dimensionless) terms.

Preferences and Evaluation of Utility Functions

Preferences are directly incorporated in the utility functions by assigning to each an appropriate weighting factor. If minimizing hazards is k times as important as minimizing costs, then weighting factors of k/(k+1) and 1/(k+1) would be assigned to the hazard utility and the cost utility, respectively. These weighting factors would reflect, or give a measure of, the preferences for a given utility function.

Evaluation of utility functions becomes more difficult when the utility function represents the preferences of a particular interest group. Past decisions can help provide empirical data that can be used for quantifying the tradeoffs and therefore the k values. The given utilities are weighted by their preferences and are summed over all the objectives. For n alternatives, the value of the i-th alternative would be determined according to

$$V_i = \frac{k}{(k+1)} U(H_i) + \frac{1}{(k+1)} U(\$_i) \tag{3.5}$$

where

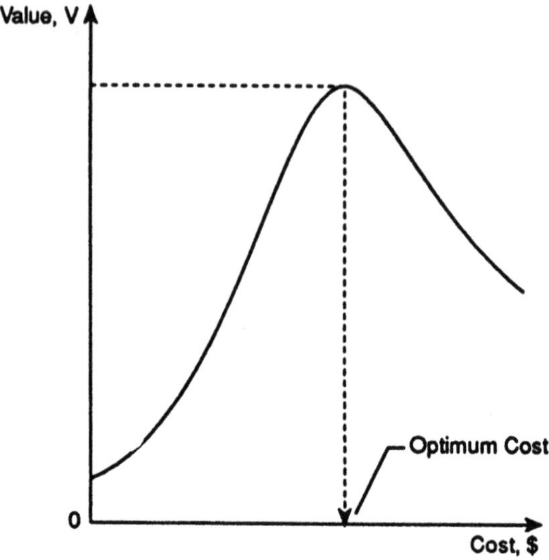

Figure 3.4 Value function for costs.

V_i = the total relative value for the ith alternative
$U(H_i)$ = the utility of hazard H for the ith alternative
$U(\$_i)$ = the utility cost, \$, associated with alternative i

In general, the largest total relative value would eventually be selected as the best alternative.

Utility Optimization

From a plot of the total relative value vs. the cost (Figure 3.4), the optimum cost can be obtained, corresponding to that giving the maximum total relative value, as the most cost-effective option for project execution. The optimum cost is equivalently obtained mathematically as follows:

$$\frac{\delta V}{(\delta \$)} = \frac{\delta}{(\delta \$)}\left[\frac{k}{(k+1)}U(H) + \frac{1}{(k+1)}U(\$)\right] = 0 \qquad (3.6)$$

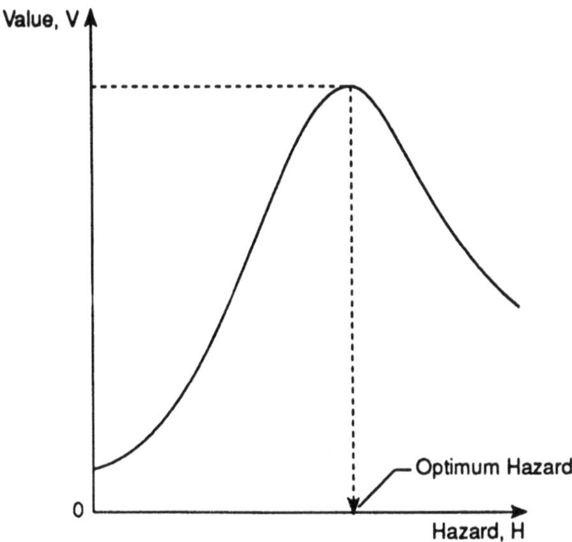

Figure 3.5 Value function for hazards.

or

$$k\frac{\delta U(H)}{(\delta \$)} = -\frac{\delta U(\$)}{(\delta \$)} \tag{3.7}$$

where

$\frac{\delta U(H)}{(\delta \$)}$ is the derivative of hazard utility relative to cost

$\frac{\delta U(\$)}{(\delta \$)}$ is the derivative of cost utility relative to cost

From this relation, ($) can be solved for to determine the optimum cost.

In a similar manner, a plot of total relative value against hazard provides a representation of the "optimum hazard" (Figure 3.5). Again, this result is evaluated in an analytical manner similar to that for cost, as

$$\frac{\delta V}{(\delta \$)} = \frac{\delta}{(\delta H)}\left[\frac{k}{(k+1)}U(H) + \frac{1}{(k+1)}U(\$)\right] = 0 \tag{3.8}$$

or

$$k\frac{\delta U(H)}{(\delta H)} = -\frac{\delta U(\$)}{(\delta H)} \tag{3.9}$$

Definitions for the derivatives are similar to those given for the cost evaluations above, except that the differentiation in Equations 3.6 and 3.7 are with respect to hazard, H. Solving for H yields the optimum value for the hazard.

Multiattribute utility theory provides the procedures for determining whether one set of impacts is more or less desirable than another set. Based on the assumption that humans are reasonable utility maximizers, the decision maker's value judgments would normally lead to the determination of satisfactory management schemes. Alternative sites for new facilities for hazardous waste treatment, storage, and disposal can be evaluated using multiattribute decision analysis methods to determine optimal solutions for corrective action alternatives. With such a formulation, an explicitly logical and justifiable solution can be assessed for the complex decisions involved in hazardous waste management problems.

3.5 UTILIZATION OF RISK-TIME CURVES

Extensive societal pressures are being exerted to lower the risks to which people are exposed during daily living, and in no area of concern are these pressures greater than in relation to hazardous waste exposures. Thus, in a problem such as the remediation of a hazardous waste site, or in the planning of a new waste site, it is inappropriate to consider only the economic costs associated with the remedial action selected or the site design costs. Instead, it is necessary to include the risk considerations, both in terms of absolute magnitude of risk and in the temporal variability of the risk. To ignore the temporal variability would be to ignore a significant part of the problem. One approach that is useful in this risk and cost tradeoff is via use of risk-time curves as described below. It is to be noted that the concept is general; however, for ease of discussion, the focus will be on the problem of selection between remediation alternatives for cleanup of a contaminated site (although any problem in which the variability of exposure risks over time is substantial could equally well be addressed by the procedure).

The use of cost-time curves in the selection between remediation alternatives has been standard practice for many years. Typically, the curves themselves have only been considered peripherally since, using a discount rate, the temporal variability of the costs is modified to a present value and/or translated into an equivalent annual cost. The equivalencing concept is utilized to allow comparisons between alternatives by removing the individual temporal variations and putting the comparisons between the alternatives onto an equal basis. Risk-time curves are of a similar nature, except that the curves indicate the temporal, changing levels in terms of risk as a function of time, for each of the remediation alternatives (McBean, 1990). The general character of the curves is best demonstrated by an example (purposely kept simple for ease of explanation). Consider the following case in which industrial wastes have been improperly disposed by placement (i.e., dumping) of hazardous material into trenches and the material covered with soil. As a result of this poor disposal practice, the surrounding groundwater is being contaminated. Remediation alternatives include the following:

1. *Do-nothing alternative* — This alternative involves no capital and operating expenditure in terms of dollars, but will result in the contamination of the local groundwater supplies. At this site, there could be an increased incidence of cancer in the local population as the chemicals spread and enter the groundwater and lakes in the immediate vicinity. The nature of the migration pathways are depicted in Figure 3.6.
2. *Excavation/incineration/relandfilling alternative* — Excavation of the refuse will result in elevated emission levels from the waste as it is uncovered, transported, stored, and handled at a staging area, and finally rehandled for treatment at an incinerator. During excavation, the potential exists for explosions to occur resulting from the mixing of incompatible wastes. In addition, incineration of the wastes and the adjacent contaminated soil removed with the wastes will result in flue gas emissions and will generate a significant quantity of fly ash and unburnable residue. Although good incineration practices will minimize any emissions from the incineration process, the incineration of thousands of tons of site material will result in release of some chemicals to the atmosphere. A schematic of the activities and the consequent migration pathways is depicted in Figure 3.7.
3. *In-place containment and groundwater pumping and treatment alternative* — Offsite contaminant migration can be prevented or minimized by using a physical barrier (a grout wall, bentonite trench, and sheet piles) and a groundwater pumping system. In this situation, there will be air emissions of volatile organics as the groundwater is treated by air stripping. In addition, there will be some groundwater contamination until the effectiveness of groundwater pumping is fully realized. A schematic depiction of the activities and the consequent migration pathways associated with this remediation alternative are depicted in Figure 3.8.

Each of the above remediation alternatives has associated capital and operating costs (measured in terms of dollars) and different elements of risk. To allow characterization of the risk-time curves, the use of pathways of migration models for contaminants must be employed. The models employed in this type of assessment must describe the mass, physical state, and the degree of containment for each chemical involved. Although such models can assume many different forms, the essential feature is that the environmental transport and fate analyses of the chemicals must be quantified as the chemicals migrate to the receptors. The mathematics for describing the pathways assessment in detail are beyond the scope to be addressed herein. Suffice it to indicate that the range of sophistication of such models is considerable. An array of mathematical models is available for the migration pathways assessment because of the variations that exist in the data available to characterize individual problems and the need for increased or

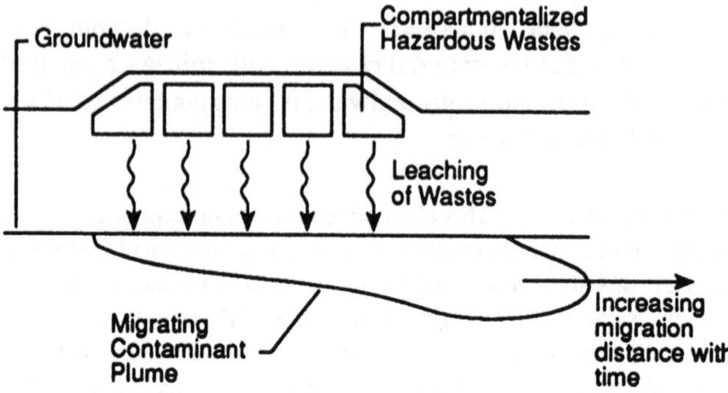

Figure 3.6 Schematic depiction of migration pathways for do-nothing alternative.

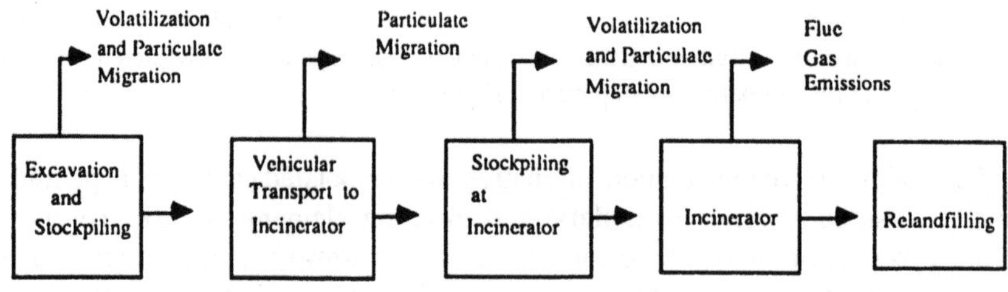

Figure 3.7 Materials handling activities and primary release mechanisms for excavation/incineration/relandfilling alternative.

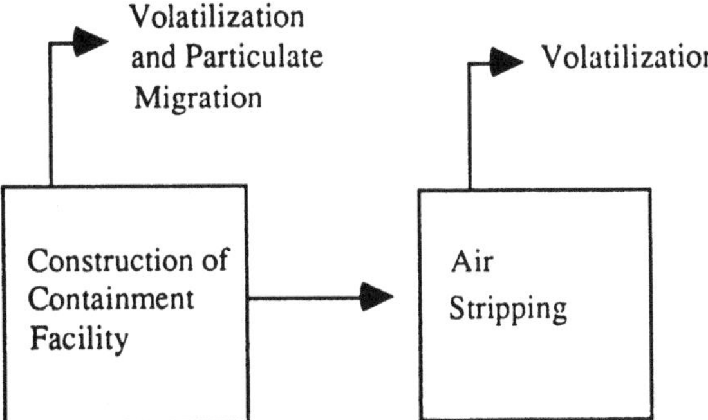

Figure 3.8 Source opportunities and primary release mechanisms for in-place containment, groundwater pumping, and treatment alternative.

decreased accuracy in characterizing the individual pathways of migration. The interested reader is referred to such books as McTernan and Kaplan (1990) for additional discussion of mathematical models of migration pathways.

The first task requires quantification of the releases of the various chemicals as a function of time for the different remediation alternatives. The second task utilizes the environmental pathways models to characterize the transfer of chemicals from the site to the receptors, thus determining the concentrations of the individual chemicals at the various receptors. These determinations must reflect the temporal variations associated with the release mechanisms, since not all of the various release mechanisms are functioning at all times. The aggregation must then take place over the array of chemicals, from all the exposure pathways, resulting in risk-time curves depicted in Figure 3.9, for the various alternatives for the remediation namely (1) do-nothing alternative, (2) excavation/incineration/relandfilling alternative, and (3) in-place containment and groundwater pumping alternative. Brief comments to the individual risk-time curves for the various remediation alternatives follows.

1. *Do-nothing alternative* — As the groundwater plume continues to migrate away form the buried refuse, the exposure risk continues to escalate as additional sources of water supply for various residents are successively contaminated.
2. *Excavation/incineration/relandfilling alternative* — This alternative has the potential to substantially reduce public risk in future generations, but increases the health risk in the short-term (over the period of 10 years of construction and implementation). In this case, mean health risk reduction to future generations will be large, but there is a large risk of transfer to the nearby residents during the 10 year period of excavation/ incineration and relandfilling activities during the excavation due to such features as the increased opportunity for volatilization, the risk of explosions during excavation, and the air emissions during incineration.
3. *In-place containment and groundwater pumping* — The health risk afforded by the proposed pump and treat alternative is elevated over the "do-nothing" alternative for the first few years due to the construction activity and the treatment emissions (volatilization) due to the air stripping and carbon adsorption processes for treatment of the pumped groundwater. The treatment emissions will continue throughout the site-life remediation, projected as 100 years.

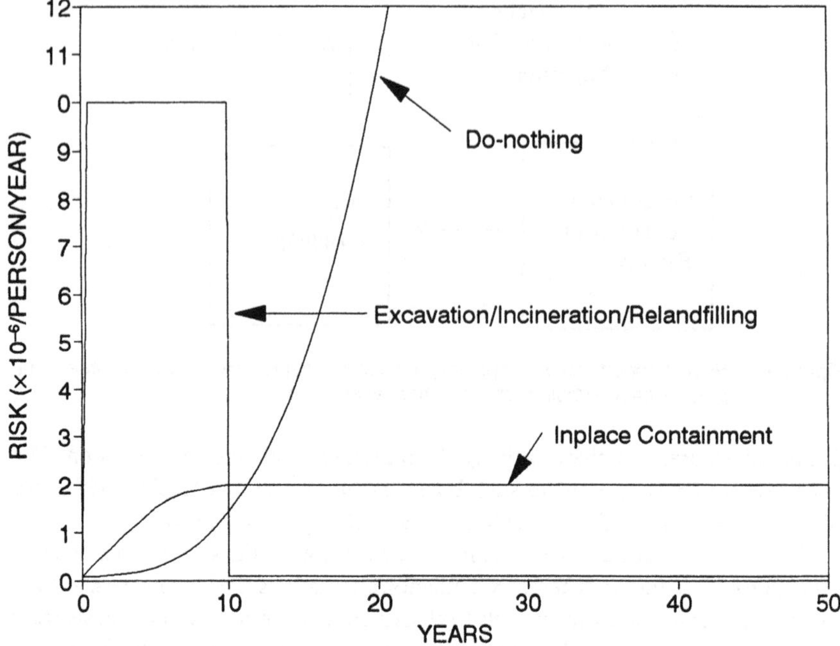

Figure 3.9 Risk-time curves for different remediation alternatives.

A significant benefit of the risk-time curves is the immediate indication by each alternative of any periods of elevated risk. It is readily apparent, for example, that the levels of risk for the in-place containment/groundwater pumping alternative does not reach the same magnitude as the excavation/incineration/relandfilling alternative, but of course, the risks associated with the former continue over a much longer time frame (Figure 3.9).

3.5.1. Equivalent-Value Assignments

Standard practice in dealing with cost-time curves utilizes the discounting of the various cost-time components back to present value, using a discount rate. A similar discounting procedure is not normal practice in risk-time curves. Instead, the normal practice usually equivalences the risk to a lifetime equivalent using no discounting. Therefore, the next task involves integration of the risk curves over the population at risk and equivalencing back to present value.

Given the information on cost- and risk-time curves, the resulting values for the three remediation alternatives, plotted as "expected cost vs expected risk," are illustrated in Figure 3.10. Acceptance of the alternative with the lower expected value of risk would imply that the excavation/incineration/relandfilling alternative would be the most desirable alternative — it has a higher cost than the in-place containment, but it has one third of the risk. On

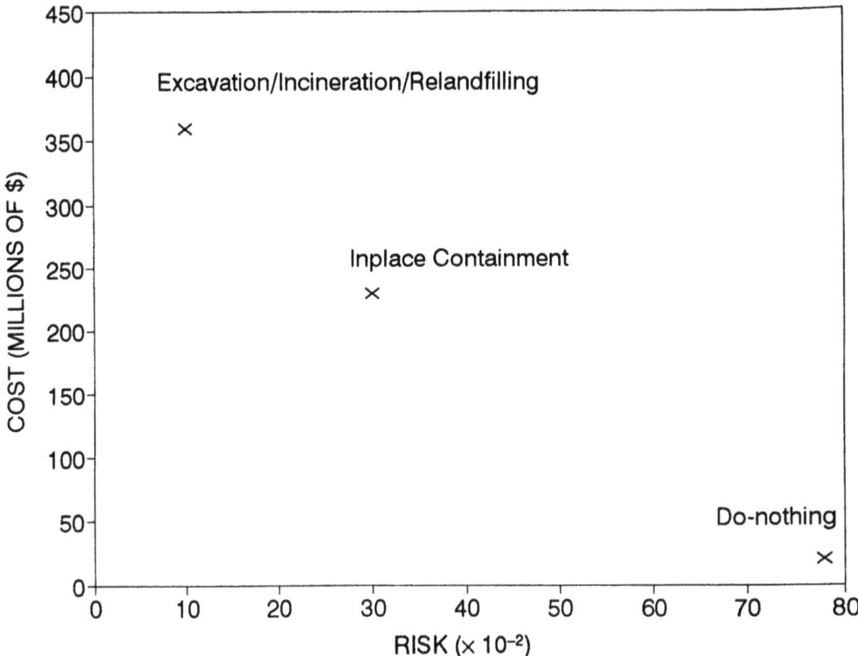

Figure 3.10 Societal risk-cost tradeoff for expected values.

the other hand, as indicated at numerous points within this text, there is substantial uncertainty associated with many data assignments in risk assessment. This is a relevant consideration in problems such as this, since the waste disposal records for old landfill sites are typically quite poor. Thus, for example, knowledge of the quantities of various chemicals buried, the specific locations relevant to the concern of mixing incompatible wastes during excavation and the resulting potential for an explosion occurring, and the ability to assign representative volatilization rates for the chemicals during stockpiling, etc. all indicate that the uncertainty in the risk-time curves may be very relevant in terms of decisionmaking. In other words, there is a range of risks reflecting the uncertainties in the data assignments utilized in generating the risk-time curves. Selection between remedial alternatives may not be adequate based on expected values alone — the decision maker may want to adopt a risk-averse approach.

To reflect the uncertainty of the risk, the plot of societal risk vs. time curve becomes as depicted in Figure 3.11; the solid lines depict the expected values of the risk-time, and the associated dashed lines indicate the uncertainty as quantified by one standard deviation away from the mean. In turn, using the risk-time curves (in the plural sense, indicating the expected value and the uncertainty aspects) and including the expected value of the costs and a measure of the uncertainty of this component, the resulting cost vs. risk tradeoffs become as depicted in Figure 3.12. The solid lines in Figure 3.12 surrounding the expected

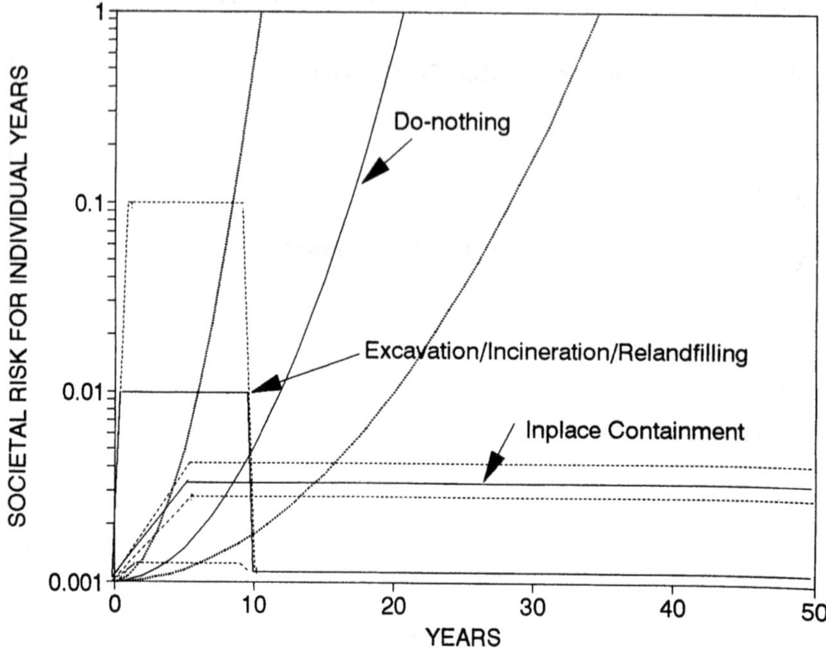

Figure 3.11 Log risk-time curves for society with different remediation alternatives and showing uncertainty.

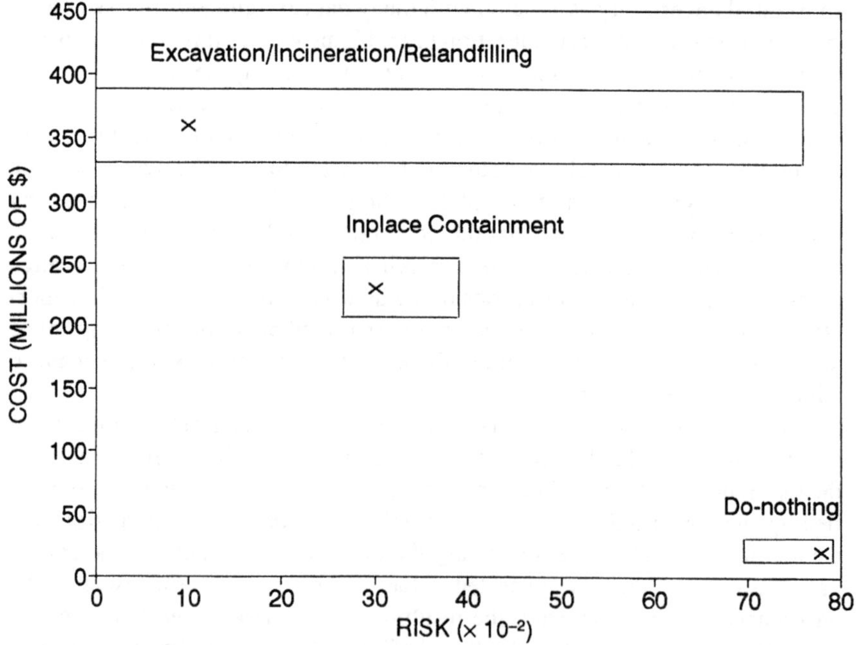

Figure 3.12 Societal risk-cost tradeoff reflecting uncertainty.

values indicate one standard deviation from the mean, for both the costs and the risk.

Several points are noteworthy. The cost uncertainty is of a smaller magnitude than the risk uncertainty. The elements of cost uncertainty relate to the projection of the future technological costs (e.g., operational costs of the treatment facility in 20 years). There is uncertainty associated with such projections, but the magnitudes are expected to be relatively small. On the other hand, the uncertainties associated with risk are substantial, as noted above. The cost uncertainty is depicted as Gaussian or normally distributed, whereas the risk uncertainty is log-Gaussian or log-normal. This is to be expected since the projections of the future costs could just as easily be on the high side as the low side. Conversely, the distribution of the risk is log-normal since it is bounded on the low side by zero and unbounded on the high side; the distribution is skewed to the right. Note that both components of uncertainty (in costs and in risks) appear symmetrical around the expected value in Figure 3.11, but the characterization of the risk is by a logarithmic axis.

As a result of including the uncertainty aspects of the risk, the selection between the alternatives is not so obvious. The excavation/incineration/relandfilling alternative has the lower expected value of risk, but also possesses the very real possibility (probability) that the risk impact is larger. In other words, the excavation/incineration/relandfilling alternative has the lower expected risk, but has a much greater probability that risks associated with this alternative are much higher. Consequently, the greater costs and the potential for greater risks might well argue for selection of the in-place containment/groundwater pumping alternative as opposed to the excavation/incineration/relandfilling alternative.

3.6 REFERENCES

Ashby, E. "What Price the Furbish Lousewort?" in Proc. 4th Conf. on Environmental Engineering Education, Toronto, (1980).

Bean, M.C. "Speaking of Risk," *ASCE Civil Eng.* 58(2):59–61 (1988).

Crouch, E. A. C. and R. Wilson. *Risk/Benefit Analysis* (Boston: Ballinger, 1982).

Daniels, S. L. "Environmental Evaluation and Regulatory Assessment of Industrial Chemicals," in 51st Annu. Conf. Water Pollution Control Federation, Anaheim, CA (1978).

Keeney, R. L. "Ethics, Decision Analysis, and Public Risk," *Risk Anal.* 4:117–129 (1984).

Keeney, R. D. and H. Raiffa. *Decisions with Multiple Objectives: Preferences and Value Tradeoffs* (New York: John Wiley & Sons, 1976).

Lifson, M. W. *Decision and Risk Analysis for Practicing Engineers* (Boston: Barnes and Noble, Cahners Books, 1972).

Massmann, J. and R. A. Freeze. "Groundwater Contamination from Waste Management Sites: The Interaction Between Risk-Based Engineering Design and Regulatory Policy. 1. Methodology. 2. Results, "*Water Resour. Res.* 23(2): 351–380 (1987).

McBean, E. "Utility of Risk-Time Curves in Selecting Remediation Alternatives at Hazardous Waste Sites," presented at 1990 Pacific Basin Conference on Hazardous Waste, Pacific Basin Consortium for Hazardous Waste Research, Honolulu, Hawaii, November 1990.

McTernan, W. and E. Kaplan. *Risk Assessment for Groundwater Pollution Control,* ASCE Monograph (New York: American Society of Chemical Engineers, 1990).

Paustenbach, D. J., Ed. *The Risk Assessment of Environmental Hazards: A Textbook of Case Studies* (New York: John Wiley & Sons, 1988).

Rowe, W. D. *Evaluation Methods for Environmental Standards* (Boca Raton, FL: CRC Press, 1983).

Starr C., and C. Whipple. "Risks of Risk Decisions," *Science* 208:1114 (1980).

U.S. EPA. Exposure Factors Handbook, EPA/600/8-89/043, U.S. Environmental Protection Agency, Washington, DC (1989).

Whipple, C. *De Minimis Risk. Contemporary Issues in Risk Analysis,* Vol. 2 (New York: Plenum Press, 1987).

CHAPTER 4

The Risk Assessment Process

The risk assessment process will generally help identify and quantify risks imposed on an individual (i.e., individual lifetime risk), the general public (i.e., population or societal risk), and/or the environment (i.e., ecological stress). Although specific forms of risk assessment may differ considerably in their detail, they share the same general logic. Whereas several schemes have been developed and/or proposed, four basic elements in a risk assessment effort are generally found in the risk analysis literature (Figure 4.1). These elements are briefly discussed below; further elaboration of these elements are given in the risk analysis literature (e.g., Rowe, 1977; NRC, 1983; U.S. EPA, 1984; Bowles et al., 1987; Hallenbeck and Cunningham, 1988; Paustenbach, 1988; U.S. EPA, 1989a; U.S. EPA, 1989b; Asante-Duah, 1990; CAPCOA, 1990; Huckle, 1991). A good risk assessment framework should consider all reasonably probable loading conditions and/or events and exposure scenarios that may be potential hazard sources. Such a procedure will help to establish the existence of risk, aid in determining the magnitude and range of the risk present, and also assist in determining risk reduction policies appropriate for the specific problem.

4.1 HAZARD IDENTIFICATION AND ACCOUNTING

Hazard identification consists of evaluating the potential adverse effects of hazardous situations to which some populations are potentially exposed. It is comprised of the *identification* and *accounting* of the initiating events with hazard-causing potentials. In a risk assessment for hazardous waste sites, this may consist of the identification of contaminant sources and the selection of the chemicals of potential concern.

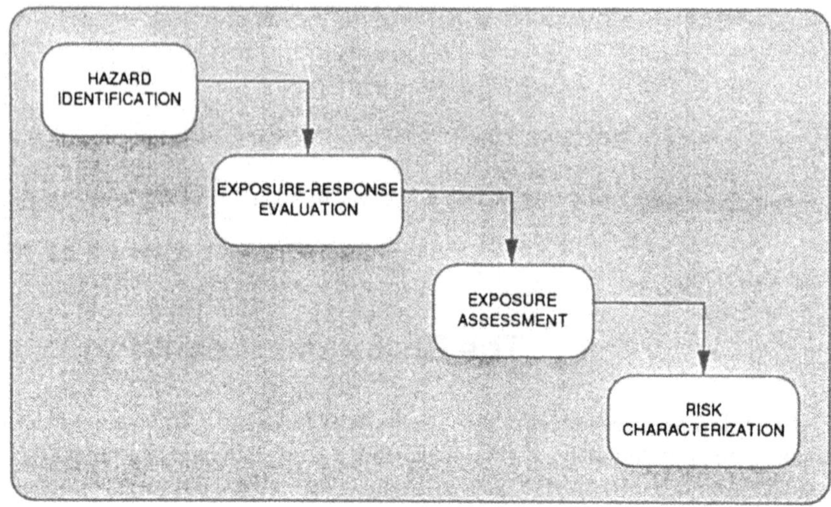

Figure 4.1 The fundamental elements of risk assessment: process model and components.

4.2 EXPOSURE-RESPONSE EVALUATION

Exposure-response evaluation involves analyzing how any adverse impacts on populations potentially at risk will change in frequency of occurrence and intensity, due to changes in levels of receptor exposures. It also consists of the determination of the system response that follows the occurrence of an initiating event. This may include a toxicity assessment and/or dose-response evaluation in a health risk assessment undertaken for a hazardous waste site problem.

4.3 EXPOSURE ASSESSMENT

Exposure assessment consists of developing realistic exposure scenarios for the specific problem. An exposure pathway analysis examines the ways by which a receptor is exposed to, or impacted by, a source of hazard. Pathways may be direct and immediate or more complex and delayed. The assessments commonly use mathematical models. For example, exposure modeling helps to relate concentrations of chemicals in the environment to exposed populations. Either of equilibrium (steady-state) or dynamic models may be used. The exposure scenario associated with a given hazardous situation may be better defined if the exposure has already occurred. In most cases, however, a risk assessment is undertaken to evaluate potential risks due to exposures that have not occurred yet, in which case hypothetical exposure scenarios are developed for this purpose. For a health risk assessment carried out for a hazardous waste program, one needs to estimate the frequency and magnitude of human exposure to the chemicals of potential concern.

If numerous potential exposure scenarios exist, or if a complex exposure scenario has to be evaluated, it may help to set up an event tree structure to cover potential outcomes and/or consequences. The event tree is a diagrammatic representation of events and consequences that can occur under various exposure scenarios. Probability values can be assigned to various branches in the tree to determine the most likely outcomes. By using such a diagram, the various exposure contingencies can be identified and organized in a systematic manner. In this case, priorities can be established for conducting the risk assessment.

4.4 RISK CHARACTERIZATION AND CONSEQUENCE DETERMINATION

Risk characterization is the final step in the risk assessment process and the first input to the risk management process. It involves the integration of the exposure-response investigation and the exposure assessment to arrive at an estimate of risk to the exposed population. Its purpose is to present the risk manager with a synopsis and synthesis of all the data that can help the manager decide what to do.

Risk characterization is the process of estimating the probable incidence of adverse impacts to potential receptors under various exposure conditions, including an elaboration of uncertainties associated with such estimates. Consequences from exposure of any receptors to hazards are determined during this stage. In a health and environmental risk assessment for hazardous waste management planning, this will consist of determining the existence, and then the probability of health and environmental effects, that may occur in potential receptors exposed at contaminant levels estimated during the exposure assessment; in this case, the process combines exposure assessment results with toxicity assessment results to estimate risks involved in a given situation.

Risk estimates are developed based on the exposure of the risk group or populations at risk. Risk characterization involves the evaluation of the significance of potential risks posed by a site and/or facility. Uncertainties and the main assumptions used to complete the whole risk assessment process are also evaluated during the risk characterization stage. Risk characterization should discuss all specific sources of uncertainty, including a consideration of background levels of exposure.

4.5 LEVELS OF DETAIL FOR RISK ASSESSMENT APPLICATIONS

Risk assessment is generally conducted to aid risk management decisions, usually aimed at programs for minimizing health and environmental risks to society. Depending on the problem situation, different degrees of detail may be required. However, the continuum of acute to chronic hazards and exposures should be fully investigated in a comprehensive assessment so that a complete spectrum of risks can be defined for subsequent risk management decisions. Irrespective of the level of detail, a well-defined protocol should be used to assess imminent risks as part of an overall policy decision on risk mitigation measures needed by society; (Figure 4.2) shows the important elements and process flow that will aid manage-

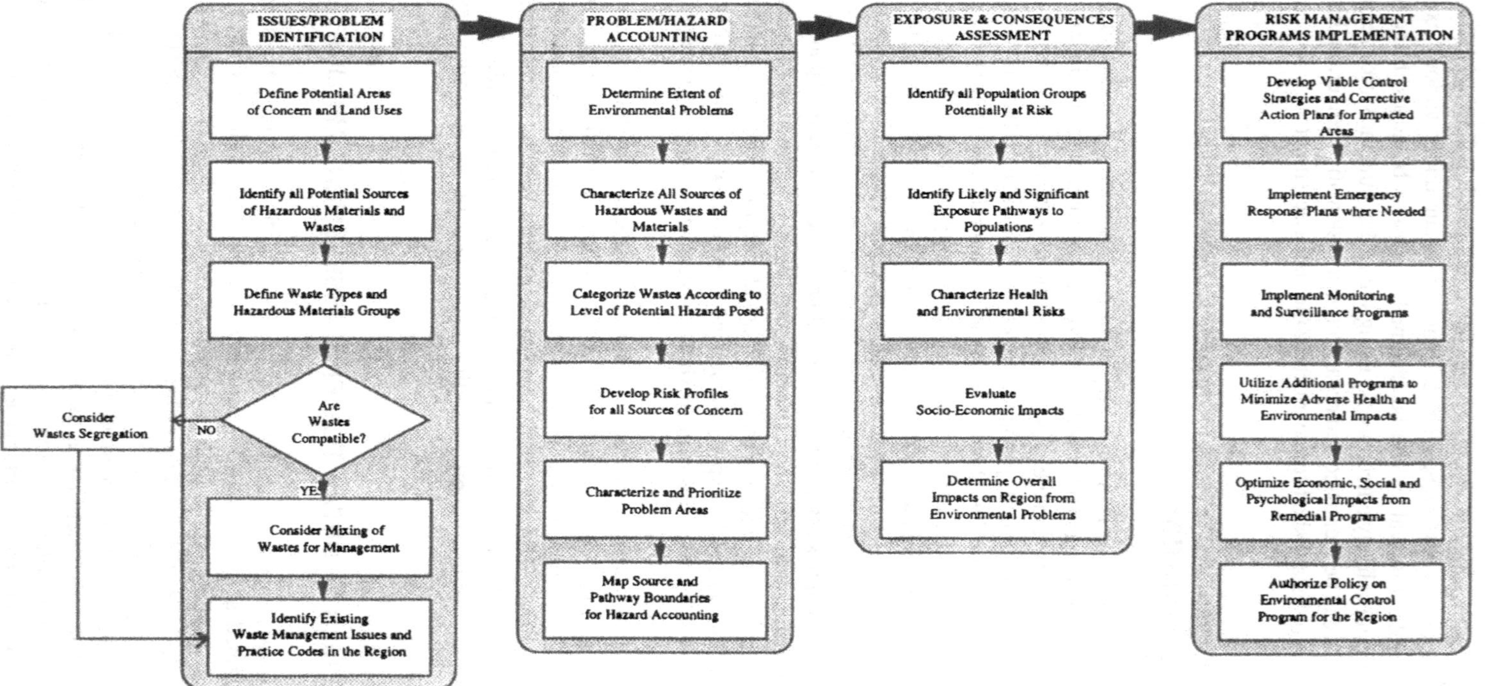

Figure 4.2 Protocol for developing environmental control programs in hazardous waste management.

ment decisions. In fact, there is considerable merit in utilizing an array of procedures in a risk assessment — simpler procedures being used to screen out those problems requiring further analyses. The decision on the level (i.e., qualitative, quantitative, or combinations thereof) at which an analysis is carried out will usually be based on the complexity of the situation and the level of risk involved, anticipated, or predicted.

4.6 REFERENCES

Asante-Duah, D. K. "Quantitative Risk Assessment as a Decision Tool for Hazardous Waste Management," in *Proc. 44th Purdue Industrial Waste Conf. (May 1989)*, (Chelsea, MI: Lewis Publishers, 1990), pp. 111–123.

Bowles, D. S., L. R. Anderson, and T. F. Glover. "Design Level Risk Assessment for Dams," *Proc. Struct. Congr. ASCE*: 210–225 (1987).

CAPCOA (California Air Pollution Control Officers Association). "Air Toxics 'Hot Spots' Program. Risk Assessment Guidelines," California Air Pollution Control Officers Association, 1990.

Hallenbeck, W. H. and Cunningham, K. M. *Quantitative Risk Assessment for Environmental and Occupational Health* (Chelsea, MI: Lewis Publishers, 1988).

Huckle, K. R. *Risk Assessment—Regulatory Need or Nightmare. Selected Papers* (London: Shell Centre, 1991).

NRC (National Research Council). *Risk Assessment in the Federal Government: Managing the Process* (Washington, DC: National Academy Press, 1983).

Paustenbach, D. J., Ed. *The Risk Assessment of Environmental Hazards: A Textbook of Case Studies* (New York: John Wiley & Sons, 1988).

Rowe, W. D. *An Anatomy of Risk* (New York: John Wiley & Sons, 1977).

U.S. EPA. "Risk Assessment and Management: Framework for Decision Making," EPA 600/9-85-002, Washington, DC (1984).

U.S. EPA. "*Risk Assessment Guidance for Superfund. Volume I: Human Health Evaluation Manual (Part A)*," EPA/540/1-89/002, Office of Emergency and Remedial Response, Washington, DC (1989a).

U.S. EPA. "*Risk Assessment Guidance for Superfund. Volume II: Environmental Evaluation Manual*," EPA/540/1-89/001, Office of Emergency and Remedial Response, Washington, DC (1989b).

CHAPTER 5

Risk Assessment Techniques and Methods of Approach

Risk assessment is a process that seeks to estimate the likelihood of occurrence of adverse effects as a result of exposures to chemical, physical, and/or biological agents in humans and ecological receptors within an ecosystem. A number of techniques are available for performing risk assessments; most of the techniques are structured around decision analysis procedures to facilitate comprehensible solutions for even complicated problems.

Risk assessment can be used both to provide a baseline estimate of existing risks attributable to an agent or hazard and to determine the potential reduction in exposure and risk given various corrective actions. Potential risks are estimated by determining the probability or likelihood of occurrence of harm, the intrinsic harmful features or properties of specified hazards, the population at risk (PAR), the exposure scenarios, and the extent of expected harm and potential effects. A number of the risk assessment approaches commonly encountered in the literature of hazardous waste management are elaborated further in this chapter.

5.1 THE HEALTH RISK ASSESSMENT METHODOLOGY

Human health risk is the likelihood or probability that a given chemical exposure or series of exposures will damage the health of exposed individuals. Health risk assessment is defined as the characterization of the potential adverse health effects of human exposures to environmental hazards (NRC, 1983). In this process, the extent to which potential receptors have been or could be exposed to selected chemical(s) is determined; the extent of exposure is then considered in relation to the type and degree of hazard posed by the chemical(s), thereby permitting an estimate to be made of the present or potential health risk to the PAR. For hazardous

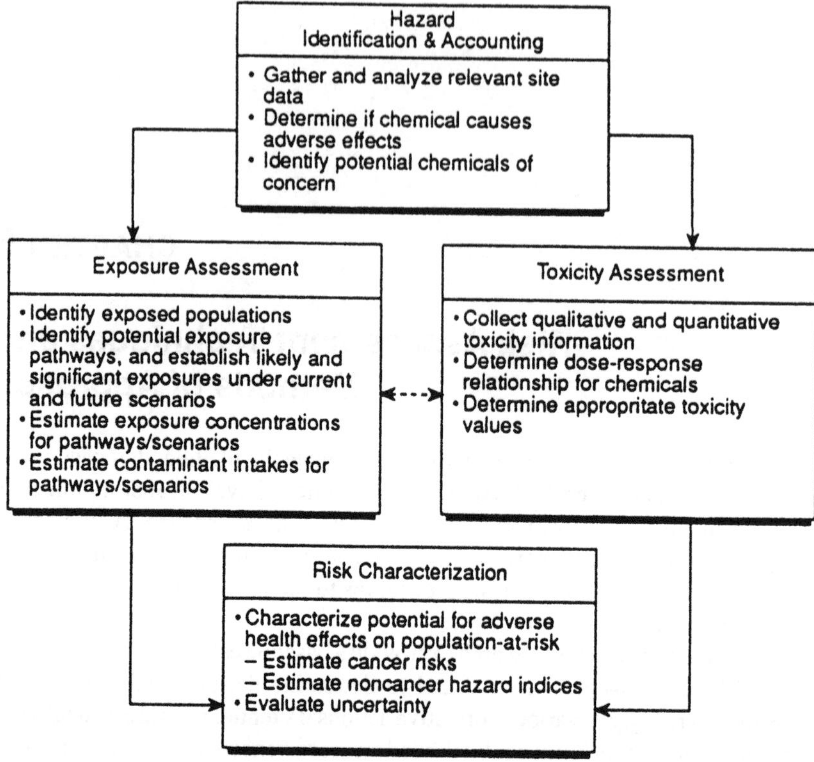

Figure 5.1 Components of the human health and chemical risk assessment process. (Adapted from U.S. EPA, 1989i.)

waste sites, a procedure generally followed for conducting a risk assessment consists of the following elements:

- Definition of the sources of contaminants
- Definition of the contaminant exposure pathways
- Identification of populations potentially at risk from contaminants
- Identification of acceptable exposure levels of contaminants
- Determination of frequency of exposure to potential receptors
- Determination of impacts or damage due to presence and exposures to chemicals

Figure 5.1 shows the basic components and steps used in carrying out a comprehensive health risk assessment. Risk assessment may be performed in response to either short-term (acute) exposures to toxic chemicals, long-term (chronic) exposures, or to combinations of these. It generally requires some level of effort in mathematical modeling, especially with respect to exposure point concentration estimation. Many sources of uncertainty surround the risk assessment process, especially because of possible incomplete exposure assessments, limited and questionable monitoring information, limitations on dose-response assessments, and/or the absence of complete toxicological profiles on some chemicals involved in the assessment.

5.1.1 Hazard Identification and Accounting

Hazard identification is the qualitative evaluation of the potential adverse health impacts of a chemical on potential receptors. The process involves identifying the chemicals of potential concern as well as the specific hazardous properties (such as persistence, bioaccumulative potentials, toxicity, and general fate and transport properties) exhibited. This primary stage of the risk assessment includes a compilation of the lists of all contaminants present at the site, the identification and selection of the chemicals of potential concern for the site, and the compilation of summary statistics for the key constituents selected for further investigation and evaluation.

Data Collection

Contaminants released to the environment are controlled by a complex set of processes including various forms of transport (e.g., intermedia transfers), transformation (e.g., biodegradation), and biological uptake (e.g., bioaccumulation or bioconcentration). Potential primary sources of contaminant release to the various environmental media include the following:

- Atmospheric contamination that may be the results of emissions of contaminated fugitive dusts and volatilization of chemicals
- Surface water contamination, resulting from contaminated runoff and overland flow of chemicals (from leaks, spills, etc.) and chemicals adsorbed to mobile sediments
- Groundwater contamination, as a result of the leaching of toxic chemicals from contaminated soils or vertical migration of chemicals from lagoons and ponds
- Soil contamination, whereby sources of surface soil contamination include intentional placement of wastes on or in the ground, or as a result of spills, lagoon failure, or contaminated runoff; chemicals can also be leached from surface soils to subsurface layers

Secondary sources will include those present in aquatic and terrestrial organisms and biota, due to earlier uptake from the primary sources. Table 5.1 summarizes the important source and target or receiving media associated with typical hazardous waste problems. All the source and impacted media should be thoroughly investigated. In fact, an adequate characterization of the site through implementation of a substantive data collection program is vital for an effective risk assessment; the types of site data and information required include (U.S. EPA, 1989i):

- Contaminant identities
- Contaminant concentrations in the key sources and media of interest
- Characteristics of sources, especially information related to release potential
- Characteristics of the physical and environmental setting that can affect the fate, transport, and persistence of the contaminants

In a typical scenario in which there is a release, contaminants may be transported via one or more media (including air, soils/sediments, surface water, and ground-

Table 5.1 Potential Release Mechanisms for Various Waste Sources and Target Media

Primary Waste and Contaminant Source	Typical Release Causes and Mechanisms	Primary Receiving or Impacted Media
Surface Impoundments (e.g., lagoons, ponds, pits)	Loading/unloading activities	Air
	Overlapping dikes and surface runoff	Soils and Sediments
	Seepage and infiltration/percolation	Surface water
	Fugitive dust generation	Groundwater
	Volatilization	
Waste management units (e.g., landfill, land treatment unit, and waste pile)	Migration of releases outside unit's runoff collection and containment system	Air Soils and sediments
	Migration of releases outside the containment area from loading and unloading operations	Surface water Groundwater
	Seepage and infiltration	Subsurface gas (in soil pores, vents, and cracks migrating through soil)
	Leachate migration	
	Fugitive dust generation	
	Volatilization	
Waste management zones (e.g., container storage area and storage tanks)	Migration of runoff outside containment area	Air Soils and sediments
	Loading/unloading areas spills	Surface water
	Leaking drums, leaks through tank shells, and leakage from cracked or corroded tanks	Groundwater
	Release from overflows	Subsurface gas (in soil pores, vents, and cracks migrating through soil)
	Leakage from coupling/uncoupling operations	
Waste treatment plants/ facilities	Effluent discharge to surface and groundwater resources	Surface water (by dissolution, dispersion, transport, etc.)
		Sediments (from adsorbed chemicals)
		Groundwater
Incinerator	Routine releases from waste handling/preparation activities	Air
	Leakage due to mechanical failure	Foliage (from particulate deposition and atmospheric fallout)

Table 5.1 (continued)

Primary Waste and Contaminant Source	Typical Release Causes and Mechanisms	Primary Receiving or Impacted Media
Incinerator	Stack emissions	Soils (from particulate deposition and atmospheric washout)
		Surface water (from particulate deposition and atmospheric washout)
Injection wells	Leakage from waste handling operations at the well head	Groundwater (by dissolution, diffusion, dispersion, etc.)
		Surface water (from groundwater recharge)

water) to potential receptors (through inhalation, dermal contact, ingestion, and/or via the food chain). Samples should be gathered and analyzed for the chemicals of concern in the appropriate media of interest. In addition, background samples are collected and evaluated to determine the possibility of a potentially contaminated site contributing to offsite contamination levels in the vicinity of the site. For instance, background soil samples may be collected near a site in areas upwind of the site; the upwind locations are determined from the wind rose (i.e., a diagram or pictorial representation that summarizes pertinent statistical information about wind speed and direction at a specified location) for the geographical region in which the site is located. These background samples would not have been significantly influenced by the subject site contamination. However, these soil samples must be obtained from media that have the same basic characteristics as the soil media at the subject site, to provide a justifiable basis for comparisons. In general, background sampling is conducted to distinguish site-related contamination from naturally occurring or other non-site-related levels of the chemicals of concern. More detailed sampling considerations and strategies for the media of concern can be found in the literature (e.g., U.S. EPA, 1988e, 1989k).

Data Evaluation

Site data relevant to the human health risk assessment are gathered and analyzed to help identify the chemicals present at a site that are to be the focus of the risk assessment process. Effective analytical protocols in the sampling and laboratory procedures are required to help minimize uncertainties associated with the data collection and evaluation aspects of the risk assessment. The applicable analytical procedures, the details of which are outside the scope of this book, should be strictly adhered to. U.S. EPA (1989i) discusses data evaluation requirements for organizing data in a suitable and usable format for a risk assessment. Anthropogenic levels, rather than naturally occurring levels, are used as a basis for evaluating background sampling data; anthropogenic levels are concentrations of chemicals that are present in the environment due to human-made, nonsite sources, such as industry and automobiles. In the context of background contamination at hazardous waste sites, the null hypothesis that there is no difference between contaminant concentrations in the background areas and the on-site chemical concentrations is tested vs. the alternative hypothesis that concentrations are higher onsite (implying a one-tailed test of significance). Broadly speaking, however, a statistically significant difference between background samples and site-related contamination should not, by itself, trigger a cleanup action; the completion of the risk assessment will ascertain the significance of the contamination based on the toxicological information integrated with exposure scenarios.

Chemicals are screened based on such parameters as toxicity, carcinogenic classifications, concentrations of detected chemicals, etc. For any site/media, contaminant concentrations are likely to vary widely across the site/media. Values used in exposure assessments and risk characterization should reflect the average concentration to which receptors potentially at risk are exposed. To represent the natural variability and heterogeneous nature for such parameters, it often is convenient to use probability distributions for the media contamination. The probability distribution can also be stated in the form of a discrete uncertainty node. For instance, consider a representation of three possible values of soil concentrations for which there are 40, 50, and 10% chances that the average soil concentrations are 100 (low), 500 (moderate), and 1000 ppm (high), respectively. This may be viewed also with respect to spatial distribution of contamination for the whole contaminated site in which 40, 50, and 10% of the site are at low, moderate, and high concentration levels. In most analyses, the nominal or average value will be used in the evaluation. However, for a more complete and comprehensive analysis, the uncertainties in the average and other pertinent variables should be explicitly defined and analyzed. Similarly, contamination levels and exposures may have temporal variations and the dynamic nature of such parameters should be incorporated in a full analysis.

Statistical procedures used for data evaluation can significantly affect the final results of the risk assessment. Over the years, there has been extensive technical literature developed regarding the "best" probability distribution to utilize in different applications. Of the many available, the Gaussian or normal distribution has been extensively utilized to describe environmental data. However, there is considerable support for the use of the lognormal distribution in describing chemical concentration data. Consequently, chemical concentration data in air, water, and soil have been described by the log-normal distribution, rather than being defined by a normal (Gaussian) distribution (Leidel and Busch, 1985; Gilbert, 1987; Rappaport and Selvin, 1987). The use of lognormal statistics for the data set $X_1, X_2, X_3, ..., X_n$ requires that the logarithmic transform of these data, $\ln(X_1), \ln(X_2), \ln(X_3), ..., \ln(X_n)$, can be expected to be normally distributed. The logarithmic transform acts to suppress outliers so that the mean is a much better representation of the central tendency of the sample distribution. In fact, the use of a normal distribution (the central tendency of which is measured by the arithmetic mean) to describe environmental contaminant distribution, rather than lognormal statistics (the central tendency of which is defined

Table 5.2 Example Statistical Analysis of Environmental Data

	Concentration Values (μg/L)	
Month/Parameter	**Original "Raw" Data, X**	**Log-Transformed Data Y = ln(X)**
January	0.049	–3.016
February	0.056	–2.882
March	0.085	–2.465
April	1.200	0.182
May	0.810	–0.211
June	0.056	–2.882
July	0.049	–3.016
August	0.048	–3.037
September	0.062	–2.781
October	0.039	–3.244
November	0.045	–3.101
December	0.056	–2.882
Mean	$X_{a\text{-mean}} = 0.213$	-2.445 ($X_{g\text{-mean}} = 0.087$)[a]
Standard Deviation	SD = 0.379	SD = 1.154
95% confidence interval	0.213 ± 0.834 (CL_x is –0.621 to 1.047 μg/L)[b]	-2.445 ± 2.539 (CL_x is 0.0068 to 1.099 μg/L)[c]

[a] Transforming the average of the Y values back to arithmetic values yields the geometric mean value of $X_{g\text{-mean}} = e^{-2.445} = 0.087$. It is recognized that the arithmetic mean of 0.213 mg/L is substantially larger than the geometric mean of 0.087 mg/L. The two large sample values in the data set very strongly bias the arithmetic mean; the logarithmic transform acts to suppress the extreme values.

[b] The development of a 95% confidence limit (refer to standard statistics texts for details of procedures) for the untransformed data gives a confidence interval of $0.213 \pm 0.379t = 0.213 \pm 0.834$ (where $t = 2.20$ obtained from the student t – distribution for $n = 12 - 1 = 11$ degrees of freedom). This indicates a nonzero probability of a negative concentration, making it nonsensical.

[c] The development of a 95% confidence limit (refer to standard statistics texts for details of procedures) for the log–transformed data gives a confidence interval of $-2.445 \pm 1.154t = -2.445 \pm 2.539$ (where $t = 2.20$ obtained from the student t–distribution for $n = 12 - 1 = 11$ degrees of freedom). Transforming these values back to the arithmetic realm gives the 95% confidence interval from 0.0068 to 1.099 mg/L, which makes more sense than provided under (b) above.

by the geometric mean) will often result in significant overestimation and may be overly conservative. As an example, consider the case for the estimation of the mean, standard deviation, and confidence interval of monthly groundwater sample concentration data as listed in Table 5.2.

The results from this example analysis illustrate the potential effects from the choice of distribution types for a given evaluation or investigation. As a result, statistical analysis procedures should be used in data evaluation that reflect the character of the distribution. The appropriateness of any distribution assumed to fit a given data set should always be checked; this is accomplished by using such procedures as the chi-square test for goodness-of-fit, as described in standard textbooks of statistics (e.g., Wonnacott and Wonnacott, 1972; Miller and Freund, 1985; Freund and Walpole, 1987).

In general, the use of arithmetic or geometric averages for estimating average concentrations tends to bias the averages. Furthermore, sparse sampling data usually are all that there is from previous site investigations; these tend to be highly variable, and arithmetic or geometric averaging would not produce representative concentration estimates. Geostatistical techniques that account for spatial variations in concentrations may be employed for estimating the average concentrations at a site required for the long-term exposure assessment (e.g., Zirschy and Harris, 1986; U.S. EPA, 1988). A technique called block kriging is frequently used to estimate soil chemical concentrations in sections of a hazardous waste site in which sparse sampling data do exist. The site is divided into blocks (or grids), and concentrations are determined within blocks by using interpolation procedures that incorporate sampling data in the vicinity of the block. The sampling data are weighted in proportion to the distance of the sampling location from the block. Also, weighted moving-average estimation techniques based on geostatistics are applicable for estimating mean contamination present at a site. Because of the uncertainty associated with any estimate of exposure concentration, the upper confidence limit (i.e., the 95% upper confidence limit) on the average is frequently used in evaluations.

Treatment of Sample Non-Detects

All laboratory analytical techniques have detection limits (DLs) below which only "less than" values may be reported; the reporting of such values provides a degree of quantification. This is important because even at or near their detection limits, the concentration levels of some particular contaminants may be of considerable importance in a risk assessment. However, uncertainty about the actual values below the DL can bias or preclude subsequent statistical analyses. One approach in the calculation of applicable statistical values in a data evaluation involves the use of a value of one half of the sample quantitation limit (SQL) (or simply called the detection limit, DL). Half the DL is usually assumed (as a proxy or estimated concentration) for nondetectable (ND) levels (instead of assuming a value of zero or neglecting such values), provided there was at least one detected value from the analytical results (and/or if there is reason to believe that the chemical is possibly present in the sample at a concentration below the SQL). This method conservatively assumes that some level of the chemical is present and arbitrarily sets that level at $^1/_2$(SQL) (or $^1/_2$[DL]) when it is an ND value. In fact, the U.S. EPA (U.S. EPA, 1989i) suggests the use of the SQL value itself if there is reason to believe that the chemical concentration is closer to this value than to one half the SQL. It should be noted that although these assignments provide a degree of quantification, they may on the other hand considerably affect subsequent analyses and evaluations. Where it is apparent that serious biases may result, more spohisticated analytical and evaluation methods may be warranted.

5.1.2 Exposure Assessment

An exposure assessment is conducted to estimate the magnitude of actual and/ or potential human exposures to chemical constituents, the frequency and duration of these exposures, and the pathways by which humans are potentially exposed to chemicals from a hazardous waste site. The exposure estimates are used to assess whether any threats exist based on existing exposure conditions at or near a potentially contaminated site. Exposure assessment involves describing the nature and size of the population exposed to a substance (i.e., the risk group, which refers to the actual or hypothetical exposed population) and the magnitude and duration of their exposure. Several characteristics of the chemicals of concern will provide an indication of the critical features of exposure, as well as information necessary to determine the distribution, uptake, residence time, magnification, and breakdown of a chemical to new chemical compounds (Hallenbeck and Cunningham, 1988). Indeed, the physical and chemical characteristics of the chemicals can also affect the intake, distribution, half-life, metabolism, and excretion of such chemicals by potential receptors. The evaluation could concern past or current exposures or exposures anticipated in the future. Several techniques may be used for the exposure assessment, including

- Modeling of anticipated future exposures
- Environmental monitoring of current exposures
- Biological monitoring to determine past exposures

The exposure assessment phase of the health risk assessment involves the characterization of the physical and exposure setting, including contaminant distributions leading from sources on the site to the points of exposure; the identification of significant migration and exposure pathways; the identification of potential receptors, or the PAR; the development of site conceptual model(s) and exposure scenarios (including the determination of current and future land uses, and the analysis of environmental fate and persistence); the estimation/modeling of exposure point concentrations for the critical pathways and environmental media; and the estimation of chemical intakes for all potential receptors and significant pathways of concern. The process is used to estimate the rates at which chemicals are absorbed by organisms through all mechanisms including ingestion, inhalation, and dermal absorption. Populations potentially at risk are defined, and concentrations of the chemicals of concern are determined in each medium to which potential receptors may be exposed. Finally, using the appropriate site-specific exposure parameter values, the intakes of the chemicals of concern are estimated.

Exposure Pathways

An exposure pathway is the potential route that contaminants take to reach potential receptors. The route and duration of exposure greatly influences the impact on the receptor. Exposure duration may be short- (acute) or long term (chronic). Exposure pathways are determined by integrating information from the initial site characterization with knowledge about potentially exposed populations and their likely behavior. Table 5.3 indicates examples of the multiple pathways typically observed from hazardous waste sites. The significance of the migration pathway in a particular application is evaluated on the basis of whether the contaminant migration could cause significant adverse human exposures and impacts.

Table 5.3 Potential Multipathway Exposure Routes for Chemical Releases from Hazardous Waste Sites

Transport/Exposure (Contaminated) Medium of Concern	Primary (Direct) Exposure Routes	Examples of Secondary (Indirect) Pathways[a]
Air	Inhalation	Mother's milk
	Dermal absorption	Poultry, meat, and eggs diet
	Crop ingestion (from direct deposition)	Dairy products
Subsurface gas	Inhalation	Mother's milk
		Poultry, meat, and eggs diet
		Dairy products
Soil	Soil ingestion	Crop ingestion (from plant uptake)
	Dermal contact	Poultry, meat, and eggs diet
	Inhalation of particulates	Dairy products
Groundwater	Inhalation of volatiles	Crop ingestion (from plant uptake @ irrigation water)
	Water ingestion	
	Dermal absorption	Poultry, meat, and eggs diet (from use of water in feed)
		Dairy products (from use of water in feed)
Surface water	Inhalation of volatiles	Fish ingestion
	Ingestion of water	Crop ingestion (from plant uptake @ irrigation water)
	Ingestion of contaminated biota	Poultry, meat, and eggs diet (from use of water in feed)
		Dairy products (from use of water in feed)
Aquatic and terrestrial biota	Ingestion @ food chain	Fish ingestion

[a] Secondary pathways of exposure are those which result from assimilation of the chemical into a food source or that reaches the ultimate potential receptor via an intermediary.

An exposure pathway is considered complete only if all of the following elements are present:

- A source of contaminant
- A mode of transport (i.e., a mechanism of chemical release to the environment)
- A contaminant release pathway (including transport media) and exposure route
- Receptor contact at potential exposure points in affected media

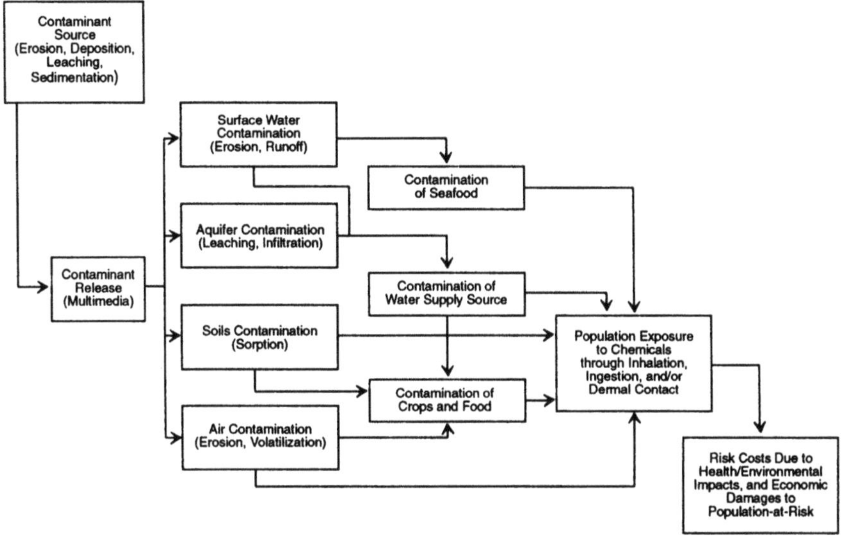

Figure 5.2 A simplified conceptual model for a typical exposure scenario. (Adapted from Asante-Duah, 1990.)

For exposure to occur, a complete pathway is necessary, and a risk assessment must address all exposure pathways. The accuracy with which exposure is characterized could be a major determinant of the ultimate validity of a risk assessment. Failure to identify an important pathway may seriously detract from the validity of any risk assessment.

Developing Exposure Scenarios

An exposure scenario is the qualitative connection between a contaminant source through one or more environmental media to some receptor population(s). The route of exposure to the population, such as inhalation, ingestion, or dermal contact, is identified as part of the exposure scenario. A schematic representation of the characterization process for developing an exposure scenario is shown, generically, in Figure 5.2. If numerous potential exposure scenarios exist, or if a complex exposure scenario has to be evaluated, it becomes helpful to use the event tree structure to cover potential outcomes and/or consequences; the event tree concept offers an efficient way to develop exposure scenarios. An example of an event tree structure is depicted in Figure 5.3. By using such an approach, the various exposure contingencies can be identified and organized in a systematic manner.

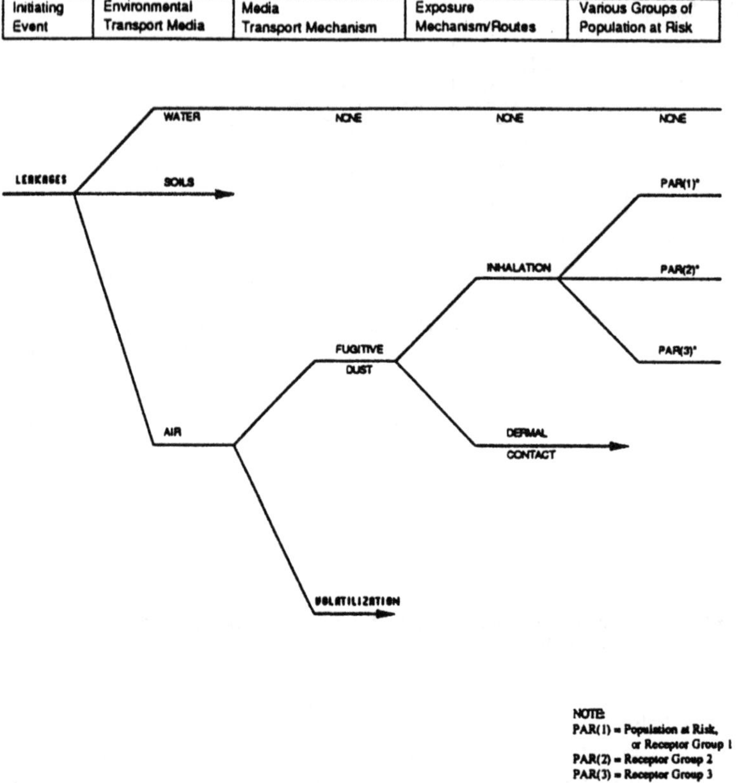

Figure 5.3 Development of exposure scenarios by use of the event tree concept.

Once developed, priorities can be established for focusing the available effort in the aspects of greatest concern. Table 5.4 illustrates the equivalent analytical protocol for developing the set of exposure scenarios; it is noteworthy that this representation is analogous to the event tree structure.

The exposure scenario associated with a given hazardous situation may be reasonably well defined if the exposure is known to have already occurred; in most cases, however, the risk assessment is being undertaken to evaluate the potential risks due to exposures that may not yet have occurred, in which case hypothetical exposure scenarios are developed for this purpose.

Spectrum of Exposure Scenarios. Exposure scenarios are derived and modeled based on the movement of chemicals in the various environmental compartmental media. A schematic overview of selected processes that includes a conceptualization of typical mechanisms that may affect contaminant migration at a waste site is shown in Figure 5.4. For instance, precipitation may infiltrate the soil onsite and leach contaminants from the wastes and soil as it migrates through the material and the unsaturated soil zone. Infiltrating water may continue its vertical migration and encounter the water table at the top of the saturated zone; the mobilized contaminants may be diluted by the available groundwater flow. Once a contaminant enters the groundwater system, it is possible for it to be transported by groundwater to a discharge point. There also is the possibility of continued vertical migration of contaminants into the bedrock aquifer system. Contaminants may also be carried by surface runoff into surface water bodies. Air releases present additional release pathways. The following *potential* exposure scenarios may be considered representative of the exposure pattern anticipated from a contaminated site:

- Direct human exposure onsite via ingestion of dirt (including pica), inhalation of airborne contaminants, and/or absorption through the skin after dermal contact with contaminated soil.
- Direct human exposure offsite via inhalation of fugitive dust, ingestion of settled dust, and/or dermal contact with chemicals adsorbed onto soil particles.
- Direct human exposure resulting from onsite use of groundwater; exposure may be via ingestion of groundwater used for municipal or local water supplies, inhalation (e.g., during showering activities), and/or dermal contact (from use of the ground water for washing and showering).
- Direct human exposure resulting from offsite use of groundwater; exposure may be via ingestion of groundwater used for domestic water supplies, inhalation (e.g., during showering activities), and/or dermal contact from use of the groundwater for washing and showering.
- Direct human exposure resulting from offsite use of surface water (that has been contaminated from surface runoff and/or ground water discharge); exposure may be via ingestion of surface water, inhalation, and/or dermal contact (from use of the surface water for washing).
- Direct human exposure resulting from off-site recreational use of surface water (that has been contaminated from surface runoff and/or groundwater discharges); exposure may be via ingestion of surface water, inhalation, and/or dermal contact.
- Indirect human exposure resulting from bioaccumulation in *river* fish that is consumed by humans; aquatic life may be exposed to contaminants as a result of runoff and/or groundwater discharges into river(s).
- Indirect human exposure resulting from ingestion of game or livestock (as a result of bioaccumulation through the food chain).
- Indirect human exposure resulting from ingestion of dairy products from cattle that consumed feed and water containing surface residues of chemicals.
- Indirect human exposure resulting from ingestion of crops with bioaccumulated chemicals deposited onto soil, directly onto edible portions of plants, or accumulated through root uptake.
- Inter-human transfers, such as ingestion of human breast milk containing chemicals absorbed by the feeding mother.

Table 5.4 Example Analysis of Potential Exposure Pathways

Potential Mechanisms of Releases	Environmental Transport Media	Transport Mechanisms	Potential Receptor Location	Exposure Routes	Potentially Exposed PAR
Infiltration Volatilization Leakages Spill	Air	Volatilization	Onsite	Inhalation	Utility workers Casual site visitors Remediation workers Facility employees
			Offsite	inhalation	Utility workers Local employees Nearby residents Nearby school
		Wind erosion (air particulates/ fugitive dust)	Onsite	Inhalation	Utility workers Casual site visitors Remediation workers Facility employees
				Dermal contact	Utility workers Casual site visitors Remediation workers Facility employees
			Offsite	Inhalation	Utility workers Local employees Nearby residents Nearby school
				Dermal contact	Utility workers Local employees Nearby residents Nearby school

Infiltration Volatilization Leakages Spill	Soils/sediments	Soil erosion	Onsite	Dermal contact	Utility workers Casual site visitors Remediation workers Facility employees
			Offsite	Dermal contact	Utility workers Local employees Nearby residents Nearby school
				Ingestion	Children eating soil
	Groundwater	Leaching to aquifier	Offsite	Dermal contact	Local employees Nearby residents Nearby school
				Ingestion	Local employees Nearby residents Nearby school
				Inhalation	Local employees Nearby residents Nearby school
	Surface water	Groundwater discharge	Offsite	Dermal contact	Local employees Nearby residents Nearby school
				Ingestion	Local employees Nearby residents Nearby school
				Inhalation	Local employees Nearby residents Nearby school

Table 5.4 (continued)

Potential Mechanisms of Releases	Environmental Transport Media	Transport Mechanisms	Potential Receptor Location	Exposure Routes	Potentially Exposed PAR
Infiltration Volatilization Leakages Spill	Surface water	Surface runoff, drainage, and ponding	Onsite	Dermal contact	Utility workers Casual site visitors Remediation workers Facility employees
			Offsite	Dermal contact	Utility workers Local employees Nearby residents Nearby school
				Ingestion	Local employees Nearby residents Nearby school
				Inhalation	Local employees Nearby residents Nearby school

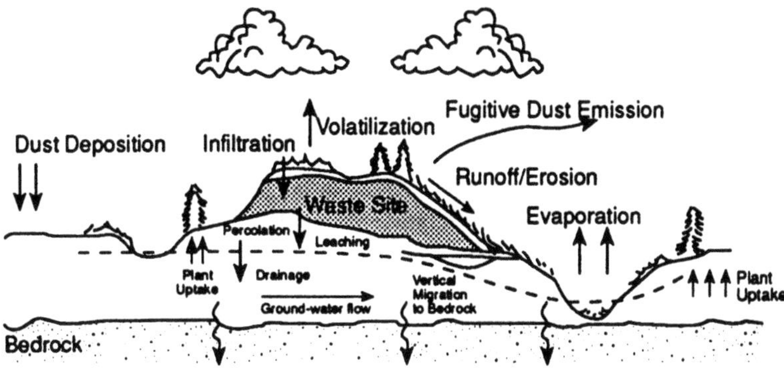

Figure 5.4 A schematic overview of typical processes affecting contaminant migration at waste sites.

Several tasks are undertaken to facilitate the development of complete exposure scenarios, including determining the sources of site contamination; identifying constituents of concern; identifying affected environmental media; delineating contaminant migration pathways; identifying potential receptors; determining potential exposure routes; constructing a conceptual model for the site; and delineating likely and significant exposure pathways. At all times, it is necessary to develop realistic exposure scenarios to support evaluation of the risks posed by a potentially contaminated site and to allow appropriate decisions regarding the need for and extent of remediation, if any.

Multimedia Pollutant Transport and Environmental Modeling

Chemical contaminants entering the environment tend to be partitioned or distributed across various environmental media and biota. The distribution of chemicals entering the environmental compartments is the result of a number of complex processes (Table 5.5). Nevertheless, the potential hazards and/or risks associated with the individual chemicals is very much dependent on the extent of multimedia exposures of potential receptors. Thus, a good prediction of chemical concentrations in the various compartmental media is essential for the completion of exposure and risk assessments.

Table 5.5 Important Transformation Processes Affecting the Fate and Migration of Hazardous Waste Constituents in the Environment

Media Category of Concern	Process	Key Factors
Soils	Biodegradation	Waste degradability Waste toxicity Acclimation of microbial community Aerobic/anaerobic conditions Soil pH Temperature Nutrient concentrations
	Photodegradation (photolytic reactions)	Solar irradiation Exposed surface area
	Hydrolysis	Functional group of chemical Soil pH and buffering capacity Temperature
	Oxidation/reduction (redox reactions)	Chemical class of contaminant Presence of oxidizing agents
	Volatilization	Partial pressure Henry's law constant Soil diffusion Temperature
	Adsorption	Effective surface area of soil Cation exchange capacity (CEC) Fraction organic content (f_{oc}) of soil Octanol/water partition coefficient (k_{ow})
	Dissolution	Solubility Soil pH and buffering capacity Complex formation
Sediments	Adsorption	Effective surface area of sediments Cation exchange capacity (CEC) Fraction organic content (f_{oc}) of sediments Octanol/water partition coefficient (k_{ow})
Surface water	Hydrolysis Volatilization Dissolution Evaporation Dilution	Functional group of chemical Temperature Solubility Water pH and buffering capacity Complex formation Partial pressure Henry's law constant Sedimentation
Groundwater	Dissolution Dilution	Solubility Water pH and buffering capacity Complex formation Advection and dispersion Molecular diffusion
Air	Deposition Photodegradation	Solar irradiation Exposed surface area

In conducting exposure assessments, transport and fate of contaminants can be predicted through use of various methods, ranging from simple mass-balance and analytical procedures to multidimensional numerical solution of coupled differential equations. In fact, numerous model classification systems exist, one of which divides the array of models into the following broad categories:

- Black-box, or regression-based models
- Analytical models
- Numerical models

The choice of which model to use is subject to numerous factors; simply moving to a more complicated model will not necessarily ensure a better solution. Since a model is a mathematical representation of a complex system, some degree of mathematical simplification must be made about the system being modeled. The simplifications may enter via the model itself being simple and/or the data input utilized in the model being approximate. Data limitations must be considered since it is not possible to obtain all of the input parameters due to the complexity (e.g., anisotropy and nonhomogeneity) of natural systems. In general, the black-box and analytical models are appropriate for well-defined systems for which extensive data are available (e.g., to develop the regression model) and/or for which the limiting assumptions are valid; for complicated systems, solutions are usually by numerical models.

Through the use of the mathematical models, a comprehensive exposure assessment is developed. That becomes an input to the risk assessment. Multimedia mathematical models for contaminant fate and transport can be used to generate information to supplement available field information gathered during site characterization. In such applications, mathematical algorithms are used to predict the potential for contaminant migration from a contaminated site to potential receptors. A thorough review of site background information will confirm the need for any modeling effort.

Using the predictions of contaminant transport, simplified exposure assessments are performed for the critical pathways to the potential receptors. The common and general types of modeling practice used in the exposure assessment relate to atmospheric, surface water, groundwater and unsaturated zone, multimedia, and food chain models. Typically, the following scenarios will be simulated and evaluated using the appropriate models for a contaminated site problem:

- Infiltration of rain water
- Erosion/surface runoff release of chemicals
- Emission of particulates and vapors
- Chemical fate and transport through the unsaturated zone
- Chemical transport through the aquifer system
- Mixing of groundwater with surface water

Mathematical models often serve as valuable tools for evaluating the behavior and fate of chemical constituents in the various environmental media. For instance, the Graphical Exposure Modeling System (GEMS), an interactive computer system developed by EPA's Office of Pesticide and Toxic Substances, provides a simple interface to environmental modeling, physiochemical property estimation, statistical analysis, and graphic display capabilities, with data manipulations which support all of the functions. Fate and transport models are provided for soils, groundwater, air, and surface water and are supported by various data sets, including demographic, hydrologic, pedologic, geologic, climatic, economic, etc. conditions. A more complete listing of selected models applicable to exposure modeling is contained in Table D.1 (Appendix D). Additional models of interest are discussed in the literature of exposure modeling (e.g., U.S. EPA, 1986e, 1987c, 1988g, 1988h).

Due to the heterogeneity in environmental compartments and natural systems, models used for exposure assessments should be adequately tested, and sensitivity runs should be carried out to help determine the most sensitive and/or critical parameters considered in the evaluation. Assistance on specific questions relating to exposure modeling (such as data requirements, appropriate models, etc.) can be obtained from EPA's Center for Exposure Assessment in Athens, GA.

Model Uses. To manage contamination problems and/or potentially contaminated sites, mathematical models may be needed to perform several functions such as for developing an initial screening of sites or problems; assessment of the probable contaminant behavior at specific sites; design of disposal schemes and assessment of performance goals; prediction of contaminant migration in various compartmental media; and the design of monitoring and corrective action plans.

Regarding exposure assessment, one of the major benefits associated with using mathematical models is that environmental concentrations useful for exposure assessment and risk characterization can be estimated for several locations and time periods of interest. Since field data frequently are limited and insufficient to accurately and completely characterize site and nearby conditions, models can be particularly useful for studying spatial and temporal variabilities, together with potential uncertainties. In addition, sensitivity analyses can be performed by varying specific parameters and using models to explore the ramifications (by monitoring changes in model outputs).

Environmental Fate. The fate of chemical compounds released into the environment forms an important basis for evaluating the exposure of biological receptors to hazardous chemicals. Environmental fate analysis is used to assess the movement of chemicals between media of concern or environmental compartments. For instance, the affinity that contaminants have for soil affects their mobility by retarding transport. Consequently, hydrophobic or cationic contaminants that are migrating in solution are subject to retardation effects. The hydrophobicity of a contaminant can greatly affect its fate and the resulting risk posed by such a chemical. This explains some of the different rates of contaminant migration occurring in the subsurface environment. Also, the phenomenon of adsorption is a major reason why the sediment zones of surface water bodies/systems may become highly contaminated with specific organic and inorganic chemicals. In general, a number of natural processes work to lessen or attenuate contaminant concentrations in the environment; the mechanisms of natural attenuation include dispersion/dilution, ion exchange, precipitation, adsorption and absorption, filtration, gaseous exchange, and biodegradation.

(I)

Soil Air <== {K_W} ==> Soil Water <== {K_{OC}} ==> Soil (organic carbon)

(II)

Soils/Sediments <== {K_{OC} } ==> Water <=== {BCF} ===> Biota (fish)

Figure 5.5 Partitioning in model environmental compartments.

The distribution of organic chemicals among environmental compartments can be defined in terms of simple equilibrium expressions (Figure 5.5), where the K_w, K_{oc}, and BCF terms refer to partitioning coefficients, to be further elaborated later on in this section (Swann and Eschenroeder, 1983). The general assumption is that all environmental compartments are well mixed to achieve equilibrium between them. With respect to inorganic chemicals, metals generally exhibit relatively low mobilities in soils (Evans, 1989). Also, relatively insignificant partitioning would be expected among environmental compartments for metals. Rather, the inorganics will tend to adsorb onto soils (which may become airborne or be transported by erosion) and sediments (that may be transported in water). Important factors affecting environmental fate and/or intermedia transfers are briefly discussed below.

Physical State — Solid wastes are generally less susceptible to release and migration than liquids. However, processes such as leaching, erosion, and/or runoff and physical transport of waste particulates can act as significant release mechanisms.

Water Solubility — The solubility of a chemical in water is the maximum amount of the chemical that will dissolve in pure water at a specified temperature, usually expressed in terms of weight per weight (e.g., ppb, ppm, mg/kg) or weight per volume (e.g., ppb, ppm, µg/L, mg/L). Lyman et al. (1990), among others, describe several different approaches to the estimation of water solubility. Solubility is an important factor affecting release and subsequent migration and fate of a chemical constituent in the surface and groundwater environments. In fact, among the various parameters affecting the fate and transport of organic chemicals in the environment, water solubility is one of the most important, especially with regards to the hydrophilic compounds. Highly soluble chemicals are easily and quickly distributed by the hydrologic system. Such chemicals generally tend to have relatively low adsorption coefficients for soils and sediments and also relatively low bioconcentration factors in aquatic biota. Furthermore, they tend to be more readily biodegradable.

Diffusion and Dispersion — Diffusive processes create mass spreading due to molecular diffusion in response to concentration gradients. Thus, diffusion coefficients are used to describe the movement of a molecule in a liquid or gas medium as a result of differences in concentration; it is used to calculate the dispersive component of chemical transport. The higher the diffusivity, the more likely a chemical is to move in response to concentration gradients. Dispersive processes create mass mixing due to heterogeneities (e.g., velocity variations). Consequently, for example, as a pulse of contaminant migrates through the soil, the peaks in concentration are decreased by spreading.

Volatilization — Volatilization is the process by which a chemical compound evaporates into the vapor phase from another environmental compartment. The volatilization of chemicals is an important mass transfer pathway. Knowledge of volatilization rates is necessary to determine the amount of chemical that enters the atmosphere and the change of pollutant concentrations in the source media. The transfer process from the source (e.g., water body, sediments, soil) to the atmosphere is dependent on the physical and chemical properties of the chemical compound in question, the presence of other pollutants, and the physical properties of the source media and the atmosphere. Lyman et al. (1990), among others, elaborate several estimation methods for evaluating this parameter.

Vapor pressure — Vapor pressure is the pressure exerted by a chemical vapor in equilibrium with its solid or liquid form at any given temperature. It is a relative measure of the volatility of a chemical in its pure state and is an important determinant of the rate of volatilization. It is used to calculate the rate of volatilization of a pure substance from a surface or in estimating a Henry's law constant (c.f.) for chemicals with low water solubility. Numerous estimation procedures exist in the technical literature; Lyman et al. (1990), among others, elaborate some estimation methods for this parameter. The higher the vapor pressure, the more likely a chemical is to exist in significant quantities in a gaseous state; thus, constituents with high vapor pressure are more likely to migrate from soil and groundwater to be transported in air.

Boiling point (B.P.) — B.P. is the temperature at which the vapor pressure of a liquid is equal to the pressure of the atmosphere on the liquid. Besides being an indicator for the physical state of a chemical, the B.P. also provides an indication of its volatility, an important parameter for hazard assessments. Other physical properties, such as critical temperature and latent heat (or enthalpy) of vaporization may be predicted by use of its normal B.P. as an input.

Henry's law constant — Henry's law constant provides a measure of the extent of chemical partitioning between air and water at equilibrium; it indicates the relative tendency of a constituent to volatilize from aqueous solution to the atmosphere based on the competition between its vapor pressure and water solubility. Contaminants with low Henry's law constant values will tend to favor the aqueous phase and volatilize to the atmosphere more slowly than constituents with high values. This parameter is important in determining the potential for intermedia transport to the air media. As an example of its application, the concentration in soil gas is related to the concentration in an underlying aquifer groundwater by

$$\frac{C_{sg}}{C_w} = \frac{H}{RT}$$

where

C_{sg} = concentration of the chemical in soil gas (mg/m^3)
C_w = concentration of the chemical in groundwater (µg/L)
H = Henry's law constant (atm-m^3/mol)

R = Gas constant (atm-m^3/mol·K)
T = Temperature (K)

Several other forms of expressing H also exist in the technical literature. The variation in H between chemicals is extensive. As a general guideline, for H values in the range of 10^{-7} to 10^{-5} (atm-m^3/mol), volatilization is low; for H between 10^{-5} and 10^{-3} (atm-m^3/mol), volatilization is not rapid but possibly significant; and for H > 10^{-3} (atm-m^3/mol), volatilization is rapid.

Water/air partition coefficient (K_w)—The distribution of a chemical between water and air is an expression representing the reciprocal of Henry's Law constant, H, viz.:

$$K_w = \frac{C_{water}}{C_{air}} = \frac{1}{H}$$

where

C_{air} = concentration of the chemical in air (mg/L)
C_{water} = concentration of the chemical in water (mg/L)

Octanol/water partition coefficient (K_{ow}) — The octanol/water partition coefficient (K_{ow}) is defined as the ratio of a chemical's concentration in the octanol phase (organic) to its concentration in the aqueous phase of a two-phase octanol/water system (Lyman et al., 1990). Thus,

$$K_{ow} = \frac{\textit{Concentration in octanol phase}}{\textit{Concentration in aqueous phase}}$$

and is dimensionless. It provides a measure of the extent of chemical partitioning between water and octanol at equilibrium; it has become an important parameter in studies of the environmental fate of organic chemicals. K_{ow} can be used to predict the magnitude of the tendency of an organic constituent to partition between the aqueous and organic phases of a two-phase system, such as surface water and aquatic organisms. The higher the value of K_{ow}, the greater the tendency of an organic constituent to adsorb to soil or waste matrices containing appreciable organic carbon or to accumulate in biota. It has been found to be related to water solubility, soil/sediment adsorption coefficients, and bioaccumulation factors for aquatic life. Chemicals with low K_{ow} (<10) values may be considered relatively hydrophilic, whereas those with high K_{ow} (>10^4) values are very hydrophobic; thus, the greater the K_{ow}, the more likely a chemical is to partition to octanol than to remain in water. The hydrophilic chemicals tend to have high water solubilities, small soil/sediment adsorption coefficients, and small bioaccumulation factors for aquatic life. High K_{ow} values are generally indicative of the ability of a chemical to accumulate in fatty tissues and therefore bioaccumulate in the food chain. It is also a key variable in the estimation of skin permeability.

K_{ow} values may be estimated by several methods discussed by Lyman et al. (1990), among others.

Organic carbon adsorption coefficient (K_{oc})—The sorption characteristics of a chemical may be normalized to obtain a sorption constant based on organic carbon which is essentially independent of any soil. The organic carbon adsorption coefficient (K_{oc}) provides a measure of the extent of partitioning of a chemical constituent between soil/sediment organic carbon and water at equilibrium. K_{oc} is the ratio of the amount of constituent adsorbed per unit weight of organic carbon in the soil or sediment to the concentration of the constituent in aqueous solution at equilibrium. Also called the organic carbon partition coefficient, K_{oc} is a measure of the tendency for organics to be adsorbed by soil and sediment and is expressed as:

$$K_{oc}[ml/g] = \frac{\mu g \ chemical \ adsorbed \ / \ g \ soil \ organic \ carbon}{\mu g \ chemical \ dissolved \ / \ ml \ of \ water}$$

The extent to which an organic constituent partitions between the solid and solution phases of a saturated or unsaturated soil, or between runoff water and sediment, is determined by the physical and chemical properties of both the constituent and the soil (or sediment). The tendency of a constituent to be adsorbed to soil is dependent on its properties and on the organic carbon content of the soil or sediment. When constituents have a high K_{oc}, they have a tendency to partition to the soil or sediment. In such cases, sediment sampling would be appropriate. This value is a measure of the hydrophobicity of a chemical. The more highly sorbed, the more hydrophobic — or the less hydrophilic — a substance. The K_{oc} is chemical specific and largely independent of the soil or sediment properties. Values of K_{oc} may range from 1 to 10^7. The higher the K_{oc}, the more likely a chemical is to bind to soil or sediment than to remain in water. Lyman et al. (1990), among others, elaborate several estimation methods for evaluating this parameter.

Soil-water partition coefficient (K_d)—The distribution of a chemical between water and adjoining soil or sediment may be described by an equilibrium expression that relates the amount of chemical sorbed to soil or sediment to the amount in water at equilibrium; this is often described by a soil/water distribution coefficient, K_d. For most environmental concentrations, K_d can be approximated by

$$K_d[ml/g] = \frac{(concentration \ of \ adsorbed \ chemical \ in \ soil, \ C_s)}{(concentration \ of \ chemical \ in \ solution \ in \ water, \ C_w)}$$

or

$$K_d[ml/g] = \frac{[(\mu g \ chemical \ / \ g \ soil)]}{[(\mu g \ chemical \ / \ g \ water)]}$$

On this basis, K_d describes the sorptive capacity of the soil and allows estimation of the concentration in one medium, given the concentration in the adjoining medium. For hydrophobic contaminants,

$$K_d = f_{oc} K_{oc}$$

where f_{oc} is the fraction of organic carbon in the soil. The mobility of contaminants in soil depends not only on properties related to the physical structure of the soil, but also on the extent to which the soil material will retain, or adsorb, the hazardous constituents. The extent to which a constituent is adsorbed depends on chemical properties of the constituent and of the soil. Therefore, the sorptive capacity must be determined with reference to a particular constituent and soil pair. The soil-water partition coefficient (K_d) is generally used to quantify soil sorption. It provides a soil- or sediment-specific measure of the extent of chemical partitioning between soil or sediment and water, unadjusted for dependence on organic carbon. K_d is the ratio of the adsorbed contaminant concentration to the dissolved concentration, at equilibrium; the higher the value of K_d, the less mobile is the contaminant, because for large values of K_d, most of the chemical remains stationary and attached to soil particles due to the high degree of sorption. Thus, the higher the K_d, the more likely a chemical is to bind to soil or sediment than to remain in water.

Bioconcentration factor (BCF) — The BCF is the ratio of the concentration of a chemical constituent in an organism or whole body (e.g., a fish) or specific tissue (e.g., fat) to the concentration in water at equilibrium, given by

$$BCF = \frac{(concentration\,in\,biota)}{(concentration\,in\,surrounding\,medium)}$$

$$= \frac{[(\mu g\,chemical\,/\,g\,biota — fish)]}{[(\mu g\,chemical\,/\,g\,medium — water)]}$$

The partitioning of a chemical between water and biota (fish) also gives a measure of the hydrophobicity of the chemical. Ranges of BCFs for various constituents and organisms can be used to predict the potential for bioaccumulation and, therefore, to determine whether sampling of the biota may be necessary. The accumulation of chemicals in aquatic organisms is of increased concern as an important environmental and health hazard. The BCF indicates the degree to which a chemical residue may accumulate in aquatic organisms, coincident with ambient concentrations of the chemical in water. It is a measure of the tendency of a chemical in water to accumulate in the tissue of an organism. Values of BCF range from 1 to over 10^6. Constituents exhibiting a BCF greater than 1.0 are potentially bioaccumulative. Generally, constituents exhibiting a BCF greater than 100 cause the greatest concern (U.S. EPA, 1987b). The concentration of the chemical in the edible portion of the organism's tissue can be estimated by multiplying the concentration of the chemical in surface water by the fish BCF for that chemical. Indeed, the BCF is an estimate of the bioaccumulation potential for biota in general, not just for fish. Thus, the average concentration in fish or biota is given by

$$C_{fish-biota}(\mu g / kg) = Concentration\,in\,water, C_{water}(\mu g / kg) \times BCF$$

This parameter is an important determinant for human intake via ingestion of aquatic foods. Lyman et al. (1990), among others, elaborate several estimation methods for evaluating this parameter.

Degradation/chemical half-lives — Degradation, whether biological, physical, or chemical, is often reported in the literature as a half-life, which is usually measured in d. It is usually expressed as the time it takes for one-half of a given quantity of a compound to be degraded. Half-lives are used as measures of persistence, since they indicate how long a chemical will remain in various environmental media; long half-lives (e.g., greater than a month or a year) are characteristic of persistent constituents. Media-specific half-lives provide a relative measure of the persistence of a chemical in a given medium, although actual values can vary greatly depending on site-specific conditions. For example, the absence of certain microorganisms at a site, or the number of microorganisms, can influence the rate of biodegradation and, therefore, the half-life. As such half-life values should be used only as general indications of chemical persistence (U.S. EPA, 1987b). Nevertheless, in general, the greater the half-life, the more persistent a chemical is likely to be.

Biodegradation — Biodegradation is one of the most important environmental processes affecting the breakdown of organic compounds. It results from the enzyme-catalyzed transformation of organic constituents, primarily from microorganisms. The ultimate fate of a constituent introduced into a surface water or other environmental system (e.g., soil) could be a constituent or compound other than the species originally released. For example, trichloroethylene (TCE) biodegrades to produce 1,2 dichloroethylene (1,2-DCE), vinyl chloride, and other compounds; from a toxicity viewpoint, 1,2-DCE is less toxic and vinyl chloride more toxic than TCE. Biodegradation potential should therefore be considered in designing monitoring programs. The biological degradation may also initiate other chemical reactions, such as oxygen depletion in microbial degradation processes, creating anaerobic conditions and the initiation of redox potential-related reactions.

Photolysis — Photolysis can be an important dissipative mechanism for specific chemical constituents in the environment. Similar to biodegradation, photolysis (or photodegradation) may cause the ultimate fate of a constituent introduced into a surface water or other environmental system (e.g., soil) to be different from the constituent originally released. Hence, photodegradation potential should also be considered in designing sampling and analysis programs.

Chemical degradation (hydrolysis and oxidation/reduction) — Similar to photodegradation and biodegradation, chemical degradation, primarily through hydrolysis (a chemical transformation process in which an organic molecule reacts with water, forming a new carbon-oxygen (C-O) bond and cleaving the C bonding with the original molecule) and oxidation/reduction (REDOX) reactions, can also act to change chemical constituent species once they are introduced to the environment. Hydrolysis of organics usually results in the introduction of a hydroxyl group (-OH) into a constituent structure. Hydrated metal ions (particularly those with a valence of 3 or greater) tend to form ions in aqueous solution, thereby enhancing species solubility. Oxidation may occur as a result of chemical oxidants being formed during photochemical processes in natural waters. Similarly, reduction of constituents may take place in some surface water environments (primarily those with low oxygen levels).

Miscellaneous equilibrium constants — Equilibrium constants are important predictors of the chemical state of a compound in solution. In general, a constituent which is dissociated (ionized) in solution will be more soluble and therefore more likely to be released to the environment and more likely to migrate in a surface water body. Many inorganic constituents, such as heavy metals and mineral acids, can occur as different ionized species depending on pH. Organic acids, such as the phenolic compounds, exhibit similar behavior. It should also be noted that ionic metallic species present in the release may have a tendency to bind to particulate matter, if present in a surface water body, and to settle out to the sediment over time and distance. In fact, heavy metals are removed in natural attenuation by ion exchange reactions, whereas trace organics are removed primarily by adsorption. Metallic species also generally exhibit bioaccumulative properties. When metallic species are present in a release, both sediment and biota sampling would be appropriate.

Examination of the physical and chemical properties of a chemical can often allow an estimate of its environmental partitioning. Qualitative analysis of the fate of a chemical can also be made by analogy with other chemicals whose fate are well documented. If the chemical under study is structurally similar to a previously studied one, some parallel can be drawn to the environmental fate of the analogue. In addition, several site characteristics may influence the environmental fate of chemicals, including the amount of ambient moisture, humidity levels, temperatures, and wind speed; geologic, hydrologic, pedologic, and watershed characteristics; topographic features of the site and vicinity; vegetative cover of site and surrounding area; and land use characteristics.

Intermedia Pollutant Transfers. Chemicals released to one environmental medium are affected by several complex processes and phenomena, facilitating transfers to other media. The potential for inter-media transfer of releases from the soil medium to other media is particularly significant; contaminated soil can be a major source of contamination to ground water, air, subsurface gas and surface water. Hazardous wastes or constituents, particularly those having a moderate to high degree of mobility, can leach from the soil to the groundwater. Volatile wastes or constituents may contribute to subsurface gas to the vadose zone and releases to air. Contaminated soils can also contribute to contamination of surface waters, especially through runoff during heavy rains or storms. Similarly, the potential for intermedia transport of constituents from other media to the soil also exists. For example, hazardous waste or constituents may be transported to the soil via atmospheric deposition through the air medium and also through releases of subsurface gas.

As pollutants are released into various environmental media, several factors contribute to their migration and transport. For instance, in the groundwater system, the solutes in the porous media will move with the mean velocity of the solvent by advective mechanism; in addition, other mechanisms governing the spread of contaminants include hydraulic dispersion and molecular diffusion (which is caused by the random Brownian motion of molecules in solution that occurs whether the solution in the porous media is stationary or has an average motion). Furthermore, the transport and concentration of the solute(s) are affected by reversible ion exchange with soil grains; the chemical degeneration with other constituents; fluid compression and expansion; and in the case of radioactive wastes, by the radioactive decay.

Estimation of Potential Receptor Exposures

Using applicable exposure models based on the physicochemical properties of the contaminants of potential concern, and conservative but realistic assumptions regarding contaminant migration and equilibrium partitioning, exposure to potential receptors may conservatively be estimated according to the following generic relationship:

$$EXP = IF \times C_m \times EDF \times CF$$

where

EXP = receptor dose or exposure (mg/kg/day)
IF = intake factor(s) (e.g., inhalation rate—m^3/day; ingestion rate—mg/day, L/day)
C_m = concentration in media (e.g., $\mu g/m^3$ or mg/L or mg/kg)
EDF = exposure duration factor(s)
CF = conversion factor

Values for IF and EDF may be obtained from standard reference manuals and documents (e.g., U.S. EPA, 1989b). These values are derived from information relating to the MEL. The MEL is used to calculate levels of a compound in an environmental media (i.e., soil, water, or air) such that its value is not likely to be exceeded during the course of specified categories of human activity and/or exposures. MEL values may be drawn from several sources, such as the guidelines for ADIs; virtually safe doses (VSDs); MCLs for drinking water; MCLGs for drinking water; threshold limit values (TLVs) for occupational exposures; the Food and Drug Administration (FDA) guidelines for concentrations in foods; no observed adverse effect levels (NOAELs); and acute toxicity values (e.g., LC_{50}, LD_{50}). The concentration in the various media, C_m, may be obtained from field measurements, estimated by simple mass balance analyses or other appropriate contaminant transport models, or may be determined from equilibrium and partitioning relations. For instance, considering a situation in which groundwater is feeding into a surface water body, the concentration of a chemical in the surface water may be related to its concentration in groundwater by the following mass balance relationship:

$$C_{sw} = C_{gw} \left\{ \frac{Q_{gw}}{\left(Q_{gw} + Q_{sw}\right)} \right\}$$

where

C_{sw} = concentration in surface water (mg/L)
C_{gw} = concentration in groundwater (mg/L)
Q_{gw} = flow rate of groundwater (cfs)
Q_{sw} = flow rate of surface water (cfs)

The site-related exposure point concentrations are determined once the exposure scenarios and potentially affected populations are identified. If the transport of compounds associated with the site is under steady-state conditions, monitoring data are generally adequate to determine potential exposure concentrations. If there are no data available, or if conditions are transient (such as pertains to a migrating plume in groundwater), models are better used to predict concentrations. Many factors — including the fate and transport processes affecting the chemicals of concern — must be considered when selecting the most appropriate model. In any case, in lieu of an established trend in historical data indicating the contrary, the site is considered to be in steady state with its surroundings.

Since exposure could be occurring over long time periods (say, up to a lifetime of about 70 years), it is important in a detailed analysis to consider whether degradation or other transformation of the chemical at the source would occur. In such cases, the chemical and its degradation properties should be reviewed. If significant degradation is likely to occur, exposure calculations become much more complicated. In that case, source contaminant levels must be calculated at frequent intervals and summed over the exposure period. Assuming first-order kinetics for instance, an approximation of the degradation effects can be obtained by multiplying the concentration by a degradation factor, DGF, defined by

$$DGF = \frac{\left(1 - e^{-kt}\right)}{kt}$$

where

k = chemical-specific degradation rate constant (d^{-1})
t = time period over which exposure occurs (d)

For a first-order decaying substance, k is estimated from the following relationship:

$$t_{1/2}[days] = \frac{0.693}{k} \text{ or } k\left[days^{-1}\right] = \frac{0.693}{t_{1/2}}$$

where $t_{1/2}$ is the half-life, which is the time after which the mass of a given substance will be one half its initial value. It should be recognized in carrying out all these manipulations, however, that in many cases when a substance undergoes degradation, it produces an end product that could be of potentially equal or greater concern (such as is the case when trichloroethylene biodegrades to vinyl chloride). Consequently, for simplicity, the decay factor will normally be ignored, except in situations where the end product is known to present no potential hazards to potential receptors.

Intake and Dose Calculation. Once exposure point concentrations in all media of concern have been estimated, the intakes and/or doses to potentially exposed populations need to be determined. Intake is defined as the amount of chemical coming into contact with the receptor's body or exchange boundaries (such as the skin, lungs, or gastrointestinal tract), and dose is the amount of chemical absorbed by the body into the bloodstream.

The absorbed dose differs significantly from the externally applied dose (called exposure or intake). Intakes and doses are normally calculated in the same step of the exposure assessment, where the former multiplied by an absorption factor yields the latter value. The methods by which each type of exposure is estimated are well documented in the literature of exposure assessment (e.g., DOE, 1987; U.S. EPA, 1988h; 1989b; 1989i; CAPCOA, 1990). The general equation for calculating chemical intakes by the PAR is expressed by the following relationship:

$$I = C \times CR \times CF \times FI \times ABS_s \times EF \times ED \times 1/BW \times 1/AT$$

where

- I = intake, adjusted for absorption (mg/kg/day)
- C = chemical concentration in media of concern (e.g., mg/kg; mg/L)
- CR = contact rate (e.g., mg soil/day; liters water/day)
- CF = conversion factor
- FI = fraction of intake from contaminated source (unitless)
- ABS_s = bioavailability/absorption factor (%)
- EF = exposure frequency (d/years)
- ED = exposure duration (years)
- BW = body weight (kg)
- AT = averaging time (period over which exposure is averaged — days)

Appendix B1 of this book contains a more detailed presentation of the exposure estimation equations. For a comprehensive evaluation, standard parameters suggested by the U.S. EPA are used, with appropriate and justifiable adjustments being made for problem-specific cases.

For each exposure pathway under consideration, an intake per event is developed. This value quantifies the amount of a chemical contacted during each exposure event, where "event" may have different meanings depending on the nature of the scenario under consideration (e.g., each d of inhalation of the chemical in air constitutes one inhalation exposure event). The quantity of a chemical absorbed into the bloodstream per event, represented by the dose, is calculated by additionally considering any pertinent physiological parameters (such as gastrointestinal absorption rates). When the level of dose (systemic absorption) from an intake is unknown, or cannot be estimated by a defensible argument, intake and dose are considered to be the same (i.e., 100% absorption into the bloodstream from contact is assumed). This approach provides a conservative estimate of the actual exposures. It assumes that the potential receptor is always in the same location, exposed to the same ambient concentration, and that there is 100% absorption on exposure. This would hardly ever represent any real-life situation, and lower exposures will be expected due to the fact that potential receptors will generally be exposed to lower or even near-zero levels for the times spent outside the "zones of influence."

Event-based intake values are converted to final intake values by multiplying the intake per event by the frequency of exposure events over the time frame being considered. Long-term (chronic) exposures are based on the number of events that are assumed to occur within an assumed 70-year lifetime (U.S. EPA, 1989i). Chronic daily intake (CDI) is the projected human intake over the long-term period and is calculated by multiplying the average, or the reasonably maximum exposure (RME) media concentrations, by the human intake and body weight factors. Subchronic daily intake (SDI), on the other hand, is the projected human intake over a short-term period such as only a portion of a lifetime (U.S. EPA, 1989i) and is also calculated by multiplying the RME (for the maximum or worst-case scenario) media concentrations by the human intake and body weight factors.

SDIs are used to evaluate subchronic noncarcinogenic effects, and CDIs are used to evaluate both carcinogenic risks and chronic noncarcinogenic effects. In general, the carcinogenic effects (and sometimes the chronic noncarcinogenic effects) from a contaminated site involve estimating the lifetime average daily dose (LADD). For noncarcinogenic effects, the average daily dose (ADD) is generally used. The ADD differs from the LADD in that the former is not averaged over a lifetime; rather, it is the average daily dose pertaining to the d of exposure. The maximum daily dose (MDD) will typically be used in estimating acute or subchronic exposures.

5.1.3 Toxicity Assessment

A toxicity assessment is conducted as part of a health risk assessment to qualitatively and quantitatively assess the potential for adverse human health effects from exposure to the chemicals of potential concern. The quantitative portion of the toxicity assessment entails identifying the relevant toxicity indices against which exposure point intakes and doses can be compared during the risk characterization stage of the overall assessment. Such assessment may include a consideration of experimental studies that uses animal data for extrapolation to humans, as well as epidemiological studies. The qualitative aspect of the assessment includes summaries of the adverse human health effects, typical environmental levels or background concentrations, toxicokinetics, toxicodynamics, and ecotoxicology associated with each chemical of potential concern.

The toxicity assessment component of the risk assessment considers the types of adverse health effects associated with chemical exposures, the relationship between magnitude of exposure and adverse effects, and related uncertainties such as the weight of evidence of the carcinogenicity of a particular chemical in humans (Appendix C). A detailed toxicity assessment for chemicals found at contaminated sites is generally accomplished in two steps (1) hazard assessment; and (2) dose-response assessment. These steps are briefly discussed below. Appendix D in this book describes selected information sources for toxicity parameters.

Typically, risk assessments rely heavily on existing toxicity information developed for specific chemicals. Where toxicity information does not exist, decisions can be made to exclude the chemical from the evaluation or to estimate toxicological data from that of similar compounds (with respect to molecular weight and structure-activity). Structure-activity analysis is a technique which can be applied to derive an estimate for the toxicity of a chemical when direct experimental or observational data are lacking.

Hazard Assessment

Hazard assessment is the process of determining whether exposure to an agent can cause an increase in the incidence of an adverse health effect (e.g., cancer, birth defects, etc.), involving characterization of the nature and strength of the evidence of causation. It involves gathering and evaluating data on the types of health injury or disease that may be produced by a chemical and on the conditions of exposure under which injury or disease is produced. Hazard assessment may also involve characterizing the behavior of a chemical within the body and the interactions it undergoes with organs, cells, or even parts of cells. Data of the latter types may be of value in answering the ultimate question of whether the forms of toxicity known to be produced by a substance in one population group or in experimental settings, are also likely to be produced in all humans.

Methods commonly used for assessing the hazardous nature of substances include (Lave, 1982)

- Case clusters
- Structural toxicology
- Laboratory study of simple test systems
- Long-term animal bioassays
- Human (epidemiological) studies

Case clusters are based on the identification of an abnormal pattern of disease. This procedure tends to be more powerful in identifying hazards, especially when the resulting condition is extremely rare; when the health condition is more common in the general population, the method is not very powerful. Since the PAR is essentially never known in detail, the case cluster method necessarily yields no conclusive evidence, only rather vague suspicions. Structural toxicology involves searching for similarities in chemical structure that might identify carcinogens. The close association between mutagens and carcinogens lead to a general presumption that mutagenic substances are also carcinogenic. Animal bioassays are laboratory experimentations, generally with rodents; statistical models are used to extrapolate from animal bioassays to humans. Epidemiology is a more scientific, systematic form of case cluster analysis with an attempt to control for confounding factors in the experimental design or statistical analysis.

Dose-Response Assessment

Dose-response assessment is the process of quantitatively evaluating the toxicity information and characterizing the relationship between the dose of the contaminant administered or received (i.e., exposure to an agent) and the incidence of adverse health effects in the exposed populations. It is the process by which the potency of the compounds is estimated by use of dose-response relationships. For carcinogens, this involves estimating the probability that an individual exposed to a given amount of chemical will contract cancer due to that exposure. Potency estimates may be given as "unit risk factor" (in $\mu g/m^3$ or ppm) or as "potency slopes" (in units of $[mg/kg/day]^{-1}$). Data are derived from animal studies or, less frequently, from studies in exposed human populations.

The risks of a substance cannot be ascertained with any degree of confidence unless dose-response relations are quantified, even if the substance is known to be toxic. There may be many different dose-response relationships for a substance if it produces different toxic effects under different conditions of exposure. Dose-response curves are functional relationships between the amounts of chemical substance and its morbidity/lethality. The response of a toxicant depends on the mechanism of its action; in the simplest case, the response, R, is directly proportional to its concentration, [C], so that

$$R = k[C]$$

where k is a rate constant. This would be the case for a pollutant that metabolizes rapidly, but even so, the response and the value of the rate constant would tend to differ for different risk groups of individuals and for unique exposures. If the toxicant accumulates in the body, the response is defined as

$$R = k[C]t^n$$

where t is the time and n is a constant. For cumulative exposures, the response would generally increase with time. Thus, the cumulative effect may show as linear until a threshold is reached, after which secondary effects begin to affect and enhance the responses. The cumulative effect may be related to what is referred to as the "body burden" (BB). The body burden is determined by the relative rates of absorption (ABS), storage (ST), elimination (ELM), and biotransformation, (BT) (Meyer, 1983):

$$BB = ABS + ST - ELM - BT$$

Each of the factors involved in the quantification of the body burden is dependent on a number of biological and physiochemical factors. In fact, the response of an individual to a given dose cannot be truly quantitatively predicted, since it depends on many extraneous factors such as general health and diet of individual receptors or the PAR.

Three major classes of mathematical extrapolation models are often used for relating dose and response in the subexperimental dose range:

1. Tolerance distribution models, including Probit, Logit, and Weibull
2. Mechanistic models, including One-hit, Multi-hit, and Multi-stage
3. Time-to-occurrence models, including Log-normal and Weibull

Indeed, other independent models — such as linear, quadratic, and linear-cum-quadratic — may also be employed for this purpose. The details of these are beyond the scope of this book and are discussed elsewhere (e.g., CDHS, 1986). From the quantitative dose-response relationship, toxicity values are derived that can be used to estimate the incidence of adverse effects occurring in humans at different exposure levels.

Quantifying Toxicological Effects: Toxicity Parameters

Chemicals that give rise to toxic endpoints other than cancer and gene mutations are often referred to as "systemic toxicants" because of their effects on the function of various organ systems; the toxic endpoints are referred to as "noncancer or systemic toxicity." Most chemicals that produce noncancer toxicity do not cause a similar degree of toxicity in all organs, but usually demonstrate major toxicity to one or two organs. These are referred to as the target organs of toxicity for the chemicals (Klaassen et al., 1986; U.S. EPA, 1989d). In addition, chemicals that cause cancer and gene mutations also commonly evoke other toxic effects (i.e., systemic toxicity).

For the purpose of health risk assessment, chemicals are usually categorized into carcinogenic and noncarcinogenic groups. Noncarcinogens operate by threshold mechanisms, i.e., manifestation of systemic effects requires a threshold level of exposure or dose to be exceeded. Systemic toxicity is generally treated as if there is an identifiable exposure threshold below which there are no observable adverse effects. Chronic noncarcinogenic health effects are assumed to exhibit a threshold level — i.e., continuous exposure to levels below the threshold produce no adverse or noticeable health effects. This characteristic distinguishes systemic endpoints from carcinogenic and mutagenic endpoints, which are often treated as nonthreshold processes. The threshold principle is not applicable for carcinogens, since no thresholds exist for this group. It is noteworthy, however, that among some professional groups, there is the belief that certain carcinogens require a threshold exposure level to be exceeded to provoke carcingenic effects.

Often, it becomes necessary to compare receptor intakes of chemicals with doses shown to cause adverse effects in humans or experimental animals. This can be achieved by estimating the MDD and LADD resulting from environmental exposures (expressed in mg/kg/day). The dose at which no effects are observed in human populations or experimental animals is referred to as the "no observed effect level" (NOEL). Where data identifying a NOEL are lacking, a "lowest observed effect level" (LOEL) may be used as the basis for determining safe threshold doses. For acute effects, short-term exposures/doses shown to produce no adverse effects are involved; this is called the "no observed adverse effect level" (NOAEL).

Traditionally, risk decisions on systemic toxicity have been made using the concept of "acceptable daily intake" (ADI) derived from an experimentally determined NOAEL. The ADI is the amount of chemical (in mg/kg body weight/day) to which a receptor can be exposed to on a daily basis over an extended period of time — usually a lifetime — without suffering a deleterious effect. A NOAEL is an experimentally determined dose at which there has been no statistically or biologically significant indication of the toxic effect of concern. In cases where a NOAEL has not been demonstrated experimentally, the term "lowest observed adverse effect level" (LOAEL) is used. For chemicals possessing carcinogenic potentials, the LADD is compared with the NOEL identified in the long-term bioassay experimental tests; for chemicals with acute effects, the MDD is compared with the NOEL observed in short-term animal studies. In assessing the chronic and subchronic effects of noncarcinogens and also noncarcinogenic effects associated with carcinogens, the experimental dose value (e.g., NOEL values) is typically divided by a safety (or uncertainty) factor to yield an RfD (or ADI).

For exposure of humans to noncarcinogenic effects of chemicals, the ADI is used as a measure of exposure considered to be without adverse effects. For carcinogenic effects, and assuming a no-threshold situation, an estimate of the excess cancer per unit dose, called the unit cancer risk (UCR), or the cancer slope factor (SF) is used. Overall, the quantitative evaluation of toxicological effects consists of the following specific steps:

- Compilation of toxicological profiles (including the intrinsic toxicological properties of the chemicals of concern, which may include their acute, subchronic, chronic, carcinogenic, and/or reproductive effects)
- Determination of appropriate toxicity indices (e.g., ADIs, RfDs, and SFs [or cancer potency factors, CPFs])

The toxicity parameters are dependent on the route of exposure; however, oral RfDs and SFs will normally be used for both ingestion and dermal exposures.

Derivation of Reference Doses (RfDs) or Acceptable Daily Intakes (ADIs)

RfD (or the ADI) is defined as the maximum amount of a chemical that the human body can absorb without experiencing chronic health effects; it is generally expressed in units of milligrams per kilogram of bodyweight per d (mg/kg/day). Although often used interchangeably with the ADI, the RfDs are based on a more rigorously defined methodology. It is an estimate of continuous daily exposure of a noncarcinogenic substance for the general human population (including sensitive subgroups) which appears to be without an appreciable risk of deleterious effects. Subchronic RfD is used to refer to cases involving only a portion of the lifetime, whereas chronic RfD refers to lifetime exposures. The RfD is a "benchmark" dose operationally derived from the NOAEL by consistent application of general "order-of-magnitude" uncertainty factors (UFs) (also called "safety factors") that reflect various types of data sets used to estimate RfDs. In addition, a modifying factor (MF) is sometimes used which is based on professional judgment of the entire data base of the chemical. More generally stated, RfDs (and ADIs) are calculated by dividing a NOEL (i.e., the highest level at which a chemical causes no observable changes in the species under investigation), a NOAEL (i.e., the highest level at which a chemical causes no observable adverse effect in the species being tested), or a LOAEL (i.e., that dose rate of chemical at which there are statistically or biologically significant increases in frequency or severity of adverse effects between the exposed and appropriate control groups), which are derived from human or animal toxicity studies by one or more UFs and MFs.

To estimate the risk of acute exposures, levels of acceptable short-term exposure may be developed, representing, for instance, the maximum 1-d exposure levels that are anticipated not to result in adverse effects in most individuals. Where available, acute ADIs can be based on EPA's 1-d drinking water health advisories. Where worker exposures are involved, OSHA's permissible exposure limits (PELs), or TLVs where no PEL has been established, serve as ARARs for acute exposures. Also, in rather rare cases where only TLV data may be all that is available, acceptable intake levels may be established/derived by correcting for continuous exposure and further dividing by a safety factor (of 100) to account for highly sensitive segments of impacted populations.

When no toxicological information exists for a chemical, concepts of a structure-activity relationship may have to be employed to derive acceptable intake levels by influence and analogy to closely related or similar compounds. In such cases, some reasonable degree of conservatism is suggested in any judgement call to be made.

Approach for Estimating RfDs (or ADIs). RfDs are typically calculated using a single exposure level and UFs that account for specific deficiencies in the toxicological data base. Both the exposure level and UFs are selected and evaluated in the context of the available chemical-specific literature. After all toxicological, epidemiologic, and supporting data have been reviewed and evaluated, a key study is selected that reflects best availabe data on the critical effect. Dose-response data points for all reported effects are examined as a component of this review. U.S. EPA (1989d) discusses specific issues of particular significance in this endeavor — including the types of response levels (ranked in order of increasing severity of toxic effects as NOEL, NOAEL, LOAEL, and frank effect level [FEL], defined as overt or gross adverse effects) considered in deriving RfDs for systemic toxicants.

The RfD (or ADI) is determined by use of the following equation:

$$Human\,dose\left(e.g., ADI\ or\ RfD\right) = \frac{Experimental\,dose\left(e.g., NOAEL\right)}{\left(UF \times MF\right)}$$

or, specifically

$$RfD = \frac{NOAEL}{\left(UF \times MF\right)}$$

The UF (also safety factor) used in calculating the RfD reflects scientific judgment regarding the various types of data used to estimate RfD values. It is used to offset the uncertainties associated with extrapolation of data, etc. Generally, the UF consists of multiples of 10, each factor representing a specific area of uncertainty inherent in the available data. For example, a factor of 10 may be introduced to account for the possible differences in responsiveness between humans and animals in prolonged exposure studies. A second factor of 10 may be used to account for variation in susceptibility among individuals in the human population. The resultant UF of 100 has been judged to be appropriate for many chemicals. For other chemicals, with data bases that are less complete (for example, those for which only the results of subchronic studies are available), an additional factor of 10 (leading to a UF of 1000) might be judged to be more appropriate. For certain other chemicals, based on well-characterized responses in sensitive humans (as in the effect of fluoride on human teeth), a UF as small as 1 might be selected (Dourson and Stara, 1983).

In general, the following guidelines are useful in selecting uncertainty and modifying factors for the derivation of RfDs (Dourson and Stara, 1983; U.S. EPA, 1986b; U.S. EPA, 1989b; U.S. EPA, 1989d):

Standard UFs —

- Use a tenfold factor when extrapolating from valid experimental results in studies using prolonged exposure to average healthy humans. This factor is intended to account for the variation in sensitivity among the members of the human population, due to heterogeneity in human populations, and is referenced as "10H." Thus, if NOAEL is based on human data, a safety factor of 10 is usually applied to the NOAEL dose to account for variations in sensitivities between individual humans.
- Use an additional tenfold factor when extrapolating from valid results of long-term studies on experimental animals when results of studies of human exposure are not available or are inadequate. This factor is intended to account for the uncertainty

involved in extrapolating from animal data to humans and is referenced as "10A." Thus, if NOAEL is based on animal data, the NOAEL dose is divided by an additional safety factor of 10, to account for differences between animals and humans.

- Use an additional tenfold factor when extrapolating from less than chronic results on experimental animals when there are no useful long-term human data. This factor is intended to account for the uncertainty involved in extrapolating from less than chronic (i.e., subchronic or acute) NOAELs to chronic NOAELs and is referenced as "10S."
- Use an additional tenfold factor when deriving an RfD from a LOAEL, instead of a NOAEL. This factor is intended to account for the uncertainty involved in extrapolating from LOAELs to NOAELs and is referenced as "10L."
- Use an additional up to tenfold factor when extrapolating from valid results in experimental animals when the data are "incomplete." This factor is intended to account for the inability of any single animal study to adequately address all possible adverse outcomes in humans and is referenced as "10D."

MF —

- Use professional judgment to determine the MF, which is an additional UF that is greater than zero and less than or equal to ten. The magnitude of the MF depends upon the professional assessment of scientific uncertainties of the study and data base not explicitly treated above; e.g., the completeness of the overall data base and the number of species tested. The default value for the MF is 1.

In general, the choice of the UF and MF values reflect the uncertainty associated with the estimation of an RfD from different human or animal toxicity data bases. For instance, if sufficient data from chronic duration exposure studies are available on the threshold region of the critical toxic effect of a chemical in a known sensitive human population, then the UF used to estimate the RfD may be set at unity (1). That is, these data are judged to be sufficiently predictive of a population subthreshold dose, so that additional UFs are not needed (U.S. EPA, 1989d).

Determination of the RfD for a Hypothetical Example

Using the NOAEL — Consider the case of a study made on 250 animals (e.g., rats) that is of subchronic duration, yielding a NOAEL dosage of 5 mg/kg/day. Then,

$$UF = 10H \times 10A \times 10S = 1000$$

In addition, there is a subjective adjustment (MF) based on the high number of animals (250) per dose group:

$$MF = 0.75$$

These factors then give UF × MF = 750, so that

$$RfD = \frac{NOAEL}{(UF \times MF)} = \frac{5}{750} = 0.007(mg / kg / day)$$

Using the LOAEL — If the NOAEL is not available, and if 25 mg/kg/day had been the lowest dose tested that showed adverse effects,

$$UF = 10H \times 10A \times 10S \times 10L = 10,000$$

Using again the subjective adjustment of MF = 0.75, one obtains

$$RfD = \frac{LOAEL}{(UF \times MF)} = \frac{25}{7500} = 0.003(mg / kg / day)$$

Quantification of Toxicological Effects of Lead as an Example. Because of the importance of chemicals such as lead in a risk assessment, and since no RfDs are available through the U.S. EPA data base systems, an approach based on the use of ADIs may be utilized to estimate the acceptable chronic intakes (AICs) for the different potential receptor groups identified for a subject site.

Surrogate measures for oral RfDs — Marcus (1986) has calculated the ADIs for infants and children to be 19 μg/day and for adults to be 48 μg/day. Standard body weights of 16, 29, and 70 kg are used for children under 6 years, children between 6 to 12 years, and adults, respectively. Based on this information, the following AICs are calculated and used as substitute for oral RfDs for lead from the contaminated site problem:

AIC for children aged < 6 years

$$= \frac{(19 \times 10^{-3}) mg / day}{16 kg} = 1.19 \times 10^{-3}\ mg / kg / day$$

AIC for children aged 6 – 12 years

$$= \frac{(19 \times 10^{-3}) mg / day}{29 kg} = 6.55 \times 10^{-4}\ mg / kg / day$$

AIC for adults

$$= \frac{(48 \times 10^{-3}) mg / day}{70 kg} = 6.86 \times 10^{-4}\ mg / kg / day$$

Surrogate measures for inhalation RfDs — The Health Effects Assessment (HEA) (from the Environmental Criteria and Assessment Office of the U.S. EPA) bases its inhalation AIC for lead on the air standard of 1.5 μg/m^3. Inhalation rates of 0.25, 0.46, and 0.83 m^3/h are used for children under 6 years (16 kg), children between 6 to 12 years (29 kg), and adults (70 kg), respectively. Based on this information, the following inhalation AICs are calculated and used as substitute for inhalation RfDs for lead from the contaminated site problem:

AIC for children aged < 6 years

$$= \frac{\left[\left(1.5\mu g / m^3\right) \times \left(0.25 m^3 / h\right) \times (24 h / day) \times \left(10^{-3} mg / \mu g\right)\right]}{16\, kg}$$

$$= 5.63 \times 10^{-4} mg / kg / day$$

AIC for children aged $6 - 12$ years

$$= \frac{\left[\left(1.5\mu g / m^3\right) \times \left(0.46 m^3 / h\right) \times (24 h / day) \times \left(10^{-3} mg / \mu g\right)\right]}{29\, kg}$$

$$= 5.71 \times 10^{-4}\, mg / kg / day$$

AIC for adults

$$= \frac{\left[\left(1.5\mu g / m^3\right) \times \left(0.83 m^3 / h\right) \times (24 h / day) \times 10^{-3} mg / \mu g\right]}{70\, kg}$$

$$= 4.29 \times 10^{-4} mg / kg / day$$

Similar procedures may be used for estimating applicable toxicity parameters for noncarcinogenic effects of other chemicals of concern.

Inter-Conversions of RfD Values. RfD values for inhalation exposure are usually reported both as a concentration in air (mg/m^3) and as a corresponding inhaled dose (in mg/kg/day). RfD values for oral exposures are reported in mg/kg/day; an oral RfD value can be converted to a corresponding concentration in drinking water as follows:

$$mg / L \text{ in water} = \frac{oral\ RfD\left(mg / kg / day\right) \times body\, weight\ (kg)}{ingestion\, rate\ \left(L / day\right)}$$

Risk Characterization Considerations. In a risk characterization process, comparison is made between the RfD and the estimated exposure dose (EED). The EED should include all sources and routes of exposure involved. If the EED is less than the RfD, the need for regulatory concern may be small. An alternative measure also considered useful to risk managers is the "margin of exposure" (MOE), which is the magnitude by which the NOAEL of the critical toxic effect exceeds the EED, where both are expressed in the same units. Suppose the EED for humans exposed to a chemical substance (with a RfD of 0.005 mg/kg/day) under a proposed use pattern is 0.02 mg/kg/day (i.e., the EED is greater than the RfD), then:

$$NOAEL = RfD \times (UF \times MF) = 0.005 \times 1000 = 5 mg / kg\text{-}day$$

and

$$MOE = \frac{NOAEL}{EED} = \frac{5\,(mg/kg/day)}{0.02\,(mg/kg/day)} = 250$$

Because the EED exceeds the RfD (and the MOE is less than the UF × MF of 1000), the risk manager will need to look carefully at the data set, the assumptions for both the RfD and the exposure estimates, and the comments of the risk assessors. In addition, the risk manager will need to weigh the benefits associated with the case and other non-risk factors in reaching a decision on the regulatory dose (RgD), defined by

$$RgD = \frac{NOAEL}{MOE}$$

The MOE may be used as a surrogate for risk; as the MOE becomes larger, the risk becomes smaller.

Determination of SFs and UCRs

The cancer SF (also cancer potency factor or potency slope) is a measure of the carcinogenic toxicity of a chemical generally required for completing a health risk assessment. Exposure to any level of a carcinogen is considered to have a finite risk of inducing cancer associated with it, i.e., carcinogenic exposure is generally not considered to have a no-effect threshold. The SF is the cancer risk (proportion affected) per unit of dose and is usually expressed in milligrams of substance per kilogram body weight per d (mg/kg/day). For instance, to estimate risks from exposures in food, one multiplies the SF (risk per mg/kg/day), the concentration of the chemical in the food (ppm), and the daily intake (mg) of that food together; the total dietary risk is found by summing risks across all foods.

For evaluating risks from chemicals found in certain other environmental sources, dose-response measures are expressed as risk per concentration unit. These measures are called the unit risk for air (inhalation) and the unit risk for drinking water (oral). The continuous lifetime exposure concentration units for air and drinking water are usually expressed in micrograms per cubic meter ($\mu g/m^3$) and micrograms per liter (μg/L), respectively. If the fraction of the agent that is absorbed from the diet for humans and animals differs, a correction factor is applied when extrapolating the animal-derived value to humans.

Scientific investigators have developed numerous models to extrapolate and estimate low-dose carcinogenic risks to humans from high-dose carcinogenic effects usually observed in experimental animal studies. Such models yield an estimate of the upper limit in lifetime risk per unit of dose (or the UCR, or unit risk, UR). The U.S. EPA generally uses the linearized multistage model to generate UCRs. This model, known to make several conservative assumptions, results in highly conservative risk estimates, yielding overestimates of actual UCR for carcinogens; in fact, the actual risks may be substantially lower than that predicted by the upper bounds of this model (Paustenbach, 1986).

Structural similarity factors, etc. can be used to estimate cancer potency units for chemicals not having one, but that are suspected to be carcinogenic. This is achieved, for instance, by estimating the geometric mean of a number of similar compounds whose UCRs are known and using this as a surrogate value for the chemical with unknown UCR.

Derivations and Conversion of Cancer Potency Slope to Unit Risk Values. The unit risk estimates the upper-bound probability of a "typical" or "average" person contracting cancer when continuously exposed to 1 μg/m^3) of the chemical over an average 70-year lifetime. Potency estimates are also given in terms of "potency slopes"; a potency slope is the probability of contracting cancer due to exposure to a given lifetime dose in units of mg/kg/day. The potency, SF, can be converted to UCR (also UR, or unit risk factor, URF) by adopting several assumptions. The most critical factor is that the endpoint of concern must be a systematic tumor, so that potential target organs experience the same blood concentration of the active carcinogen regardless of the method of administration. This implies an assumption of equivalent absorption by the various routes of administration. The basis for these conversions is the assumption that at low doses, the dose-response curve is linear, so that

$$P(d) = SF \times \{dose\}$$

where

P(d) = response (probability) as a function of dose
SF = potency slope factor (mg/kg/day)$^{-1}$
{dose} = amount of chemical intake (mg/kg/day)

Inhalation potency factor—Risks associated with unit chemical concentration in air is estimated as follows:

$$\begin{aligned} &risk\ per\ \mu g / m^3 (air) \\ &= slope\ factor(risk\ per\ mg / kg / day) \times \frac{1}{body\ weight(kg)} \\ &\times inhalation\ rate(m^3 / day) \times 10^{-3} (mg / \mu g) \end{aligned}$$

Thus, the inhalation potency can be converted to a UCR value by applying the following conversion factor:

$$\{(kg\text{-}day)/mg\} \times \{1/70\,kg\} \times \{20\,m^3/day\} \times \{1\,mg/1000\,\mu g\} = 2.86 \times 10^{-4}$$

Thus, the lifetime excess cancer risk from inhaling 1 μg/m^3 concentration for a full lifetime is

$$UCR(\mu g / m^3)^{-1} = (2.86 \times 10^{-4}) \times SF$$

Alternatively, the potency, SF, can be derived from the unit risk as follows:

$$SF = (3.5 \times 10^3) \times UCR$$

The assumptions used involve a 70-kg body weight and an average inhalation rate of 20 m^3/day.

Risk-specific concentrations in air — Risk-specific concentrations of chemicals in air is estimated from the unit risk in air as follows:

$$Air\ concentration, \mu g / m^3 = \frac{specified\ risk\ level(R) \times body\ weight(BW)}{SF \times inhalation\ rate \times 10^{-3}}$$

$$= \frac{specified\ risk\ level(R)}{UCR} = \frac{1 \times 10^{-6}}{UCR(\mu g / m^3)^{-1}}$$

The assumptions used involves a specified risk level of 10^{-6}, a 70-kg body weight, and an average inhalation rate of 20 m^3/day.

Oral potency factor — Risks associated with unit chemical concentration in water is estimated as follows:

$$risk\ per\ \mu g / L(water)$$

$$= slope\ factor(risk\ per\ mg / kg / day) \times \frac{1}{body\ weight(kg)} \times ingestion\ rate$$

$$(L / day) \times 10^{-3}(mg / \mu g)$$

Thus, the ingestion potency can be converted to a UCR value by applying the following conversion factor:

$$\{(kg\text{-}day)/mg\} \times \{1/70\,kg\} \times \{2\ L/day\} \times \{1\,mg/1000\,\mu g\} = 2.86 \times 10^{-5}$$

Thus, the lifetime excess cancer risk from ingesting 1 μg/L concentration for a full lifetime is

$$UCR(ug / L)^{-1} = (2.86 \times 10^{-5}) \times SF$$

Alternatively, the potency, SF, can be derived from the unit risk as follows:

$$SF = (3.5 \times 10^4) \times UCR$$

The assumptions used involve a 70-kg body weight and an average water ingestion rate of 2 L/day.

Risk-specific concentrations in water — Risk-specific concentrations of chemicals in drinking water can be estimated from the oral slope factor; the water concentration corrected for an upper-bound increased lifetime risk of R is given by

$$mg / L \, in \, water = \frac{(specified\, risk\, level, R \times body\, weight, BW)}{(slope\, factor, SF \times ingestion\, rate)}$$

$$= \frac{specified\, risk\, level\,(R)}{UCR(oral)}$$

or

$$= \frac{1 \times 10^{-6} \times 70\, kg}{slope\, factor(mg / kg / day)^{-1} \times 2\ L / day} = \frac{3.5 \times 10^{-5}}{SF}$$

The assumptions used involve a specified risk level of 10^{-6}, a 70-kg body weight, and an average water ingestion rate of 2 L/day.

5.1.4 Risk Characterization

Risk characterization is the process of estimating the probable incidence of adverse impacts to potential receptors under various exposure conditions, including an elaboration of uncertainties associated with such estimates. It is the final step in the risk assessment process and the first input to the risk management process. Its purpose is to present the risk manager with a synopsis and synthesis of all the data that should contribute to a conclusion with regards to the nature and extent of the risk. The risk characterization involves the integration of the exposure and toxicity assessments to arrive at an estimate of risk to the exposed population, both qualitatively and quantitatively. The exposure estimates and toxicity values used in the risk characterization should either both be expressed as absorbed doses or both expressed as administered doses (or intakes).

Risk characterization involves the quantitative estimation of the actual and potential risks and/or hazards due to exposure to each key chemical constituent, and also the possible additive effects of exposure to mixtures of the chemicals of concern. During risk characterization, chemical-specific toxicity information is compared against both measured contaminant exposure levels and, in some cases, those levels predicted through fate and transport modeling to determine whether current or future levels at or near a site under investigation are of potential concern. The risks to potentially exposed populations from exposure and subsequent intake of the chemicals of potential concern are characterized by the calculation of noncarcinogenic hazard quotients and indices and/or carcinogenic risks. These parameters are then compared with applicable standards for risk decisions in hazardous waste management.

An adequate characterization of risks and hazards at a potentially contaminated site allows a site remediation process to be better focused. Cleanup criteria can be developed based on the "acceptable" level of risks to potential receptors. Exposures resulting in the greatest risk can be identified and site mitigation measures selected to address these issues. In this sense, the risk assessment process integrates the information obtained during a remedial investigation (RI) into a coherent set of goals for the feasibility study (FS) phase of the site investigation for a potentially contaminated site.

Aggregate Effects of Chemical Mixtures

There are numerous complexities and inherent uncertainties involved in the analysis of contaminated site problems. The wastes found at contaminated sites tend to be heterogeneous and variable mixtures that may contain several distinct compounds, distributed over wide spatial regions and several compartmental media. The toxicology of complex mixtures is not well understood, further complicating the problem. Large uncertainties exist regarding the potential for these compounds to cause various health and environmental effects. Nonetheless, there is the need to assess the cumulative health risks for several chemicals measured or predicted in any environmental medium (U.S. EPA, 1986b). The method of approach assumes additivity of effects for carcinogens when evaluating chemical mixtures or multiple carcinogens. Any carcinogens which are not included in the quantitative analysis due to lack of potency values should be identified and discussed qualitatively.

For multiple pollutant exposures to noncarcinogens and noncarcinogenic effects of carcinogens, constituents should be grouped by the same mode of toxicological action (i.e., those which induce the same toxicological endpoint, such as liver toxicity). Toxicological endpoints that will normally be considered in a hazard index with respect to chronic toxicity include cardiovascular systems (CVS); central nervous system (CNS); immune system; reproductive system (including teratogenic and developmental effects); kidney; liver; and respiratory system. Cumulative risk is evaluated through the use of a hazard index that is generated for each health "endpoint." Chemicals with the same endpoint should be included in a hazard index calculation. Strictly speaking, constituents should not be grouped together unless the toxicological endpoint is known to be the same. If any calculated hazard index exceeds unity, then the health-based criterion for the chemical mixture has been exceeded and the need for interim corrective measures must be addressed. The risk assessment process must address the multiple endpoints or effects and also the uncertainties in the dose-response functions for each effect. Generally, the risk assessment is facility specific and the calculated risks should be combined for pollutants originating from a given facility or group in the case-study affecting same receptor groups.

Adjustments for Absorption Efficiency

Absorption adjustments may be necessary in the risk characterization stage to ensure that the site exposure estimate and the toxicity value for comparison are both expressed as absorbed doses or both expressed as intakes. Adjustments may be necessary to match the exposure estimate with the toxicity value if one is based on an absorbed dose and the other is based on an intake (i.e., administered dose).

Adjustments may also be necessary for different vehicles of exposure (e.g., water, food, or soil). Furthermore, adjustments may be necessary for different absorption efficiencies, depending on the medium of exposure. In the absence of reliable information, 100% absorption is normally used for most chemicals; for metals, approximately 10% absorption may be considered as a reasonable upper bound for other than the inhalation exposure route. Adjustment procedures are discussed in the literature (e.g., U.S. EPA, 1989i).

Absorption factors should not be used to modify exposure estimates in those cases where absorption is inherently factored into the toxicity/risk parameters used for the risk characterization. Thus, "correction" for fractional absorption is appropriate only for those values derived from experiments/studies based on absorbed dose. Consequently, no "correction" due to incomplete absorption is appropriate when these standards are used. Correction for fractional absorption is appropriate in two cases in particular:

- Interaction with environmental media or other contaminants may alter absorption from that expected for the pure compound.
- Assessment of exposure via a different route of contact from what was utilized in the experimental studies establishing the SFs and RfDs.

Absorbed dose should be used in risk characterization only if the applicable toxicity parameter (e.g., SF or RfD) has been adjusted for absorption; otherwise, simply use intake (undjusted for absorption) for the calculation of risk levels.

Estimation of Carcinogenic Risks

The risk of contracting cancer can be estimated by combining information about the carcinogenic potency of a chemical and exposure to the substance. For potential carcinogens, risks are estimated as the incremental probability of an individual contracting cancer over a lifetime as a result of exposure to the potential carcinogen (i.e., the excess or incremental individual lifetime cancer risk). The carcinogenic risks are estimated by multiplying the cancer SF, which is the upper 95% confidence limit of the probability of a carcinogenic response per unit intake over a lifetime of exposure, by the estimated intakes — yielding incremental risk values. The carcinogenic effects of the chemicals of concern are calculated according to the following relationship (U.S. EPA, 1989i):

$$Risk, CR = CDI \times SF$$

where

CR = probability of an individual developing cancer (dimensionless)
CDI = chronic daily intake for long-term exposure (i.e., averaged over 70 year lifetime) (mg/kg/day)
SF = slope factor (1/[mg/kg/day])

This represents the linear low-dose cancer risk model and is valid only at low risk levels (i.e., below estimated risks of 0.01). For sites where chemical intakes are high (i.e., potential risks above 0.01), the one-hit model is used; the one-hit equation for high carcinogenic risk levels is given by the following relationship (U.S. EPA, 1989i):

$$Risk, CR = 1 - \exp(-CDI \times SF)$$

where the terms are same as defined above. On the other hand, the acceptable incremental cancer risk for a chemical is estimated by the following relationship:

$$Acceptable\ incremental\ cancer\ risk = virtually\ safe\ dose(VSD) \times slope\ factor(SF)$$

where VSD represents an acceptable chemical dose or intake (in mg/kg/day).

Aggregate Effects of Multiple Carcinogenic Chemicals. The aggregate cancer risk equation for multiple chemicals is obtained by summing the risks calculated for the individual chemicals. Thus, for multiple compounds,

$$Total\ risk = \sum_{i=1}^{n}\left(CDI_i \times SF_i\right)$$

for the linear low-dose model for low risk levels, or

$$Total\ risk = \sum_{i=1}^{n}\left(1 - \exp\left(-CDI_i \times SF_i\right)\right)$$

for the one-hit model used at high carcinogenic risk levels,

where

CDI_i = chronic daily intake for the ith contaminant
SF_i = slope factor for the ith contaminant
n = total number of carcinogens

Aggregate Effects of Multiple Carcinogenic Chemicals and Multiple Exposure Routes. For multiple compounds and multiple pathways, the overall total cancer risk for all exposure pathways and all contaminants considered in the risk evaluation will be

$$Overall\ total\ risk = \sum_{j=1}^{p}\sum_{i=1}^{n}\left(CDI_{ij} \times SF_{ij}\right)$$

for the linear low-dose model for low risk levels, or

$$Overall\ total\ risk = \sum_{j=1}^{p}\sum_{i=1}^{n}\left(1 - \exp\left(-CDI_{ij} \times SF_{ij}\right)\right)$$

for the one-hit model used at high carcinogenic risk levels,

where

CDI_{ij} = chronic daily intake for the ith contaminant and j^{th} pathway
SF_{ij} = slope factor for the ith contaminant and j^{th} pathway
n = total number of carcinogens
p = total number of pathways or exposure routes

The CDIs are estimated from the equations given previously for chemical intakes, whereas the SF values are obtained from various sources or databases, including the Integrated Risk Information System (IRIS) and the Health Effects Assessment Summary Tables (HEAST), available through the U.S. EPA, or are derived from fundamental toxicological data.

As a rule of thumb, incremental risks of between 10^{-4} and 10^{-7} are generally perceived as acceptable levels for the protection of human health and the environment, with 10^{-6} used as point of departure. Due to the realization that people may be exposed to the same constituents from sources unrelated to a specific site, it is preferred that the estimated carcinogenic risk $<< 10^{-6}$.

Population Excess Cancer Burden. The two important parameters or measures for describing carcinogenic effects are the individual cancer risk and the estimated number of cancer cases — the cancer burden. The unit risk factor multiplied by the environmental concentration, or the potency slope multiplied by the CDI as discussed above, gives the estimated individual cancer risk (i.e., the added lifetime probability that an exposed individual would contract cancer due to the source in question). The individual cancer risk from simultaneous exposure to several carcinogens is assumed to be the sum of the individual cancer risks from each individual chemical. The risk experienced by the individual receiving the greatest exposure is referred to as the "maximum individual risk." The number of cancer cases due to a specific source of emission can be estimated by multiplying the individual risk experienced by a group of people by the number of people in that group. Thus, if 10 million people experience an estimated cancer risk of 10^{-6} over their lifetimes, it would be estimated that 10 (i.e., 10 million $\times$ 10^{-6}) additional cancer cases could occur. The number of cancer incidents in each receptor area can be added to estimate the number of cancer incidents over an entire region. Thus, the excess cancer burden, B_{gi}, is given by

$$B_{gi} = R_{gi} \times P_g$$

where

B_{gi} = population excess cancer burden for i^{th} chemical for exposed group, G
P_g = exposed population group (i.e., the number of persons)
R_{gi} = excess lifetime cancer risk for i^{th} chemical for the exposed population group, G

Assuming cancer burden from each carcinogen is additive, then the total population group excess cancer burden is

$$B_g = \sum_{i=1}^{N} B_{gi} = \sum_{i=1}^{N} R_{gi} \times P_g$$

and

$$Total\ population\ burden, B = \sum_{g=1}^{G} B_g = \sum_{g=1}^{G} \left\{ \sum_{i=1}^{N} B_{gi} \right\} = \sum_{g=1}^{G} \left\{ \sum_{i=1}^{N} R_{gi} \times P_g \right\}$$

Where possible, cancer risk estimates should be expressed in terms of both individual and population risk. For the population risk, the individual upper-bound estimate of excess lifetime cancer risk for an average exposure scenario is multiplied by the size of the potentially exposed population.

Calculation of Noncarcinogenic Hazards

The overall potential noncarcinogenic effects posed by the chemicals of concern is usually expressed by the hazard index (HI). The noncarcinogenic effects of the chemicals of concern are calculated according to the following relationship (U.S. EPA, 1989i):

$$Hazard\ quotient, HQ = \frac{E}{RfD}$$

where

E = chemical exposure level or intake (mg/kg/day)
RfD = reference dose (mg/kg/day)

Aggregate of Multiple Noncarcinogenic Effects of all Chemicals. The sum total of the hazard quotients for all the chemicals of concern (affecting the same organ) gives the hazard index for a given exposure pathway. The applicable relationship is

$$Total\ hazard\ index, HI = \sum_{i=1}^{n} \frac{E_i}{RfD_i}$$

where

E_i = exposure level (or intake) for the ith contaminant
RfD_i = acceptable intake level (or reference dose) for i^{th} contaminant
n = total number of noncarcinogens

Aggregate of Multiple Noncarcinogenic Effects of all Chemicals and Multiple Exposure Routes. For multiple compounds and multiple pathways, the overall total noncancer risk for all exposure pathways and all contaminants considered in the risk evaluation will be

$$Overall\ total\ hazard\ index = \sum_{j=1}^{p}\sum_{i=1}^{n}\frac{E_{ij}}{RfD_{ij}}$$

where

E_{ij} = exposure level (or intake) for the ith contaminant and jth pathway
RfD_{ij} = acceptable intake level (or reference dose) for ith contaminant and jth pathway

The E values are estimated from the equations given previously for chemical intakes, whereas the RfD values are obtained from databases such as IRIS and HEAST, available through the U.S. EPA, or are derived from fundamental toxicological information. RfDs have been established by the U.S. EPA as thresholds of exposure to toxic substances below which there should be no adverse health impact. These thresholds have been established on a substance-specific basis for oral and inhalation exposures, taking into account evidence from both human epidemiologic and laboratory toxicologic studies.

In accordance with the U.S. EPA guidelines on the interpretation of hazard indices, for any given chemical there may be potential for adverse health effects if the hazard index exceeds unity (1). The "acceptable level" itself (i.e., the RfD) incorporates a large margin of safety, so that it is possible that no toxic effects may occur even if the "acceptable level" is exceeded. However, in interpreting the results, a reference value of HI less than or equal to 1 should be taken as the acceptable reference or standard. For HI values greater than unity (i.e., HI > 1), the higher the value, the greater is the likelihood of adverse noncarcinogenic health impacts. Indeed, if HI > 1.0, it may be necessary to segregate chemicals by organ-specific toxicity and recalculate the values, since strict additivity without consideration for target organ toxicities could overestimate potential hazards (U.S. EPA, 1989i); consequently, the HI is calculated after putting chemicals into groups with same physiologic endpoints. On the other hand, due to the realization that people may be exposed to the same constituents from sources unrelated to a specific site, it is preferred that the estimated noncarcinogenic hazard index be << 1.

Distinction Between Chronic and Subchronic Noncarcinogenic Effects. The chronic noncancer HI is represented by the following modification to the general equation presented above:

$$Overall\ chronic\ hazard\ index = \sum_{j=1}^{p}\sum_{i=1}^{n}\frac{CDI_{ij}}{RfD_{ij}}$$

where

CDI_{ij} = chronic daily intake for the ith contaminant and jth pathway
RfD_{ij} = chronic reference dose for ith contaminant and jth pathway

The subchronic noncancer hazard index is represented by the following modification to the general equation presented above:

$$\text{Overall subchronic hazard index} = \sum_{j=1}^{p}\sum_{i=1}^{n}\frac{SDI_{ij}}{RfD_{sij}}$$

where

SDI_{ij} = subchronic daily intake for the ith contaminant and j^{th} pathway
RfD_{sij} = subchronic reference dose for i^{th} contaminant and j^{th} pathway

Appropriate chronic and subchronic toxicity parameters and intakes are used for completion of such estimates.

Uncertainties Recognition

Considerable uncertainty is inherent in the overall risk assessment process. Uncertainties arise due to the use of several assumptions and inferences to complete a risk assessment. For instance, health risk assessment involves extrapolations and inferences to predict the occurrence of adverse health effects under certain conditions of exposure to chemicals in the environment, based on knowledge of the adverse effects that occur under a different set of exposure conditions (e.g., different dose levels and species). Because of these types of extrapolation and projections, there is uncertainty in the conclusions that are arrived at, due in part to the several assumptions that are part of this process. The following pertinent limitations and uncertainties relate to several components of the health risk assessment process:

- Uncertainties in extrapolations relevant to toxicity information (including inherent limitations in toxicity data — arising for several reasons such as differences in the general knowledge of the toxic effects of different chemicals and uncertainties in interspecies and intraspecies extrapolation) and differences of exposure conditions (with respect actual scenarios). Some chemicals have been extensively studied under a variety of exposure conditions in several species, including humans; others may have limited investigations done on them. Because data that specifically identify the hazards to humans associated with exposure to the various chemicals of concern under the conditions of likely human exposure do not exist, it is necessary to infer those hazards by extrapolating from data obtained under other conditions of exposure, generally in experimental animals. This introduces three types of uncertainties: that related to extrapolating from one species to another (i.e., uncertainties in interspecies extrapolation), those relating to extrapolation from a high-dose region curve to a low-dose region (i.e., uncertainties in intraspecies extrapolation), and those related to extrapolating from one set of exposure conditions to another (i.e., uncertainties due to differences in exposure conditions).
- Representativeness of sampling data (including limitations in determining exposure concentrations and modeling) to data of the actual population being sampled. Uncertainties arise from random and systematic errors in the type of measurement and sampling techniques used. For instance, it is critical that sample detection limits are

lower than both the applicable standards or criteria and the concentration which may present health risks; however, this often becomes a source of uncertainty in sample analysis. Professional judgment is also frequently used to fill data gaps, based on engineering and scientific assumptions, which also has some inherent uncertainties associated with it.

- Limitations in model form, including how close to reality the model function and output are, together with model imperfections. Exposure scenarios and constituent transport models contribute uncertainty to risk assessments; transport models typically oversimplify reality, contributing to uncertainty. The natural variability in environmental and exposure-related parameters causes variability in exposure factors and, therefore, in exposure estimates developed on this basis.
- Considerable uncertainty associated with the toxicity of chemical mixtures. That is, the effects of combining two chemicals may be synergistic (effect when outcome of combining two chemicals is greater than the sum of the inputs), antagonistic (effect when the outcome is less than the sum of the two inputs), or under potentiation (i.e., when one chemical has no toxic effect, but when combined with another chemical that is toxic produces a much more toxic effect). Indeed, chemicals present in a mixture can interact to yield a new chemical or one can interfere with the absorption, distribution, metabolism, or excretion of another. Notwithstanding all these, in general, risk assessments assume toxicity to be additive.

In general, uncertainty is difficult to quantify, or at best, the quantification of uncertainty is itself uncertain. Thus, the risk levels generated in a risk assessment are useful only as a yardstick and decision-making tool for prioritization of problems, rather than being construed as actual expected rates of disease, or adversarial impacts in exposed populations. It is used only as an estimate of risks, based on current level of knowledge coupled with several assumptions. Quantitative descriptions of uncertainty, which could take into account random and systematic sources of uncertainty in potency, exposure, intakes, etc., would help present the spectrum of possible true values of risk estimates, together with the probability (or likelihood) associated with each point in the spectrum.

Model Uncertainties. Because of the various limitations and uncertainties, the results of a risk assessment cannot be considered an absolutely accurate determination of risks. Most of the techniques used for compensating for the uncertainties (such as the use of large safety factors, conservative assumptions, and extrapolation models) are designed to err on the side of safety. For these reasons, many regulatory agencies tend to use the so-called linearized multistage model for conservatism. In fact, several models have been proposed for the quantitative extrapolations of carcinogenic effects to low dose levels. However, among these models, the U.S. EPA recommends a linearized multistage model (U.S. EPA, 1986a). The linearized multistage model conservatively assumes linearity at low doses. Alternative models that are generally less conservative do exist which do not assume a linear relationship. There is often no sound basis, in a biological sense, for choosing one model over another. When applied to the same data, the various models can produce a wide range of risk estimates. The model recommended by the U.S. EPA produces among the highest estimates of risk and thus provides a greater margin of protection for human health. Moreover, this model does not provide a "best estimate" or point estimate of risk, but rather an upper-bound probability that the actual risk will be less than the predicted risk 95% of the time. However, given that no single model will apply for all chemicals, it is important to identify risk models on a case-by-case basis. In fact, Huckle (1991) suggests a presentation of the best estimate of risk (or range, with an added margin of safety) from two or three appropriate models, or a single value based on "weight-of-evidence," rather than using simply the linearized multistage model. Exceptions may occur, however, for cases of poorly studied chemicals.

Uncertainties in Uncertainty Adjustments. Experimental studies to determine the carcinogenic effects due to low exposure levels usually encountered in the environment generally are not feasible. This is because such effects are not readily apparent in the relatively short time frame over which it is generally possible to conduct such a study. Consequently, various mathematical models are used to extrapolate from the high doses used in animal studies to the doses encountered in exposure to ambient environmental concentrations. Extrapolating from a high dose (of animal studies) to a low dose (for human effects) introduces a level of uncertainty which could be significantly large. For instance, NOAELs and SFs from animal studies are usually divided by a factor of 10 to account for extrapolation from animals to humans and by an additional factor of 10 to account for variability in human responses. Given the recognized differences among species in responses to toxic insult, and between strains of the same species, it is apparent that additional uncertainties will be introduced when quantitative extrapolations and adjustments are made in the dose-response evaluation.

Potential for Risk Underestimation. It is always possible that a chemical whose toxic properties have not been thoroughly tested may be more toxic than originally believed or anticipated. For instance, a chemical not tested for carcinogenicity or teratogenicity may in fact display those effects. Furthermore, a limitation of analysis for selected "indicator chemicals" may have some limiting (even if insignificant) effects. The following factors, among others, can typically underestimate health impacts associated with chemicals evaluated in a risk assessment:

- Lack of potency data for some carcinogenic chemicals
- Risks due to compounds formed in environmental media (such as transformation products) that are not quantified
- All risks are assumed to be additive, although certain combinations of exposure may have synergistic (greater than additive) effects

Potentials for Overestimating Risks. A number of factors may cause an analysis to overestimate risks, including

- Many unit risk and potency factors are often considered plausible upper-bound estimates of carcinogenic potency, whereas the true potency of the chemical could be considerably lower.

- Exposure estimates are often very conservative.
- Possible antagonistic effects, for chemicals in which the combined presence reduces toxic impacts, are not accounted for.

5.1.5 Potential Applications

In the course of typical investigations of potentially contaminated sites, efforts are made to adequately characterize the site so that appropriate corrective actions can be implemented. Generally, risk assessment techniques can be employed to better develop the site characterization, site assessment, and corrective action plans. The scope of applications for the health risk assessment methodology discussed may vary greatly; the following specific applications are identified as part of the more common uses:

- Preliminary screening for potential problems (incorporating an analysis of baseline risks, and a consistent process to document potential public health and environmental threats from potentially contaminated sites)
- Evaluation and ranking of potential liabilities from hazardous waste facilities and properties
- Corrective measures evaluation and selection of remedial alternatives (i.e., risks posed by alternative remedial actions can be assessed before implementation)
- Prioritization of hazardous waste sites for remedial action (i.e., this helps to prioritize cleanup actions by providing consistent data for the rank-ordering of potentially contaminated sites)
- Development of target cleanup criteria for potentially contaminated sites (i.e., this provides the basis to determine levels of chemicals that can remain at a site or in environmental media without impacting public health and the environment)
- Site selection in hazardous waste management for siting of hazardous waste management facilities, including disposal sites
- Field sampling design and identification of data needs and/or data gaps

Risk assessment provides a logical, rational, and methodologically consistent approach to making cost-effective decisions. It is therefore almost imperative to make risk assessment an integral part of all investigations for potentially contaminated sites and environmental media, except that the level of detail will be case specific, ranging from qualitative through semiquantitative to detailed quantitative analyses.

5.2 METHODS OF AIR IMPACTS ASSESSMENT

CERCLA (1980) and SARA (1986) mandate the characterization of all contaminant migration pathways from hazardous wastes into the environment and an evaluation of the resulting health and environmental impacts. Furthermore, there is increasing concern that air emissions from hazardous waste sites may present a significant source of human exposure to toxic or hazardous substances. In fact, significantly low-level air emissions could pose significant threats if toxic or carcinogenic compounds are present at potentially contaminated sites, even under baseline or undisturbed conditions. Furthermore, emissions during remedial actions — especially ones involving excavation — may be much higher than baseline conditions. Emissions from RCRA and similar facilities may also pose significant threats to an impact zone. The emissions of critical concern relate to volatile organic chemicals (VOCs), semi-VOCs, particulate matter, and other chemicals associated with wind-borne particulates such as metals, PCBs, dioxins, etc. Volatile chemicals may be released into the gaseous phase from such sources as landfills, surface impoundments, contaminated surface waters, open/ruptured tanks or containers, etc. Also, there is the potential for subsurface gas movements into underground structures such as pipes and basements and eventually into indoor air. Additionally, toxic chemicals adsorbed to soils may be transported to the ambient air as particulate matter or fugitive dust.

Once released to the ambient air, a contaminant is subject to simultaneous transport and diffusion processes in the atmosphere; these conditions are significantly affected by meteorological, topographical, and source factors. Additional fundamental atmospheric processes (other than atmospheric transport and diffusion) that affect airborne contaminants include transformation, deposition, and depletion. The extent to which all these atmospheric processes act on the contaminant of concern determines the magnitude, composition, and duration of the release; the route of human exposure; and the impact of the release on the environment. Several methods exist for estimating air emissions (CAPCOA, 1990; CDHS, 1986; U.S. EPA, 1990b), including

- Direct emissions measurement
- Indirect emissions measurement
- Air monitoring and/or modeling
- Emissions (predictive) modeling

In all cases, site-specific data should be used whenever possible to increase the accuracy of the emission rate estimates. In fact, the combined approach of environmental fate analysis and field monitoring should provide an efficient and cost-effective strategy for investigating the air pathways impacts on potential receptors under varying meterological conditions.

Air Emissions Classification

Hazardous waste site air emissions may be classified as either point or area sources. Point sources include vents (e.g., landfill gas vents) and stacks (e.g., incinerator and air stripper releases); area sources are generally associated with fugitive emissions (e.g., from landfills, lagoons, and contaminated surface areas). Fugitives (associated with area sources) are released at ground level and disperse there, with less influence of winds and turbulence; point sources, generally, come from a stack and are emitted with an upward velocity, often at a height significantly above ground level. Thus, point sources are more readily diluted by mixing and diffusion, further to being at greater heights, so that ground-level concentrations are reduced. This means that area-source fugitives may be up to 100 times as hazardous on a mass per time basis compared to point sources and stacks. It is therefore important that the risk assessor describe the source type adequately in assessing potential air impacts. For instance, during the implementation of corrective action programs for potentially contaminated sites, air emissions may occur from several processes, such as thermal destruction; air stripping of groundwater; in situ venting; soils handling (e.g., excavation and transportation); and stabilization/solidification. Protocols for estimating emission levels for contaminants (fugitive, particulate, stack, VOCs, metals, etc.) from these sources are available in the literature (e.g., CAPCOA, 1990; CDHS, 1986; Mackay and Leinonen, 1975; Mackay and Yenn, 1983; Thibodeaux and Hwang, 1982; U.S. EPA, 1989e, f, g; U.S. EPA, 1990b).

Categories of Air Contaminant Emissions

Air contaminant emissions from hazardous waste sites fall into two basic categories/classes — gas phase emissions and particulate matter emissions. The emission mechanisms associated with gas phase and particulate matter releases are quite different. Gas phase emissions primarily involve organic compounds, but may also include certain metals (such as mercury), and such emissions may be released through several mechanisms such as volatilization, biodegradation, photodecomposition, hydrolysis, and combustion (U.S. EPA, 1989e). Particulate matter emissions at hazardous waste sites can be released through wind erosion, mechanical disturbances, and combustion. Hazardous chemicals, such as metals, can also be adsorbed onto particulate matter and thereby transported with the inert materials. For airborne particulates, the particle size distribution plays an important role in inhalation exposure. Large particles tend to settle out of the air more rapidly than small particles, but may be important in terms of noninhalation exposure. Very small particles (< 2.5 to 10 μm diameter) are considered to be respirable and thus present a greater health hazard than the larger particles. In addition to gas phase and particulate emissions, aerosols are an added mechanism — particularly for combustion processes. Aerosol processes can lead to chemical transformations, especially on surfaces of particulates. They are important because the usual control devices (i.e., baghouses, precipitators, etc.) do not remove aerosols with any degree of efficiency.

5.2.1 Air Pathway Analyses

An air pathway analysis (APA) is a systematic approach which involves the application of modeling and monitoring methods to estimate emission rates and concentrations of air contaminants. It can be used to assess actual or potential receptor exposures to atmospheric contamination. Thus, an APA is basically an exposure assessment for the air pathway that can be used to provide input to an overall risk assessment for a hazardous waste site or facility. A first step in assessing air impacts associated with a hazardous waste site is to evaluate site-specific characteristics and the chemical contaminants present at the site to determine if transport of hazardous chemicals to the air is of potential concern. The main components of an APA are

- Characterization of air emission sources
- Determination of the effects of atmospheric processes
- Evaluation of receptor exposure potential (for receptors of interest for various exposure periods)
- Estimation of receptor intakes and doses

The purpose of the exposure assessment is to estimate the extent of potential receptor exposures to the identified chemicals of concern. This involves emission quantification, modeling of environmental transport, evaluation of environmental fate, identification of potential exposure routes, identification of populations potentially at risk from exposures, and the estimation of both short- and long-term exposure levels.

Assessing the Potentials for Air Emissions

Hazardous waste sites may pose potential public health risks due to possible airborne soil particulate matter laden with toxic chemicals and/or volatile emissions from volatile or semivolatile chemical species generated from the site. The most important chemical parameters to consider in the evaluation of volatile air emissions are the vapor pressure and the Henry's law constant. Vapor pressure is a useful screening indicator of the potential for a chemical to volatilize from the media in which it exists. The Henry's law constant is particularly important in estimating the tendency of a chemical to volatilize from a surface impoundment or water; it also indicates the tendency of a chemical to partition between the soil and gas phase from soil water in the vadose zone or groundwater.

Several site-specific factors (such as site integrity, presence of cracks or fissures, soil organic content, soil moisture, microbial activity, ambient temperature, and the site area) influence volatilization, emission rates, and ambient chemical concentrations at hazardous waste sites and vicinity. Persistent chemicals which readily adsorb to soil (i.e., high K_d or K_{oc} values) are resistant to biodegradation, are not readily volatilized, and are most likely to remain in surface soils; hazardous waste sites with surface contamination may generate contaminated airborne particulates of potential concern. In particular, particles of < 10 μm diameter are usually considered respirable and subject to air quality control and standards. Details of the exposure assessment process and the assumptions involved, as well as guidelines and criteria for assessing the potential for particulate and volatile emissions from hazardous waste sites, can be found in the literature relating to air toxics programs (e.g., CDHS, 1986; CAPCOA, 1990).

Modes of Exposure

Contaminants released to the atmosphere are subjected to a variety of physical and chemical influences — including transformation, deposition, and depletion processes — which are but secondary to transport and diffusion processes, albeit important in several situations. Although deposition depletes concentrations of the contaminants in air, it increases the concentrations on vegetations and in soils and surface water bodies. Furthermore, deposited contaminants are subject to some degree of resuspension, especially through wind erosion for wind speeds exceeding 10 to 15 mi/h (or about 15 to 25 km/h). Consequently, the primary modes of exposure to toxic chemicals released to the atmosphere are direct inhalation, ingestion of vegetation that is contaminated as a result of deposition of particles, ingestion of contaminated dairy and meat products from animals eating contaminated crops, and dermal contact. In particular, particulate emissions can cause human exposures in a variety of ways, including

- Inhalation of respirable particulates
- Deposition on soils, leading to human exposure via dermal absorption or ingestion
- Deposition on crops or pasture lands and introduction into the human food chain
- Deposition on waterways, uptake through aquatic organisms, and eventual human consumption

The air quality evaluation can be used to support the preparation and analysis of public health and environmental assessment plans for various remedial alternatives. Atmospheric dispersion and emission source modeling can be used with appropriate air sampling data as input to an atmospheric exposure assessment.

Dispersion Modeling

Atmospheric dispersion modeling is an APA approach that can provide calculated contaminant concentrations at potential receptor locations of interest based on emission rate and meteorological data. Thus, atmospheric dispersion modeling (for Superfund and other activities) has become an integral part of the planning and decision-making process for the protection of public health and the environment. The following information, at a minimum, need to be collected and reviewed to support the air modeling program design:

- Source data (including contaminant toxicity factors, offsite sources, etc.)
- Receptor data (including identification of sensitive receptors, local land use, etc.)
- Environmental data (including dispersion data, climatology, topographic maps, soil and vegetation, etc.)
- Previous APA data (including ARARs summary, air monitoring, emission rate monitoring and modeling, dispersion modeling, etc.).

Naturally, the accuracy of the model predictions depends on the accuracy and representativeness of the input data. In general, model input data will include emissions and release parameters, meteorological data, and receptor locations. Previous air quality data that address calculated air concentrations of contaminants known to exist at the site can provide insight into existing levels of air toxic chemicals of interest. Chemical-specific information will help in identifying the indicator compounds to be modeled and the modeling methodologies to employ. Existing air monitoring data (if any) for the site area can be used in designing receptor grid and selecting chemicals to be modeled. This can also provide insight to background concentrations.

Table D.1 (Appendix D4) contains a listing of selected models that may be utilized for the execution of an APA; this list is by no means complete and exhaustive, but demonstrates the range and variations in the available models. There are many levels of complexity implicit in the models used in air pathways assessments; a brief description of the mathematical structure for such models is presented in Appendix D5.

5.2.2 Risk Characterization

The process for characterizing risks from air emissions follows the same steps as previously described under Section 5.1, once the respirable and/or inhalation concentrations are determined by appropriate field measurements and/or modeling practices. A number of relevant general assumptions are made in the assessment; important among these are

- Air dispersion and particulate deposition modeling of emissions adequately represent the fate and transport of chemical emission to ground level.
- The composition of emission products found at ground level is identical to the composition found at source, but concentrations are different.
- Conservatively assumed that the potential receptors are exposed to the maximum annual average ground-level concentrations from the emission sources for 24 h/day, throughout an assumed 70-year lifetime.
- Conservatively assumed that there is no losses of chemicals through transformation and other processes (such as volatilization or photodegradation).

In the estimation of potential risks from fugitive dust inhalation, an estimate of respirable (<10 μm aerodynamic diameter, denoted by the symbol PM-10 or PM_{10}) fraction and concentrations are required. Amount of nonrespirable (>10 μm aerodynamic diameter) concentrations may also be needed to estimate deposition of wind-blown emissions which will eventually reach potential receptors via other routes such as ingestion and dermal exposures.

5.2.3 Potential Applications

The air pathways methods of analysis find several applications, including the following:

- Estimation of VOC emission rates from landfills and ponds. A number of models exist that can be used for estimating the emissions of VOCs from landforms (represented as ponds, sludges, landfills, etc.) (e.g., Mackay and Leinonen, 1975; Thibodeaux and Hwang, 1982; U.S. EPA, 1986).
- Particulate inhalation exposure from fugitive dust. Several models are available to calculate intakes and/or doses as a result of the inhalation of fugitive dust (e.g., U.S. EPA, 1989e; CAPCOA, 1990).
- Population exposure to vapor emissions from cracked concrete foundations. A risk characterization scenario involving exposure of populations to vapor emissions from cracked concrete foundations/floors can be conducted. Applicable estimation procedures exist in the literature (e.g., Lyman et al., 1990).
- Decisions on alternative corrective measures. The potential use of risk assessment in the evaluation of alternative corrective measures involving excavation processes at a

hazardous waste site considers the levels of air emissions from available remedial options.

Once the emission rates are determined, and the exposure scenarios are defined, decisions can be arrived at with respect to potential air impacts for specified activities.

5.3 METHODS OF APPROACH IN ECOLOGICAL RISK ASSESSMENT

Ecological (or environmental) risk assessment (ERA) is a qualitative and/or quantitative appraisal of the actual or potential effects of hazardous waste sites on plants and animals other than humans and domesticated species (U.S. EPA, 1989j). Although environmental risks are associated with exposure of ecological receptors other than humans, such exposures could also affect human populations through the food chain or through changes in the ecosystem.

A contaminant entering the environment will cause adverse effects only if a number of conditions are satisfied (U.S. EPA, 1989j):

- Contaminant exists in a form and concentration sufficient to cause harm
- Contaminant comes in contact with organisms or environmental media with which it can interact
- The interaction that takes place is detrimental to life functions

An ERA must normally be conducted to assess unreasonable risks from hazardous substances to the environment. This type of assessment is directed more at the area of ecotoxicology (i.e., the study of the adverse effects of toxic materials on nonhuman organisms) and nontarget species (including endangered and threatened species). An ecological assessment seeks to determine the nature, magnitude, and transience or permanence of observed or expected effects of contaminants introduced into an ecosystem. Several consequential factors of concern are involved, in particular,

- Loss of habitat
- Reduction in population size
- Changes in community structure
- Changes in ecosystem structure and function

The process used to evaluate environmental or ecological risks parallels that for the evaluation of human health risks (Section 5.1). In both cases, potential risks are determined by the integration of information on chemical exposures with toxicological data for the contaminants of potential concern.

5.3.1 Objectives of an ERA

The objectives of an ERA include identifying and estimating the potential ecological impacts associated with the presence of chemicals of concern at a potentially contaminated site, with specific focus on determining the following:

- Biological and ecological characteristics of the study area
- Types, forms, amounts, distribution, and concentration of the contaminants of concern
- Migration pathways to and exposure of ecological receptors to pollutants
- Habitats potentially affected and populations potentially exposed to contaminants
- Actual and/or potential ecological effects/impacts and overall nature of risks

The assessment focuses on the impacts of the chemicals to the terrestrial and aquatic flora and fauna that inhabit the site and vicinity. However, ecological assessments may identify new or unexpected exposure pathways which could potentially affect human populations. On the other hand, human health risks in most situations are more substantial than ecological risks; considering that the mitigative actions taken to alleviate risks to human health are often sufficient to mitigate potential ecological risks at the same time, extensive ecological investigations is usually not required for most contaminated site problems.

5.3.2 General Considerations

In the design of an ERA program for potentially contaminated sites, common elements of populations, communities, and ecosystems should be defined; this forms the basis for developing consistent but site-specific characterization strategies for the individual sites. The following considerations should be included in the general activities conducted as part of the ecological investigation:

- Definition and role of ERA within the context of contaminated site assessment and remediation
- The concept of acceptable ecological risk
- Evaluation and selection of appropriate ecological endpoints at the population, community, and ecosystem levels
- Acute and chronic risk and secondary hazards or toxicity tests
- Evaluation of exposure and biomarkers of exposure
- Basis of the strategy adopted for the ERA that is appropriate for the case-specific problem
- Considerations in field sampling design; data analysis and evaluation; and ecological monitoring programs
- Application of ERA to the site remediation plans (i.e., for derivation of site-specific remediation objectives)

The collection and review of the existing data base on terrestrial and aquatic ecosystems, wetlands and floodplains, threatened and endangered species, soils, and other topics relevant to the study should form a prime basis for identifying any information gaps in conducting an ERA at contaminated sites.

General Types of Data Requirements

Unlike endangerment assessment (EA) for human populations, endangerment assessments for ecological species (including endangered species) often lack significant amount of critical and credible data needed to conduct a comprehensive quantitative risk assessment. Nonetheless, the pertinent data requirements should be identified and categorized in as far is practical. For instance, survey information on soil types, vegetation cover, and residential migratory wildlife may be required for terrestrial habitats, whereas comparative information needed for freshwater and marine habitats will most likely include survey data on abundance, distribution, and kinds of populations of plants (phytoplankton, algae, and higher plant forms) and animals (fish, macro- and microinvertebrates) living in the water column and in or on the bottom. A complete tabulation or schedule for all the ecosystems types relevant to the region or locality should be compiled. The biological and ecological information collected should include a general identification of the flora and fauna associated with the site and its vicinity, with particular emphasis placed on sensitive environments, especially endangered species and their habitats and those species consumed by humans or found in human food chains.

Types of Ecosystems

The types of ecosystems vary with climatic, topographical, geological, chemical, and biotic factors. Each ecosystem type has unique combinations of physical, chemical, and biological characteristics and thus may respond to contamination in its own unique way. The physical and chemical structure of an ecosystem may determine how contaminants affect its resident species, and the biological interactions may determine where and how the contaminants move in the environment and which species are exposed to particular concentrations. The following general ecosystems will normally be investigated as part of the ERA process:

- Terrestrial ecosystems (to be categorized according to the vegetation types that dominate the plant community and terrestrial animals)
- Wetlands (which are areas in which topography and hydrology create a zone of transition between terrestrial and aquatic environments)
- Freshwater ecosystems (in which environment, the dynamics of water temperature, and movement of water can significantly affect the availability and toxicity of contaminants)
- Marine ecosystems (which are of primary importance because of their vast size and critical ecological functions)
- Estuaries (which support a multitude of diverse communities and are more productive than their marine or freshwater sources—further to being important breeding grounds for numerous fish, shellfish, and species of birds)

The ecosystem types pertaining to the case-specific study should be defined and integrated into the overall ERA. For instance, an evaluation of the condition of aquatic communities may proceed from two directions.

The first direction will consist of examining the structure of the lower trophic levels as an indication of the overall health of the aquatic ecosystem. This approach emphasizes the base of the aquatic food chain and may involve studies of plankton (microscopic flora and fauna), periphyton (including bacteria, yeast, molds, algae, and protozoa), macrophyton (aquatic plants), and benthic macroinvertebrates (e.g., insects, annelid worms, mollusks, flatworms, roundworms, and crustaceans). These lower levels of the aquatic community are studied to determine whether they exhibit any evidence of stress. If the community appears to have been disturbed, the goal will be to characterize the source(s) of the stress and, specifically, to focus on the degree to which the release of waste constituents has caused the disturbance or possibly exacerbated an existing problem. An example of the latter would be the further depletion of already low dissolved oxygen levels in the hypolimnion of a lake or impoundment through the introduction of waste with a high COD and specific gravity. Benthic macroinvertebrates are commonly used in studies of aquatic communities. These organisms usually occupy a position near the base of the food chain. Just as importantly, however, their range within the aquatic environment is restricted, so that their community structure may be referenced to a particular stream reach or portion of lake substrate. By comparison, fish are generally mobile within the aquatic environment, and evidence of stress or contaminant load may not be amenable to interpretation with reference to specific releases. The presence or absence of particular benthic macroinvertebrate species, sometimes referred to as "indicator species," may provide evidence of a response to environmental stress. A "species diversity index" provides a quantitative measure of the degree of stress within the aquatic community and is an example of a common basis for interpretation of the results of studies of aquatic biological communities. Measures of species diversity are most useful for comparison of streams with similar hydrologic characteristics or for the analysis of trends over time within a single stream.

The second approach to evaluating the condition of an aquatic community focuses on a particular group or species, possibly because of its commercial or recreational importance or because a substantial historic data base already exists; it is done through selective sampling of specific organisms, most commonly fish, and evaluation of standard "condition factors" (e.g., length, weight, girth). In many cases, receiving water bodies are recreational fisheries, monitored by state or federal agencies. In such cases, it is common to find some historical record of the condition of the fish population, and it may be possible to correlate operational records at the waste management facility with alterations in the status of the fish population. Additional detail regarding the application of other measures of community structure can be found in the literature of ecological assessments and related subjects (e.g., Curns and Dickson, 1973; U.S. EPA, 1973; USGS, 1977).

Nature and Ecological Effects of Contaminants

Although a contaminant may cause illness and/or death to individual organisms, its effects on the structure and function of ecological assemblages or interlinkages may be measured in terms quite different from those used to describe individual effects. A discussion that includes a wider spectrum of ecological effects on individual organisms as well as the ecological interlinkages is therefore an important part of a comprehensive ERA. Furthermore, the biological, chemical, and environmental factors perceived to influence the ecological effects of contaminants should be identified and described. The variety of environmental variables (such as temperature, pH, salinity, water hardness, soil composition) that can influence the nature and extent of effects of a contaminant on an ecological receptor should be identified and annotated. Biological factors such as species susceptibility to contaminants, characteristics affecting population abundance and distribution, and movement of chemicals in food chains should also be evaluated.

Selection of Target Species

It generally is not feasible to evaluate each species that may be present in a potentially contaminated site and vicinity. Consequently, selected target or indicator species will normally be chosen for analysis. By using reasonably conservative assumptions in the overall assessment, it is hoped that adequate protection of these indicator species will provide protection for all other environmental species as well. In fact, not every organism may be suitable for use as target species in the evaluation of contaminant impacts on ecological systems. Thus, specific general considerations and assumptions should be applied in selecting target species. The following criteria should be considered in the selection of target species for the ERA of contaminated sites:

- Species that are threatened, endangered, rare, or of special concern
- Species that are valuable for several purposes of interest to human populations (i.e., of economic and societal values)
- Species critical to the structure and function of the particular ecosystem in which they inhabit
- Species that serve as indicators of important changes in the ecosystem
- Relevance of species in the site and vicinity

The presence of threatened or endangered species and/or habitats critical to their survival should be documented. The location of such species should be determined. Similarly, sensitive sport or commercial species and habitats essential for their reproduction and survival should be identified. Information on these should be obtained from appropriate federal, provincial, state, regional, local, and/or private institutions and other organizations. It is important to consider both the effects of chemical pollutants on the endangered population as well as on the habitats critical to their survival.

Ecological Endpoints

The development of an ERA requires the identification of one or several ecological assessment endpoints. These endpoints define the environmental resources which are to be protected and which, if found to be impacted, determine the need for corrective actions. Appropriate site-specific endpoints are an important component to the implementation of corrective actions that is based on the ERA results.

5.3.3 Exposure Assessment

A chemical may be released into the environment and then be subject to physical dispersal into the air, water, soils, and/or sediments. The chemical may then be

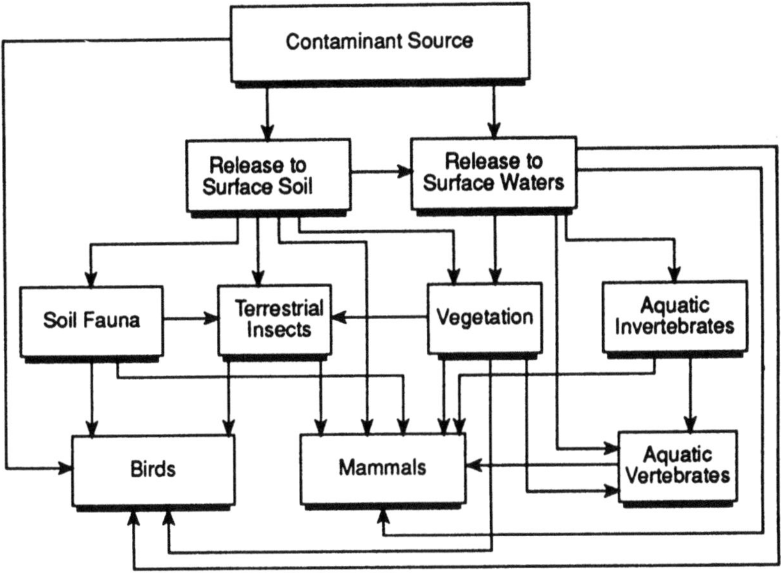

Figure 5.6 A simplified example conceptual diagram of pollutant transport pathways of chemicals through an ecosystem.

transported spatially and into the biota and perhaps chemically or otherwise modified or transformed and degraded by both abiotic processes (such as photolysis, hydrolysis, etc.) and/or microorganisms present in the environment. The resulting transformation products may have different environmental behavior patterns and toxicological properties from the parent chemical. Nonetheless, it is the exposure scenarios that determine the potential for any adversarial impacts. Thus, the nature of the target organisms (e.g., birds, fish, etc.) must be identified together with the nature of exposure (such as acute, chronic, or intermittent), so that a complete ERA can be conducted.

The objectives of the exposure assessment are to identify potential ecological receptors, determine exposure routes for receptors, and estimate the degree of contact and/or intakes of the chemicals of concern by the potential receptors. The exposure routes are selected based on the behavior patterns and/or ecological niches of the target species and communities. The amount of exposure is based on the maximum plausible exposure concentrations of the chemicals as obtained from data evaluation (including monitoring and modeling information). A food chain (also, food web) should be constructed for the target species in order to develop appropriate exposure scenarios. The total daily intake (in mg/kg/day) of the target species are then calculated by summing the amounts of constituents ingested and absorbed from all sources (e.g., soil, vegetation, surface water, fish tissue, and other target species), and also that absorbed through inhalation and dermal contacts.

Exposure in crops consist mainly of root uptake and also possible absorption from deposited contaminants on plant foliage/leaves. Figure 5.6 is a simplified and generic conceptual representation of typical interlinkages due to consumption, uptake, and absorption by elements in the food chain.

Calculation of Chemical Intakes by Potential Receptors

Procedures for the estimation of chemical concentrations in vegetation (from deposition, uptake, etc.) or in animal products and others (due to animal exposures via inhalation, soil ingestion, etc.) are presented in the literature of endangerment assessment (e.g., CAPCOA, 1990). Typically, wildlife or game daily chemical exposure and the resulting body burden may be estimated as follows:

$$E = C \times F \times \left[\sum_i D_{di} D_{ci} D_{ui} \right] \times BAF \times (1/BW)$$

where

- E = exposure or average daily dose (μg/kg/day)
- C = chemical concentration in media (e.g., soil) averaged over appropriate exposure period (70 years if to be subsequently consumed by humans, etc.)
- F = food consumption rate (g/day)
- D_{di} = component i of diet
- D_{ci} = consumption factor of i
- D_{ui} = bioaccumulation or uptake coefficient for i
- BAF = bioavailability factor
- BW = body weight (kg)

For example, a deer's daily exposure to a chemical (e.g., dioxin) in an ecological setup and the resulting body burden need to be estimated before a hunter's oral exposure from eating deer meat can be estimated. In this case, the deer's average daily uptake of dioxins will be given by:

$$E = C \times F \times \{G_d G_c G_u + I_d I_c I_b\} \times BAF \times (1/BW)$$

where

- G_d = grass component of diet
- G_c = grass consumption factor
- G_u = soil chemical uptake coefficient for grass
- I_d = rodent component of diet
- I_c = rodent consumption factor
- I_b = rodent bioavailabilty factor

Subsequently, the steady-state (or average) body burden level of chemicals in the deer, at which intake equals excretion, can be determined as a function of the exposure, *viz*:

$$S = \frac{\{k \times E \times T_{1/2}\}}{t}$$

where

- S = steady-state body burden level (μg/kg)
- k = chemical-specific kinetic rate constant (d^{-1})
- E = exposure, average daily dose (μg/kg/day)
- $T_{1/2}$ = chemical half-life (d)
- t = time interval between exposures (taken to be one d for daily continuous exposure) (d)

Consequently, the potential human intake is calculated as follows:

$$E_{deer} = (S) \times (R_c) \times (ED) \times (BAF) \times (1/BW) \times (1/LT) \times (CF)$$

where

- E_{deer} = exposure, LADD (μg/kg/day)
- S = steady-state chemical level in deer (μg/g)
- R_c = consumption rate (kg/day)
- ED = exposure duration (d)
- BAF = bioavailability factor
- BW = body weight (kg)
- LT = lifetime (d)
- CF = conversion factor (=10^3 g/kg)

Similar calculations can be performed for other game animal species. The details of the calculations involved in such analyses can be found in the literature elsewhere (e.g., Paustenbach, 1988).

5.3.4 Ecotoxicity Assessment

Similar to what is done for the human health endangerment assessment, the scientific literature is reviewed to obtain ecotoxicity information for the chemicals of potential concern relevant to the ecological receptors identified during the exposure assessment. Subsequently, critical toxicity values of the contaminants of concern are derived for the target receptor species and ecological communities of concern.

5.3.5 Risk Characterization

Environmental effects include changes in aquatic and terrestrial natural resources brought about by exposure to chemical substances. Knowledge of such effects may be important in analyzing chemical migration pathways and potential human exposures; however, knowledge of environmental effects is also important in analyzing non-human risks of a chemical release. Risk characterization steps similar to that discussed under the human endangerment assessment (Section 5.1) is followed in attempts to characterize ecological risks. The doses determined for the ecological receptors and community during the exposure assessment are integrated with the applicable and appropriate toxicity values and information derived in the ecotoxicity assessment to arrive at plausible risk estimates. This process includes a discussion of uncertainties.

In characterizing risks or threats to environmental receptors, the following determinations are made:

- The probability or likelihood of an adverse effect occuring
- The temporal character of each effect (transient vs. reversible vs. permanent)
- The magnitude of each effect
- Receptor populations or habitats that will be affected

Ecological risk characterization entails both temporal and spatial components. Beyond criteria exceedances, risk characterization is most likely to be a weight-of-evidence judgment.

Uncertainty Analysis

All risk estimates are dependent on numerous assumptions and consideration of the many uncertainties that are inherent in the ERA process. In any evaluation of the level of risk associated with a site, it is necessary to address the level of confidence or the uncertainty associated with the estimated risk. Uncertainties are associated with both toxicity information (such as ecotoxicity values and site-specific dose-response assessments) and exposure assessment information. Consequently, factors that may significantly increase the uncertainty of the ERA should be identified and addressed in a qualitative and, where possible, quantitative manner. Three common qualitatively distinct sources of uncertainty to be evaluated are the inherent variability, parameter uncertainty, and model errors.

5.3.6 Potential Applications

The ERA allows for the identification of habitats and organisms that may be affected by the chemicals of potential concern present at a contaminated site. For instance, drainage waters from agricultural lands often contain elevated levels of several agricultural chemicals (pesticides, herbicides, etc.). An important scientific consideration for decision makers studying contamination related to agricultural chemical usage is fish and wildlife habitat. In addition, when the presence of such chemicals of concern results in direct or indirect exposure to humans, a public health concern also arises that needs to be evaluated together with the potential ecological impacts.

5.4 PROCEDURES FOR DEVELOPING RISK-BASED CRITERIA IN REMEDIAL ACTION PLANNING

An important goal in EA (viz, health and environmental risk assessment) is to provide a framework for developing the risk information necessary to assist in the

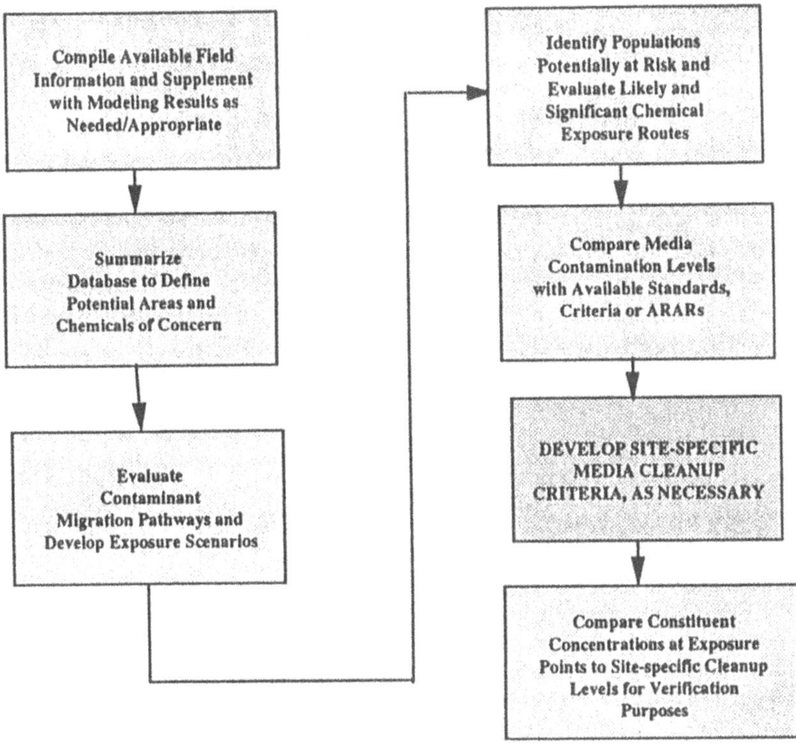

Figure 5.7 Process flow for developing site remediation information base within a risk assessment protocol.

decision-making process relating to the remediation of potentially contaminated sites (Figure 5.7). The ultimate goal in corrective actions at hazardous waste sites is to protect public health and the environment. An important consideration in developing a remedial action plan (RAP) for a contaminated site is the level of cleanup to be achieved—i.e., "how clean is clean?" This could be the driving force behind remediation costs. It is therefore prudent to allocate adequate resources to develop the cleanup criteria.

The cleanup level is a site-specific criterion that a remedial action would have to satisfy in order to keep exposure to potential receptors at or below an "action level" (AL). The AL is the concentration of a chemical in a particular medium that, when exceeded, presents significant risk of adverse impact to potential receptors (CDHS, 1986). The ALs tend to drive the cleanup process for a contaminated site; but then the ALs may not always result in "acceptable" risk levels due to the nature of the critical exposure scenarios, receptors, and other conditions that are specific to the candidate site. It is therefore recommended to recalculate the residual risks that will remain if remediation is carried out to the ALs and, if necessary, to develop more stringent and health-protective levels that will meet the "acceptable" risk level criteria.

The assessment of health and environmental risks play an important role in the remedial investigation/feasibility study (RI/FS), in the corrective action planning, and also in the risk mitigation and risk management strategies for potentially contaminated sites.

5.4.1 The General Protocol for Deriving Health and Environmental Criteria

Cleanup decisions can be developed by back-modeling acceptable soil concentrations (ASCs) or the maximum acceptable concentration (MAC) based on chemical intake/dose that represents acceptable MELs. In general, once the degree of risk or hazard resulting from existing levels of contamination has been determined, then by working backwards with information based on dilution/attenuation potentials, degradability and degradation products, distribution coefficients, and/or mass balances, media criteria may be established for a potentially contaminated site.

A site-specific cleanup limit (called a "recommended site cleanup limit," RSCL) can be applied to contaminated sites to determine the level of cleanup warranted. The RSCL is derived from a consideration of the acceptable MEL along with probable exposure levels and the type of exposure scenarios. In fact, the type of exposure scenarios envisioned may drive the RSCL. The RSCLs derived for the various pathways from defined exposure scenarios will aid in developing cleanup options such that public health and/or the environment are not jeopardized by any residual contamination. Criteria set by the use of this method should aid in the selection of appropriate remedial technologies that will meet performance goals. Depending on what chemical constituents are identified to be removed and the levels of cleanup required, one remedial technology may be determined as being more efficient than an alternative one.

5.4.2 Derivation of Site-Specific Risk-Based Cleanup Levels

This section identifies procedures for estimating site-specific media cleanup levels for potentially contaminated sites. The general approach used is based on the assumption that a contaminant deposited in the soil will be distributed among the environmental compartmental media in accordance with its physical and chemical properties, as well as the properties of the media. The cleanup criteria developed will be site-specific and dependent on the exposure scenarios identified for the particular problem.

Risk appraisal protocols may be used for developing cleanup criteria in the restoration of contaminated sites. Once the degree of risk or hazard due to current levels of contamination has been determined, the extent to which contaminant levels at the source must be reduced to *not* exceed some "acceptable" risk level can be obtained by "back-modeling" from an acceptable exposure limit of concentration at the most critical potential receptor location. This type of calculation is most appropriate for determining acceptable site cleanup levels that will adequately protect the soil, groundwater, surface water, and atmospheric pathways. Factors to consider in setting such threshold values include the following:

- Degree and type of risk involved
- Actual or potential intended use of site
- Exposure pathways (from source to receptors)
- PAR
- Site characteristics affecting exposure
- Variability in exposure scenarios

A stochastic approach that tries to represent uncertainties may be introduced by use of distribution functions to describe appropriate parameters used in the estimation or back-modeling process. The use of distribution functions allows for the introduction of elements of fuzziness of the data used or to estimate the range and distribution of data as necessary.

Site Cleanup Limits Based on Attenuation-Dilution Factors

Several authors (e.g., Dawson and Sanning, 1982; Brown, 1986; U.S. EPA, 1987; Santos and Sullivan, 1988) have given various relations for computing cleanup limits that either account for attenuation alone or dilution only. It is suggested here that the cleanup level, RSCL, be derived from the following relationship that accounts for both total attenuation and net dilution concurrently:

$$RSCL = \{Std\} \times \pi_m \{AF\} \times \pi_m \{DF\}$$

where

RSCL = cleanup level in soil (mg/kg)
Std = receiving media criteria or standard to be met
$\pi_m^{\{AF\}}$ = cumulative attenuation factor, defining the loss of contaminant during transport — i.e., product of the intermedia distribution constants
$\pi_m^{\{DF\}}$ = cumulative dilution factor during transport, i.e., product of ratios of (concentration at source) to (concentration at receiving media)

For instance, in the case of leachate transport from soil into a river, define

$$\pi_m \{DF\} = DF_1 \times DF_2$$

where DF_1 represents the dilution factor for leachate entering into groundwater (the aquifer dilution factor) and DF_2 from aquifer into river (the river dilution factor). These are represented by

$$DF_1 = \frac{(total\, aquifer\, flow)}{(flow\, through\, contaminated\, site)}$$

$$DF_2 = \frac{(river\, low\, flow)}{(aquifer\, flow)}$$

For a single intermedia transfer, the soil cleanup level may be estimated by a more simplified relationship. For instance, for leachate transport from soil into groundwater that is a source of drinking water supply, the RSCL is given by

$$RSCL = DWS \times AF \times DF$$

where

DWS = drinking water standard
AF = attenuation of contaminant in soil (approximate typical values in range 10 to 100)
DF = dilution of contaminant by groundwater (approximate typical values in range 1 to 10)

Soil cleanup levels may be established such that groundwater quality standards are met by using such simplistic relationships. These approaches represent reasonably conservative ways of setting cleanup criteria. Such conservative ways imply being protective of public health by ensuring that risks are not underestimated.

Estimation of MACs by Mass Balance Analyses

Simple mass balance analytical relationships can be applied between various environmental compartments by coupling current contaminant loadings with required or anticipated loadings together with available media standards in order to derive cleanup limits. A back-calculation to obtain the required cleanup criteria can then be carried out according to the following simple relationship:

$$C_{max} = \frac{C_{std}}{C_r} \times C_s$$

where

C_{max} = maximum acceptable source concentration
C_{std} = receiving media criteria for receptors
C_r = receiving media concentration
C_s = source concentration prior to cleanup

Consider, for instance, a situation where standards exist that should be met for a creek adjoining a hazardous waste site. The chronic water quality standards that should be met are compiled. Then, by performing back-calculations, based on contaminant concentrations in the creek as a result of the current constituents loading from the site, a conservative estimate is made as to what the MAC should be at this site. Such concentration limit ensures that the creek is not adversely impacted, based on the pathway defined in the exposure scenario. Based on the appropriate maximum acceptable soil concentration value on site, C_{max}, the site may be cleaned up to such levels as *not* to impact the surface water quality. This then becomes the RSCL for the subject site.

The Media AL as a Cleanup Limit

The media ALs are calculated for both systemic toxicants and carcinogens; the more stringent of the two, where both exist, is selected as the site cleanup limit, so as to achieve levels more protective of the public health and the environment. This would represent the maximum acceptable contaminant levels for site cleanup. The following criteria, assuming dose additivity, must be met by a cleanup action:

$$\sum_{j=1}^{p}\sum_{i=1}^{n}\frac{C_{\max ij}}{AL_{ij}} < 1$$

where i is a medium number and j is contaminant index.

Risk-specific doses (RSDs) for carcinogens — Health-based criteria for carcinogens, known as RSDs (also, virtually safe dose, VSD), may be developed for site-specific cases. An RSD is the daily dose of a carcinogenic chemical that, over a lifetime, will result in an incidence of cancer equal to a specified risk level. This yields environmental concentrations that, under site-specific intake assumptions, correspond to cumulative lifetime cancer risks of 10^{-6} for Class A and B carcinogens, or 10^{-5} for Class C carcinogens. Alternative risk levels ranging from 10^{-7} to 10^{-4} may be appropriate in some cases, such as when sensitive populations (e.g., nursing homes or schools) are potentially exposed or when potential for exposures are rather remote.

The RSD (or VSD) concentrations are calculated based on the appropriate risk level, and this becomes the AL for the specific problem. The governing equation for calculating ALs for carcinogenic constituents present at a potentially contaminated site is given by

$$C_{\max} = \frac{(R \times BW \times LT \times CF)}{(SF \times I \times A \times ED)}$$

where

C_{max} = AL (equal to the RSD or VSD) in medium of concern (e.g., soil @ mg/kg)
R = specified (acceptable) risk level (dimensionless)
BW = body weight (kg)
LT = assumed lifetime (years)
CF = conversion factor (=10^6 mg/kg for soil ingestion exposure; 1.00 for water ingestions)
SF = cancer SF (1/[mg/kg/day])
I = intake assumption (e.g., soil ingestion rate @ g/day)
A = absorption factor (dimensionless)
ED = exposure duration (years)

Each of the carcinogens is assigned 100% of the acceptable/estimated excess carcinogenic risk (of 1×10^{-6}) in calculating the health-protective concentration levels. Estimated excess carcinogenic risk is not allocated among the carcinogens because each individual carcinogen may have different mode of biological action and target organs. This may be different for noncarcinogenic effects.

For instance, allowable exposure due to ingestion of 2 L of water containing methylene chloride (with oral SF = 7.5×10^{-3} [mg/kg/day]$^{-1}$) by a 70-kg weight adult over a 70-year lifetime is given by

$$C_{methchl} = \frac{\left[10^{-6} \times 70 \times 70 \times 1\right]}{\left[0.0075 \times 2 \times 1 \times 70\right]} = 0.005 mg / L$$

That is, the RSD allowable exposure level for methylene chloride with an excess lifetime cancer risk level of 10^{-6} is estimated to be 5 μg/L.

Next, human exposure levels for ingestion of both water and fish is determined from the following modified equation:

$$RSD = \frac{\left[R \times BW \times LT \times CF\right]}{\left[SF\left(I + (DIA \times BCF)\right) \times A \times ED\right]}$$

$$= \frac{\left[10^{-6} \times 70 \times 70 \times 1\right]}{\left[0.0075 \times \left(2 + (0.0065 \times 0.91)\right) \times 1 \times 70\right]} = 0.005 mg / L$$

assuming an average daily consumption of aquatic organisms, DIA, of 6.5 g/day and a bioconcentration factor (BCF) of 0.91 L/kg. Thus, the allowable exposure levels for drinking water and eating aquatic organisms contaminated with methylene chloride is also approximately 5 μg/L. Indeed, EPA guidance would probably require reducing $C_{methchl}$ to about 20% of the value calculated, in view of the fact that there are other sources of exposure (e.g., air, food, etc.); this should generally be factored into the overall assessment and, in particular, in decision-making and risk management processes.

ALs for noncarcinogenic effects — For health-based criteria for noncarcinogenic effects of both carcinogens and noncarcinogens, known as RfDs (also ADI), there is an estimate of the threshold exposure limit below which no adverse effects are anticipated. The governing equation for calculating ALs for noncarcinogenic effects of constituents present at a potentially contaminated site is given by

$$C_{max} = \frac{(RfD \times BW \times CF)}{(I \times A)}$$

where

C_{max} = AL in medium of concern (e.g., soil @ mg/kg)
RfD = reference dose (mg/kg/day)
BW = body weight (kg)

CF = conversion factor (=10^6 mg/kg for soil ingestion exposure; 1.00 for water ingestions)
I = intake assumption (e.g., soil ingestion rate @ mg/day)
A = absorption factor (dimensionless)

The right hand side may be multiplied by a percentage factor to account for contribution to HI by each noncarcinogenic chemical subgroup.

For example, in an alternate concentration limit (ACL) application, limits for human exposure may be based on the RfDs. These are calculated by assuming the ingestion of 2 L/day of drinking water by a 70-kg adult. Thus, the allowable exposure concentration is calculated as follows:

$$C_{rfd} = \frac{[RfD mg / kg - day \times BW\, kg]}{[DW\, L / day \times A]}$$

Consider ethylbenzene with an RfD of 0.1 mg/kg/day. Then, the allowable surface water exposure concentration (based on surface water ingestion) is estimated at

$$C_{ebz} = \frac{[0.1 mg / kg - day \times 70 kg]}{[2\, L / day \times 1]} = 3500 \mu g / L$$

For exposure through ingestion of contaminated fish (quantified by using the BCF, assuming an adult will eat, say, 6.5 g of fish per d, then the maximum allowable surface water concentration for exposure by drinking water and fish ingestion (also referred to as the drinking water equivalent level, DWEL) is

$$C_{ebz}[mg / L] = \frac{[RfD\, mg / kg - day \times BW\, kg]}{[2\, L / day + (0.0065 kg \times BCF\, L / kg)] \times 1}$$

The BCF for ethylbenzene is estimated at approximately 37.5 L/kg, so that

$$C_{ebz}[mg / L] = \frac{[0.1 mg / kg - day \times 70 kg]}{[2\, L / day + (0.0065 kg \times 37.5\, L / kg)]} = 3120 \mu g / L$$

For substances that are both carcinogenic and systemically toxic, the lower of the RSD or RfD criterion should be used.

ACLs

The hazardous waste regulations under RCRA requires owners and operators of hazardous waste facilities to prevent potential migration of hazardous chemicals into groundwater. To this end, groundwater protection standards may be established by the use of ACLs rather than background levels or other specified standards such as MCLs. However, to obtain an ACL, permit applicants considering to establish and use ACLs must demonstrate that the hazardous constituents detected in the groundwater will not pose any substantial present or potential hazard to human health or the environment at the specified ACL levels (U.S. EPA, 1987).

ACL demonstrations tend to be site-specific, and may be a simple manipulation of such parameters as the MCLs, RfDs, RSDs, or other risk-specific levels. General descriptions of the geologic and hydrogeologic/hydrologic conditions at the facility should be part of all ACL demonstrations. In general, ACLs based on attenuation mechanisms may not be acceptable unless there is an in-depth evaluation and discussion of all pertinent processes involved. A detailed health risk assessment may be required of ACL demonstrations, and this is intended to determine allowable concentrations at the point of exposure (i.e., the point at which potential exposure to chemicals is assumed and where allowable exposure is met) for chemicals for which ACLs are requested; this may not be a requirement if the point of exposure is established at the point of compliance (i.e., the point on the downgradient side of the regulated unit where the groundwater protection standard is met, and where the ACL is set) and either MCLs or regulatory agency-approved allowable dose levels are used at this point. If there is a likely pathway for other ecological receptors to become exposed to chemicals, then environmental toxicity factors should be examined, including critical habitats and endangered or threatened species.

Miscellaneous Criteria Estimation Methods

Cleanup criteria may be derived from various other analytical relations, such as from media equilibrium partitioning coefficients. For instance, based on the assumption that the distribution of contaminants among various compartments in sediment is controlled by continuous equilibrium exchanges, chemical-specific partition coefficients (*viz*, sediment-water equilibrium partitioning) are determined (with respect to organic carbon content of sediments) which can be used to predict contaminant concentrations in sediment and/or water. Similarly, chemical-specific partition coefficients (*viz*, sediment-biota partitioning coefficient) are determined and used to predict distribution of the contaminant between sediment and benthic organism and/or interstitial water and benthic organism; BCFs are assumed constant and independent of organism or sediment. Other equilibrium relationships discussed in Section 5.1.2 and elsewhere can be employed for criteria estimations.

5.4.3 Health-Protective Cleanup Limits and Goals for Corrective Action

It is proposed here that where the computed "residual" risks following remediation will exceed allowable limits, then the RSCL be estimated in the same way as the soil AL or by mass balance analyses or from attenuation-dilution concepts, but with a consideration for the aggregation of the total number of chemicals present at the subject site. Thus, assuming each compound contributes proportionately to the HI or cancer risk, the RSCL is estimated according to the following simplistic relationship:

$$RSCL = \frac{C_{max}}{N}$$

where

C_{max} = AL in medium of concern (e.g., soil @ mg/kg)
N = number of chemical contributors to overall hazard index or cancer risk, as appropriate

Alternatively, the acceptable risk level may be apportioned between the chemical contributors to the overall target risk or hazard, assuming each constituent contributes equally or proportionately to the total acceptable risk or hazard. The risk fraction obtained for each compound can then be used to obtain the acceptable limit, RSCL, again working from the relationships established previously for the computation of the media ALs. In this case, the RSCL may be estimated by proportionately aggregating — or rather disaggregating — the target cancer risk (for carcinogens) or noncancer HI (for noncarcinogenic effects) between the chemicals of potential concern. This is carried out according to the following approximate relationships:

$$RSCL = \frac{(\% \times R \times BW \times LT \times CF)}{(SF \times I \times A \times ED)}$$

for carcinogens, and

$$RSCL = \frac{(\% \times RfD \times BW \times CF)}{(I \times A)}$$

for noncarcinogenic effects for chemicals having the same toxicological endpoints. All the terms are the same as defined previously, and % represents the proportionate contribution from a specific constituent to the target/acceptable risk level (for carcinogens) or HI (for noncarcinogenic effects of chemicals with the same physiological endpoint). One may also choose to use weighting factors (based on, say, absorption and carcinogenic class) such that Class A carcinogens are given twice as much weight as Class B, etc. in carrying out the apportionment of chemical contributions to target risk levels.

To further account for the various exposure routes, the following approximate relationship may be used for carcinogens:

$$RSCL = \frac{(\% \times R \times BW \times LT \times CF)}{(SF_\alpha \times I_\alpha \times A \times ED)} + \frac{(\% \times R \times BW \times LT \times CF)}{(SF_\beta \times I_\beta \times A \times ED)}$$

where α, β represent the fraction of total soil exposure due to inhalation and oral contact (i.e., ingestion + dermal absorption), respectively; SF_α, SF_β are the inhalation and oral SFs, respectively. Similarly, for noncarcinogenic effects involving chemicals that have the same toxicological endpoints,

$$RSCL = \frac{\left(\% \times RfD_\alpha \times BW \times CF\right)}{\left(I_\alpha \times A\right)} + \frac{\left(\% \times RfD_\beta \times BW \times CF\right)}{\left(I_\beta \times A\right)}$$

where α, β represent the fraction of total soil exposure due to inhalation and oral contact (i.e., ingestion + dermal absorption), respectively; RfD_α, RfD_β are the inhalation and oral RfDs, respectively.

Table 5.6 summarizes the steps for the computational process involved in calculating the RSCL. The assumption used here for allocating estimated excess carcinogenic risk is that all carcinogens have the same mode of biological actions and target organs; otherwise, excess carcinogenic risk is not allocated among carcinogens, but rather each assumes the same value in the computational efforts. Similarly, for the noncarcinogenic effects, the total HI is apportioned only between chemicals with the same toxicological endpoint.

A more comprehensive approach to partitioning or combining risks would involve more complicated mathematical manipulations, such as by the use of linear programming algorithms; for carcinogens, the algorithm assures that the sum of all risks from the chemicals involved over all pathways equal to or less than the set target risk (say, 1×10^{-6}), and for noncarcinogenic effects, that the sum of all hazard quotients over all pathways for chemicals with the same toxicological endpoints equal to or less than the HI criterion of 1.0.

Incorporating Decay/Degradation Rates into the Estimation of Cleanup Criteria

Since exposure scenarios used in calculating the RSCL consider the fact that exposures could be occuring over long time periods (up to an assumed lifetime of 70 years), it is important to consider whether degradation or other transformation of the chemical at the source could occur, this being done in a more detailed analysis. In such cases, the chemical and biological degradation properties of the contaminant should be reviewed. Assuming first-order kinetics as an example, an approximation of the degradation effects can be achieved by multiplying the concentration by the following factor:

$$\frac{\left(1 - e^{-kt}\right)}{kt}$$

where

k = chemical-specific degradation rate constant (d^{-1})
t = time period over which exposure occurs (d)

Table 5.6 Computational Steps for Estimating RSCLs

Steps for Calculating Cleanup Levels for Carcinogens

(1) Apportion total acceptable risk, TAR, among all N carcinogens as follows:

$$Target\ specified\ risk\ for\ chemical\ i,\ R_{\mathrm{i}} = \frac{TAR}{N};\left(e.g.,= \frac{10^{-6}}{N}\right)$$

(2) Calculate target CDI (mg/kg/day) for each chemical

$$Target\ CDI\ for\ chemical\ i,\ \mathrm{TCDI}_{\mathrm{i}} = \frac{R_i}{SF_i} = \frac{(1)}{SF_i}$$

(3) Calculate intake factor, IF, for all chemicals as follows:

$$Intake\ factor,\ IF_i = \frac{\left(I \times A_i \times ED\right)}{(BW \times LT)} \times CF$$

(4) Calculate RSCL (mg/kg) for each chemical

$$RSCL\ for\ chemical\ i, RSCL_i = \frac{TCDI_i}{IF_i} = \frac{(2)}{(3)}$$

Steps for Calculating Cleanup Levels of Noncarcinogenic effects

(1) Apportion total acceptable hazard intex, THI, among all N chemicals as follows:

$$Target\ specified\ risk\ for\ chemical\ i, HI_i = \frac{THI}{N};\left(e.g.,= \frac{1}{N}\right)$$

(2) Calculate target CDI (mg/kg/day) for each chemical

$$Target\ CDI\ for\ chemical\ i, TCDI_i = HI_i \times RfD_i = (1) \times RfD_i$$

(3) Calculate intake factor, IF, for all chemicals as follows:

$$Intake\ factor, IF_i = \frac{(I \times A_i)}{BW} \times CF$$

(4) Calculate RSCL (mg/kg) for each chemical

$$RSCL\ for\ chemical\ i, RSCL_i = \frac{TCDI_i}{IF_i} = \frac{(2)}{(3)}$$

For a first-order decaying substance, k is estimated from the following relationship:

$$t_{1/2}[days] = \frac{0.693}{k} \quad or \quad k\left[days^{-1}\right] = \frac{0.693}{t_{1/2}}$$

where $t_{1/2}$ is the half-life, which is the time after which the mass of a given substance will be one-half its initial value.

Thus, in the derivation of a modified cleanup limit, an adjusted RSCL (i.e., $RSCL_a$) — which is based on the original RSCL, a degradation rate coefficient, and a specified exposure duration — will be calculated to represent the true cleanup limit. Consequently,

$$RSCL_a = \frac{RSCL}{degradation\ factor(DGF)} = RSCL \times \frac{kt}{\left(1 - e^{-kt}\right)}$$

where

$RSCL_a$ = adjusted RSCL (i.e., remedial goals based on the RSCL and first-order degradation rate, in mg/kg, mg/L, etc.)

RSCL = risk-specific concentration (i.e., allowable chemical level protective of both carcinogenic and noncarcinogenic effects for exposure to chemicals, in mg/kg, mg/L, etc.)

$$DGF = \frac{\left(1 - e^{-kt}\right)}{kt}$$

The applicable relationship makes the following assumptions:

- First-order degradation/decay during complete exposure period
- Decay/degradation is initiated at time, t = 0 years
- The RSCL is the average allowable concentration over the exposure period

If significant degradation is likely to occur, RSCL calculations become much more complicated. In that case, predicated source contaminant levels must be calculated at frequent intervals and summed over the exposure period.

5.4.4 Potential Applications

A hierarchical approach will generally be utilized in the determination of health and environmental criteria for case-specific problems, as illustrated by Figure 5.8. The commonly available criteria used in developing RAPs have previously been discussed in this text. The site remediation risk assessment protocols find applications in developing RAPs for various contaminated sites and media and for related risk management programs. By utilizing methodologies that determine cleanup criteria based on risk assessment methodologies, corrective action programs can be carried out in a more cost-effective and efficient manner.

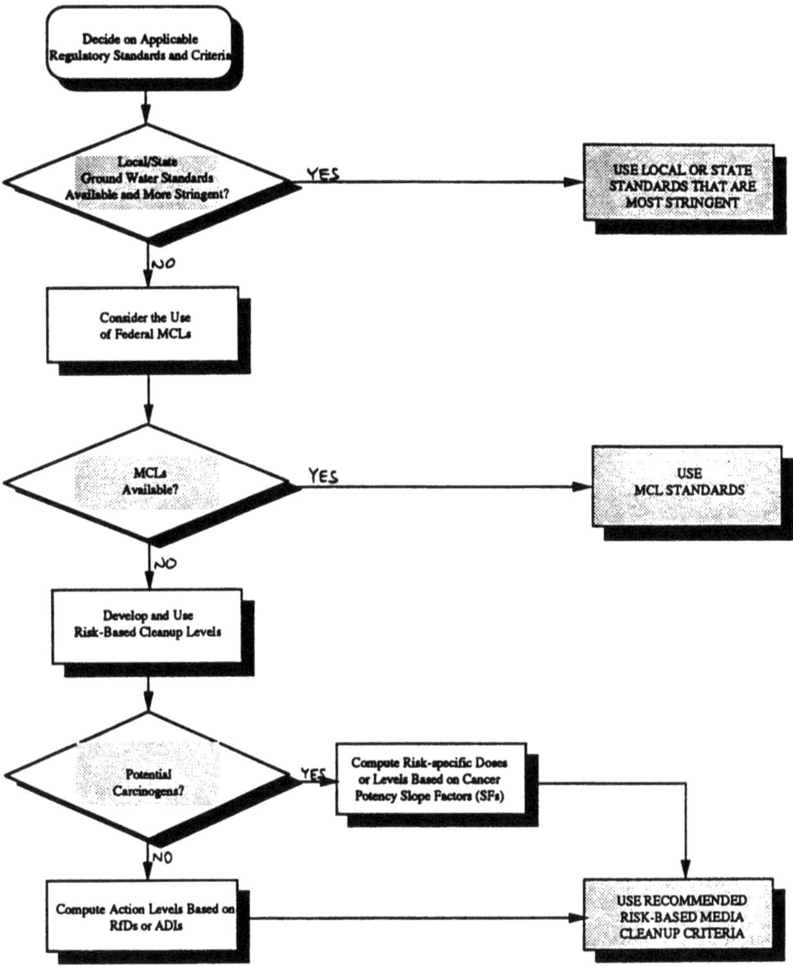

Figure 5.8 Illustrative example of a hierarchical approach to setting media cleanup criteria.

5.5 PROBABILISTIC RISK ASSESSMENT (PRA)

PRA is a method used to quantify the frequency of occurrence, the degree of system response, and the magnitude of consequences of accident events or system failures at hazardous facilities, industrial facilities, and other technological systems (Figure 5.9). Although its most extensive use has been in the nuclear industry, PRA has also been successfully used to evaluate safety for a number of industrial facilities, as well as for undertakings such as the transport of hazardous materials. A PRA may use fault tree or event tree analysis. With the PRA method, risk is defined in terms of frequency and magnitude of consequences or the failure probability (Henley and Kumamoto, 1981; Glickman and Gough, 1990), *viz*

$$Risk\left[\frac{consequence}{time}\right] = frequency\left[\frac{events}{unit\ time}\right] \times magnitude\left[\frac{consequence}{event}\right]$$

$$\text{Partial Failure Probability, Pr\{Fail\}} = \text{Pr\{Hazard\}} \times \text{Pr\{Outcome\}} \times \text{Pr\{Consequence\}}$$

$$\text{Total Failure Probability, TFP} = \sum \text{Pr\{Fail\}}$$

where:

Pr{Hazard} is the probability of an accident or failure;

Pr{Outcome} is the conditional probability of system response and outcome due to the failure event;

Pr{Consequence} is the conditional probability of adverse effects from accident sequence(s).

Figure 5.9 Risk definition in a probabilistic risk assessment.

Subsequently, the product of frequency and magnitude or the failure probability are summed over all incident sequences or failure pathways to yield total failure probability:

$$\textit{Total failure probability} = \sum_i \left\{ \textit{frequency}\left[\frac{\textit{events}}{\textit{unit time}}\right] \times \textit{magnitude}\left[\frac{\textit{consequence}}{\textit{event}}\right]\right\}$$

One major objective of risk analysis and/or assessment is to assign a probability (or frequency) to possible consequences of system failure. The concept of probability of failure required in the risk evaluation is usually defined by using the likelihood of structural breach and/or an accident event. Estimation of the applicable probability values are achieved by the use of reliability theory and/or expert judgements or by use of stochastic simulations and historical information. A Bayesian approach may always be employed to update estimates on the basis of additional information acquired in time. Through probabilistic modeling and analyses, uncertainties can be assessed properly and their effects on a given decision accounted for systematically. In this manner the risk associated with given decision alternatives may be delineated and then appropriate corrective measures taken accordingly.

5.5.1 Some Basic Concepts in Probability Theory

Quantitative probability estimates in risk analyses are often based on available statistical data and to some extent on expert judgement. In this regards, several concepts of probability may be required to assess the available information to be utilized in the evaluation process. A summary of the notations and theorems pertaining to some probability definitions and concepts commonly used in a PRA are given below. A more detailed review can be found in several standard textbooks of statistics and probability theory (e.g., Siddall, 1983; Larsen and Marx, 1985; Miller and Freund, 1985; Freund and Walpole, 1987).

- *Unconditional probability, Pr{A}*, is the fraction of items resulting in event A among the complete set of all items.
- *Conditional probability, Pr{A/B}*, is the probability of occurrence of event A, given that event B has already occurred. This is the proportion/fraction of items resulting in event A among the total set of items that give rise to event B. It is expressed by

$$\Pr\{A / B\} = \frac{\Pr\{A \cap B\}}{\Pr\{B\}}$$

where Pr{ — } is the probability of the specified event and {A ∩ B} denotes the intersection of events A and B.

- *Joint probability, Pr{A & B}*, is the fraction of items giving rise to the simultaneous occurrence of events A and B among the complete set of all items.

Thus,

$$\Pr\{A \cap B\} = \Pr\{B\} \times \Pr\{A / B\}$$

- *Independence.* Event A is said to be independent of event B if, and only if, *Pr{A/B} = Pr{A}*. This means the probability of event A is unaffected by the occurrence of event B and vice versa, so that

$$\Pr\{A \& B\} = \Pr\{A \cap B\} = \Pr\{A\} \times \Pr\{B\}$$

- *Mutually exclusive events* are events that cannot occur simultaneously. Thus,

$$\Pr\{A \cap B\} = 0$$

where Pr{ — } is the probability of the specified event and {A ∩ B} denotes the intersection of events A and B (i.e., the elements common to both events A and B).

- *Inclusive probability* is defined as

$$\Pr\{A \text{ or } B\} = \Pr\{A \cup B\} = \Pr\{A\} + \Pr\{B\} - \Pr\{A \cap B\}$$

where Pr{ — } is the probability of the specified event, and $\{A \cup B\}$ denotes the union of events A and B.

- *Total probability theorem,* states that given a set of mutually exclusive collectively exhaustive events $B_1, B_2, \ldots, B_n$, the probability Pr{A} of another event A can be expressed by

$$\Pr\{A\} = \Pr\{A \cap B_1\} + \Pr\{A \cap B_2\} + \ldots + \Pr\{A \cap B_n\}$$
$$= \sum_i \Pr\{A \cap B_i\} = \sum_i \Pr\{A / B_i\} \times \Pr\{B_i\}$$

- *Bayes' theorem,* which follows from the total probability theorem, states that the conditional probability of B_j given the event A is given by

$$\Pr\{B_j / A\} = \Pr\{B_j \cap A\} / \Pr\{A\} = \Pr\{A \cap B_j\} / \Pr\{A\}$$
$$= \frac{\Pr\{A / B_j\} \Pr\{B_j\}}{\sum_i \left[\Pr\{A / B_i\} \times \Pr\{B_i\}\right]}$$

Thus, Bayes' theorem allows updating of prior probabilities, $\Pr\{B_j\}$, to yield posterior probabilities, $\Pr\{B_j/A\}$, given new information A.

5.5.2 Risk Characterization Using Logic Trees

Decision/logic trees provide an approach of analyzing a sequence of uncertain events or conditions at hazardous waste facilities that could potentially lead to adverse consequences. Two types of decision trees can usually be used to identify sets of events leading to system failure; these commonly used logic trees are

- Event trees (which use deductive logic)
- Fault trees (that use inductive logic)

Event and fault trees are useful techniques for identification and quantification of accidents and failures of technological systems, including failure scenarios at hazardous waste facilities. In many situations, fault tree analysis is used to supplement event tree modeling by using the former to establish the appropriate probabilities of the event tree branches. The technique can indeed help designers anticipate risk in order to correct problems at the design stage rather than through retrofit technologies. Event tree concepts can also aid in developing exposure and accident scenarios in its applications to hazardous waste management issues.

PRA Concepts in Event Tree Formulation

The event tree is a diagram that illustrates the chronological ordering of event scenarios in a problem requiring decision analytical evaluation. Each event pathway is shown by a branch of the event tree. An event tree uses deductive logic, starting

HAZARD --> SYSTEM RESPONSE --> OUTCOME --> EXPOSURE --> CONSEQUENCE

Figure 5.10 The risk pathway concept. (Adapted from Bowles et al., 1987.)

with an initiating event and then using forward logic to enumerate all possible sequences of subsequent events that will help determine other possible outcomes and consequences. In probabilistic evaluations, the event tree structure requires that each event level be defined by its probability, which is conditional on preceding events in the tree structure.

The event tree model can be used to display the paths of the events or actions that relate to the safety or potential for failure of hazardous waste facilities and also the anticipated consequences for the various pathways. An event tree provides a diagrammatic representation of event sequences that begin with an initiating event or hazard and terminate in one or more undesirable consequences. Events identified as part of a failure scenario can be displayed in a tree structure that represents a sequence of events in progression, displaying branching points where several possibilities can be anticipated that can lead to an event at the top. This technique basically is an algorithm in which it is possible to assign probabilities to each of the events. Then, by simply multiplying or adding probabilities, the overall chance of failure can be calculated for a given period of time.

In general, risk can be modeled as a chain of interconnected events, according to the pathway concept illustrated in Figure 5.10. This methodology models risk as a chain of interconnected events through the use of event tree analysis (ETA). The approach allows for a systematic consideration of all potential loading conditions that may be brought to bear on a system, the potential exposure scenarios following system breach, and the consequences of all potential exposures to any population/ receptors at risk. This can be systematically developed using an event tree.

Figure 5.11 illustrates the logic used in constructing an event tree for a typical exposure scenario involving contaminant release to various environmental compartments. The event tree concepts can be used to simplify the exposure assessment in particular by transforming and representing complex situations with simplistic but adequate and manageable scenarios. This structure consequently provides a systematic approach for building a technically defensible information base for decisions on potential hazards by providing a mechanism for tackling environmental problems in a logical and comprehensive manner.

ETA can be used for hazard identification and for estimation of probability for a sequence of events leading to hazardous situations. The event tree approach will allow the evaluation of a range of possible scenarios, rather than the overly conservative approach of making an assessment for the "worst-case scenario" only; the latter will generally not be representative of the true situation that prevails. The event tree concept offers a more efficient way to perform a probabilistic risk evaluation when necessary.

The Pathway Probability (PWP) Concept. Where appropriate, in a probabilistic risk analysis, the probability of a consequence due to the occurence of a hazardous situation is defined by a so-called PWP, which is the product of an initiating probability value and the conditional probabilities of subsequent events.

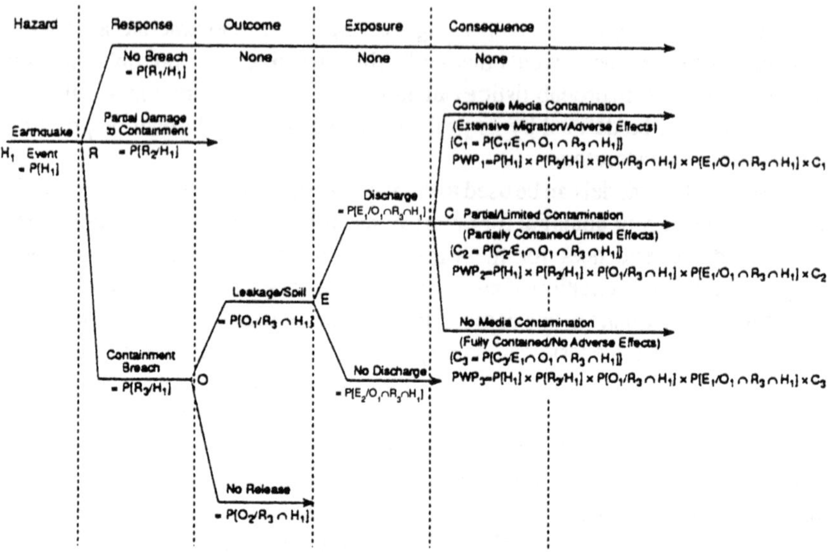

Figure 5.11 Schematic representation and illustrative example of the event tree model as used in probabilistic risk assessment.

The consequence probability, defined by the PWP, is given by the following relationship:

$$
\begin{aligned}
&PWP \\
&\quad = \Pr\{H \cap R \cap O \cap E \cap C\} \, or \, \Pr\{H.R.O.E.C\} \\
&\quad = \Pr\{H\} \times \Pr(R/H) \times \Pr\{O/H \cap R\} \times \Pr\{E/H \cap R \cap O\} \\
&\quad \times \Pr\{C/H \cap R \cap O \cap E\}
\end{aligned}
$$

where

Pr{H} is the probability of a specific hazard (H) of an initiating event occurring
Pr{R/H} is the conditional probability of system response (R), given H
Pr{O/H ∩ R} is the conditional probability of an outcome (O), given H and R
Pr{E/H ∩ R ∩ O} is the conditional probability of exposure (E), given H, R, and O
Pr{C/H ∩ R ∩ O ∩ E} is the conditional probability of a specific consequence (C), given H, R, O, and E

The initiating event is that cause or sequence of actions that create an effect or response. Such event will generally lead to follow-up conditions or consequential impacts. The fundamental assumption here is that the preceding event is the cause of subsequent events, i.e., the current event is dependent on the previous. In the case of mutual independence, this relationship becomes:

$$P\{H.R.O.E.C\} = P\{H\} \times P\{R\} \times P\{O\} \times P\{E\} \times P\{C\}$$

To determine the desired probability requires that one knows the conditional probabilities along a specific risk path, or in the case of mutually independent events, the *a priori* probability of occurrence for each event (i.e., for mutually independent events, only the probability of occurrence for each event would be required for a similar evaluation).

Risk Costs (RCs) and Impacts Assessment. Where applicable, the cost associated with the probability of failure (i.e., risk cost, RC) is estimated based on anticipated economic or environmental consequences (EC). This parameter is computed according to the following relationship:

$$Partial\ RC, C_i = \sum_j \left(PWP_i \times EC_{ij}\right), for\ i\text{-}th\ pathway$$

$$Total\ RC, C = \sum_{i=1}^{N} C_i, for\ all\ existing\ N\ pathways$$

$$\sum_{i=1}^{N}\left\{\sum_{j=1}^{IZ}\left(PWP_i \times EC_{ij}\right), for\ all\ existing\ IZ\ impact\ zones\right\}$$

EC_{ij} is the economic and/or environmental damages in impact zone j associated with the ith pathway.

The potential life impacts, which may include potential life loss, depends on the exposed population (i.e., PAR), and is computed to represent total risks as follows:

$$Life\ impacted\ /\ event = PAR_{ij} \times exposure\ probability\left(EP_{ij}\right)$$

$$Life\ impacted, LI_i = \sum_j \left(PAR_{ij} \times EP_{ij} \times PWP_i\right), for\ i\text{-}th\ pathway$$

$$Total\ LI, LI = \sum_i LI_i, for\ all\ existing\ N\ pathways$$

$$\sum_{i=1}^{N}\left\{\sum_{j=1}^{IZ}\left(PAR_{ij} \times EP_{ij} \times PWP_i\right)\right\}, for\ all\ existing\ IZ\ impact\ zones$$

PAR_{ij} represents the number of people in impact zone j associated with the ith pathway and EP_{ij} is the associated exposure probability. The average individual risk will be given by

$$Individual\ risk = \frac{\{Total\ LI\}}{\left\{\sum_j PAR_j\right\}}$$

Results from the event tree model may in general be put into a spreadsheet format for better comprehension. Such a formulation also allows for easy comparison of alternative remedial actions. The effect of each remedial alternative in reducing the risks associated with corrective actions for the technological system can be evaluated and compared.

The Fault Tree Model

A fault tree represents the combination and sequences of events which could cause specific system failure. It is traced back from a particular system failure event (called the top event) and spreads down through lower level events until it reaches the basic failure events. A fault tree has a branching structure defined by logic gates located at branch intersections. The logic gates define the causal relationship between lower level events and higher level events. Fault trees can be used to estimate the probability of occurrence of the top event, given estimates of the probabilities of occurrence of the basic events. Fault trees may be used to map all relevant possibilities and to determine the probability of the final outcome. To accomplish this latter goal, the probabilities of all component stages, as well as their logical connections, must be completely specified.

The basic concepts of fault tree construction and analysis are well documented elsewhere in literature (e.g., Lees, 1980; Dhillon and Singh, 1981; CMA, 1985). Figure 5.12 shows commonly used fault tree symbols; more details can be found in the PRA literature. A circle, diamond, or "house," represents a primary event (i.e., an event that is not developed further and has no inputs). The two basic types of fault tree logic gates are the "OR gate" and the "AND gate." Used in combination with the "NOT operator" (commonly shown as a dot above the gate), these gates can be used to define other specialized fault tree gates.

Fault Tree Analysis (FTA). FTA is a technique used to predict the expected probability of failure of a system. It seeks to relate the occurrence of an undesirable top event to one or more antecedent basic events. The FTA always starts with the definition of the undesirable event of which the probability is yet to be determined. The tree is developed to lower levels until the lowest events — called primary faults — are reached. Once all the primary event probabilities are assigned, the computation of the probability of the top event becomes an issue of Boolean algebra manipulations, for which there are several computer codes for completion.

In an FTA, an undesired state of a system is specified and the system is then analyzed in the context of its environment and operation to determine all the feasible and credible ways in which the undesirable event could possibly occur. The fault tree approach is a deductive process whereby the top event is postulated and the possible pathways for that event to occur are systematically derived. The fault tree is essentially qualitative, but is very often quantified due to its binary logic and adaptability to Boolean expressions.

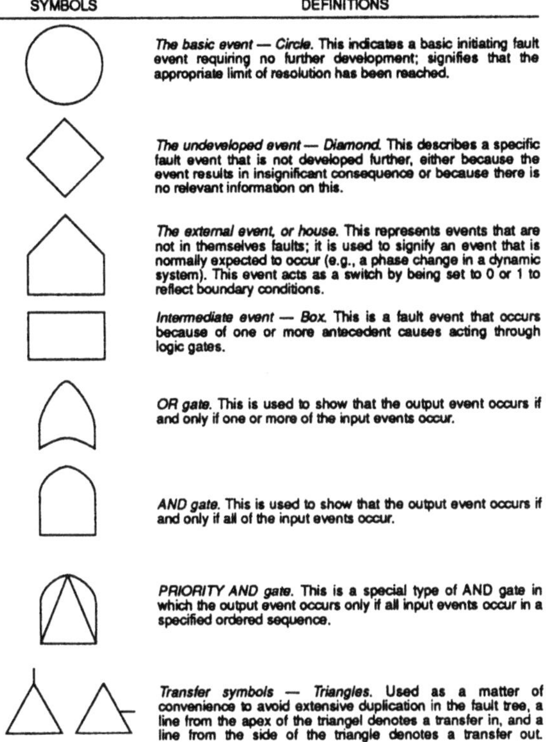

SYMBOLS
DEFINITIONS
The basic event — Circle. This indicates a basic initiating fault event requiring no further development; signifies that the appropriate limit of resolution has been reached.
The undeveloped event — Diamond. This describes a specific fault event that is not developed further, either because the event results in insignificant consequence or because there is no relevant information on this.
The external event, or house. This represents events that are not in themselves faults; it is used to signify an event that is normally expected to occur (e.g., a phase change in a dynamic system). This event acts as a switch by being set to 0 or 1 to reflect boundary conditions.
Intermediate event — Box. This is a fault event that occurs because of one or more antecedent causes acting through logic gates.
OR gate. This is used to show that the output event occurs if and only if one or more of the input events occur.
AND gate. This is used to show that the output event occurs if and only if all of the input events occur.
PRIORITY AND gate. This is a special type of AND gate in which the output event occurs only if all input events occur in a specified ordered sequence.
Transfer symbols — Triangles. Used as a matter of convenience to avoid extensive duplication in the fault tree, a line from the apex of the triangel denotes a transfer in, and a line from the side of the triangle denotes a transfer out.

Figure 5.12 Selected common fault tree symbols.

FTA may be used for hazard identification, although it is primarily used in risk assessment as a tool to provide an estimate of failure probabilities. A schematic representation of a fault tree is given in Figure 5.13; this illustrates a typical fault tree structure, but not necessarily the level of complexity in a typical fault tree. By assigning probability values to the basic events, the calculation of the probability of the top event can be achieved by performing algebraic manipulations based on some basic probability principles.

5.5.3 Other PRA Techniques and Tools

Event trees and fault trees are not the only analytical tools available for performing a PRA. Although event tree and fault tree analyses are the most powerful methods in PRA, other relatively simpler and also more complex methods are available. Several so-called system analysis methods exist that can be used in addition to, or in support of, the event and fault tree approaches. Pertinent techniques include the following:

- *Failure modes and effects analysis (FMEA),* which identifies failure modes for the components of concern and traces their effects on other components, subsystems, and systems. This approach provides an orderly examination of the hazardous conditions in a system and is simple to apply. It includes an assessment of criticality and

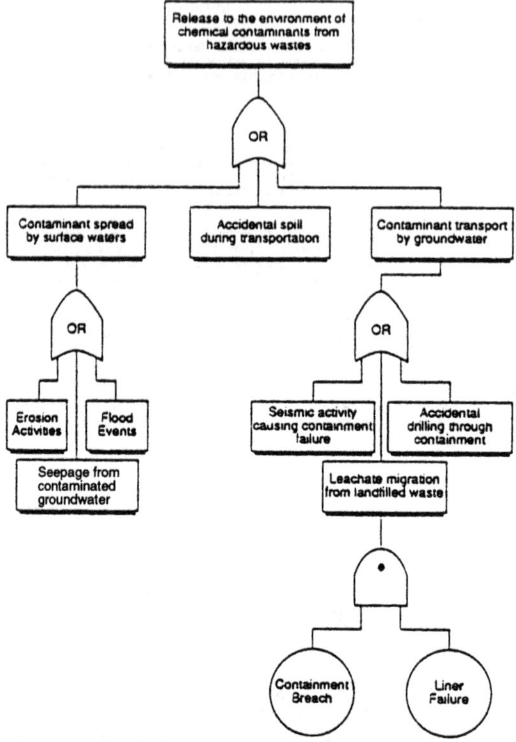

Figure 5.13 Schematic representation and illustrative example of a fault tree. The tree shown indicates the various ways in which hazardous waste chemicals may be released into the environment. To read the tree, start with the bottom row of possible initiating events.

probability of occurrence of each potential component failure mode. It is an inductive analysis that systematically details, on a component-by-component basis, all possible failure modes and identifies their resulting effects on the system.

- *Reliability Block Diagrams (RBDs),* are models generated by an inductive process whereby a given system, divided into blocks representing distinct elements, is represented according to system-success pathways and scenarios.
- *Hazard analysis (HAZAN),* or hazard quantification, is limited to the identification of hazards and considerations of strategies to employ to avoid the hazards. It involves the estimation of the expected frequencies or probabilities of events with adverse or potentially adverse consequences.
- *Hazard and operability study (HAZOP),* is a systematic, inductive technique for identifying hazards and operability problems through an entire system using guide words to identify deviations leading to hazardous situations. After the serious hazards have been identified via a HAZOP study or some other qualitative approach, a quantitative examination would be performed. HAZOP highlights specific deviations for which mitigative measures need to be developed.

This listing is by no means exhaustive, since variations or even completely different ones may be encountered in the literature and elsewhere (e.g., Zogg, 1987). Whichever method of analysis is to be used, it should optimally fulfill the requirements of the specific task. The various methods of analysis all have their strengths and weaknesses, depending on how and where they are used. The two basic kinds — inductive and deductive methods — however, can be differentiated. A thorough and comprehensive hazard analysis (particularly systematic approach to hazard identification) generally is a prerequisite for effective risk management.

5.5.4 Potential Applications of PRA Techniques

The potential for failures of hazardous waste facilities, and the inherent uncertainties associated with such facilities, all pose some degree of hazard; failures are the result of facility breach, followed by the release and migration of potentially harmful chemical contaminants through the environment. Failures may range from design flaws and deficiencies or faults, to operational and traffic accidents, to natural and man-made disasters. PRA offers techniques used to identify possible hazards and potential consequences. Such analyses can be used to improve design or operation to reduce risks. For instance, probabilistic techniques may be applied to the assessment of the structural integrity of hazardous waste facilities and containments or for the evaluation of accident scenarios in the transportation of hazardous materials. Such approaches would be based on quantitative measures of the probability that a facility or containment failure will occur or that an accident will occur during transportation and/or equipment operation.

PRA techniques utilizing the event tree model can be used in the safety evaluation of various engineered structures. The event tree concept aids in the identification and evaluation of possible failure cases which fully represent the spectrum of possible failure pathways for hazardous waste facilities. This will aid in the development of a structured risk assessment framework that will facilitate systematic decision making associated with such failures.

The PRA method aids in estimating the probabilities of events with adverse consequences or the potential to cause adverse consequences. Evaluation of risks associated with hazardous waste facilities is generally made from a broad spectrum of perspectives, including those for the public, the regulatory agencies (encompassing federal, state, and local mandates), owner/designer/operator (which may be public or private), and the insurer (who is the potential liability bearer); a risk analysis must address, in as far as possible, all these different perspectives concurrently. The PRA technique generally will serve to achieve this important objective in a logical way. The method will find several specific applications, including the following evaluations:

- Hazardous wastes storage facility design (for instance, the probability of failure for using only one liner vs. using multiple liners in a hazardous waste facility design may be evaluated and compared)
- Hazardous wastes containment and facility failure investigations (e.g., probabilities of failure are used for assessing potential economic losses and health impacts from failure or accident events involving a hazardous waste containment)
- Transportation risks evaluation (e.g., transportation risks can be analyzed by examining several variables, including the road network, loading/unloading accidents, and traffic density)

- General risk management and risk prevention programs development (e.g., risk management and prevention programs are instituted to ensure reasonable safety of industrial equipment and facilities)

Overall, risk assessment techniques provide a structured and systematic framework for evaluating the safety of hazardous waste facilities. The approach will provide an effective way to build the comprehensive and defensible information base necessary for dealing with potential hazards from hazardous waste facilities.

Investigating Hazardous Materials Facility/Containment Design and Operations

PRA provides an analytical technique for integrating diverse aspects of facility design and operation to aid in the evaluation of risks associated with a given waste facility. Safety aspects of the design of a hazardous waste facility or a hazardous materials storage facility can be evaluated by use of PRA methods. In this case, the risk evaluation will address the consequences associated with the probability of failure. Incremental risks due to failure, as a result of modifications in design criteria, can also be assessed. For instance, the probability of failure for using only one liner vs. the failure probability in using multiple liners in a hazardous waste facility design may be evaluated and compared using PRA techniques. Also, potential risk reduction by inclusion of an early monitoring system, etc. in the design of a hazardous waste facility may be evaluated by the use of risk assessment techniques and concepts.

The adequacy of facility design and operation is assessed by identifying those sequences of potential events that dominate risk, and also by establishing which sectors and features of the facility contribute most to the occurrence of accident scenarios. Thus, a PRA may help reveal the features of a facility and system that requires greater attention and thus provide a better focused plan in safety improvement programs. The information base developed in a PRA identifies dominant accident scenarios and facility aspects contributing most significantly to risk. Such information could be used in developing emergency response plans. Also, the information developed during the assessment could help in making management decisions on the allocation of limited resources for safety improvements — by directing attention to the features and scenarios dominating the facility risk.

The analysis of hazardous waste facilities of all types are often interrelated. To account for the interrelation of the different types of facilities, a modular approach to performing a hazardous wastes facility risk analysis is recommended. The modules will consist of an initiating hazards module, a system response/outcome module, and an exposure/consequences module (Asante-Duah et al., 1991). Each of these modules is applicable in evaluating the processes involved in the design, operation, maintenance, and management of hazardous waste facilities, including specific activities such as cleanup/remedial actions, storage and disposal, treatment, and transportation. The use of modular approach also allows an intramodule comparison of the various causative events, responses, and impacts.

Risks associated with the failure of hazardous materials containments and other facilities may be evaluated by use of PRA concepts, such as by using the event tree model. Probabilities of failure and conditional probabilities of system responses, outcomes, and consequences can be estimated and used for assessing potential socio-economic losses as well as health and environmental impacts from a failure or an accident event involving such facilities.

Analysis of Hazardous Materials Transportation Risks

Transport risk is of major concern, particularly with regards to long hauling of hazardous wastes. Transport problems can be a key issue for hazardous waste movement, not only from the point of view of economics and risk assessment, but also from the social and psychological perspectives as well. This is due in part to historical records of spills of virgin chemicals, oils, etc. that have occurred during transportation, and indeed, other transportation accidents (e.g., Saccomanno et al., 1989; Davidson, 1990).

Transportation of contaminated material (e.g., excavated soils, etc.) could, in the event of a spill, result in environmental and public health risks. Transportation risks can be analyzed for a system by examining several variables, including the road network, loading/unloading accidents, and traffic density. In the transport of hazardous materials, an accident during transporation will not neccessarily cause a release. Therefore, transportation risks would be estimated as the product of the probability of an accident and the conditional probability of release from a given accident.

The PRA approach may find important uses in assessing risks from transportation of hazardous materials (Theodore et al., 1989). The risks of transporting hazardous wastes may be defined in terms of the accident probability; the spill or release probabilities in an accident situation; the hazard classes for different damage scenarios (including the hazard areas for impacted zones); and the expected consequences on populations (i.e., the PAR) and environment within the accident/impacted corridor. For instance, in the investigation of potential spill incidents (due to any accidental spill as a result of mishandling, loading/unloading mishaps, vehicle accident, etc.), the number of expected spill incidents, N_s, may be estimated according to the following relationship:

$$N_{spill}\{by\ mode\ of\ transport\} =$$

$$[spill\ incidents\ /\ vehicle\ mile] \times [route\ miles\ /\ trip] \times [number\ of\ vehicle\ trips]$$

or

$$N_{spill}\{by\ mode\ of\ transport\} = [spill\ incidents\ /\ vehicle\ ton\text{-}mile]$$

$$\times [route\ miles\ /\ trip] \times [number\ of\ vehicle\ trips]$$

$$\times [tons\ hauled\ /\ trip]$$

The accident-induced releases of hazardous materials can be analyzed using fault and/or event tree methods of approach. Different levels of risks may be associated with different shipments; this will and would depend on the waste properties/ category, the spill/release scenarios, and the overall transportation environment. Serious consideration would have to be given to transportation routes and associated potential risks and costs when planning for waste management facilities in a region. In particular, a decision to operate a regional hazardous waste facility should seriously consider the implications of transporting wastes within the region.

Quasi-PRA Applications

In several situations, risk is estimated without due consideration being given to the probability of causative events. Thus, the assumption is that failure has already occurred or has an absolute chance of occurring. This scenario whereby risks are projected and based on the certainty assumption, that breach or failure has already occurred may not be realistic in all situations. It will therefore be more pragmatic to estimate actual risks, given an estimated probability of the causative event or the incidence of failure. For instance, "true" human health or environmental risks may be estimated as follows:

$$\begin{aligned} \textit{Actual risk} = & \{\textit{probability of failure incidence or causative event}\} \\ & \times \{\textit{estimated health or environmental risk value}\} \end{aligned}$$

This will represent the true risks imposed by a hazardous situation. This may be applicable under many different circumstances, and will likely result in better risk management programs.

5.6 REFERENCES

Ang, A. H.-S., and W. H. Tang. *Probability Concepts in Engineering Planning and Design*, Vol. II (New York: John Wiley & Sons, 1984).

Apostolakis, G. "Mathematical Methods of Probabilistic Safety Analysis," University of California Report No. UCLA-ENG-7464 (Los Angeles: UCLA, 1974).

Asante-Duah, D. K. "Quantitative Risk Assessment as a Decision Tool for Hazardous Waste Management," in *Proc. 44th Purdue Industrial Waste Conf. (May 1989)*, (Chelsea, MI: Lewis Publishers, 1990), pp. 111–123.

Asante-Duah, D. K., D. S. Bowles, and L. R. Anderson. "Framework for the Risk Analysis of Hazardous Waste Facilities," in Proc. of Sixth Int. Conf. on Applications of Statistics and Probability in Civil Engineering, CERRA/ICASP 6, Mexico (1991).

Bowles, D. S., L. R. Anderson, and T. F. Glover. "Design Level Risk Assessment for Dams," in *Proc. Struct. Congr. ASCE*: 210–25, Florida.

Brown, H. S. "A Critical Review of Current Approaches to Determining 'How Clean is Clean' at Hazardous Waste Sites," *Hazard. Wastes Hazard. Mater.* 3(3):233–260 (1986).

Calabrese, E. J. *Principles of Animal Extrapolation* (New York: John Wiley & Sons, 1984).
CAPCOA (California Air Pollution Control Officers Association). "Air Toxics Assessment Manual," California Air Pollution Control Officers Association, Draft Manual, August 1987 (ammended 1989), California (1989).
CAPCOA (California Air Pollution Control Officers Association). "Air Toxics 'Hot Spots' Program. Risk Assessment Guidelines," California Air Pollution Control Officers Association, California (1990).
Casarett, L. J. and Doull, J. *Toxicology: The Basic Science of Poisons* (New York: MacMillan Publishing, 1975).
CDHS (California Department of Health Services). "The California Site Mitigation Decision Tree Manual," California Department of Health Services, Toxic Substances Control Division, Sacramento (1986).
Chrostowski, P. C., L. J. Pearsall, and C. Shaw. "Risk Assessment as a Management Tool for Inactive Hazardous Materials Disposal Sites," *Environ. Manage.* 9(5):433–442 (1985).
CMA (Chemical Manufacturers Association). *Risk Analysis in the Chemical Industry* (Rockville, MD: Institutes, 1985).
Davidson, A. *In the Wake of the Exxon Valdez: The Devastating Impact of the Alaska Oil Spill* (Vancouver: Douglas & McIntyre, 1990).
Dawson, G. W. and D. Sanning. "Exposure-Response Analysis for Setting Site Restoration Criteria," *in Proc. Natl. Conf. on Management of Uncontrolled Hazardous Waste Sites*, Washington, DC (1982).
DOE (U.S. Department of Energy). "The Remedial Action Priority System (RAPS): Mathematical Formulations," U.S. Dept. of Energy, Office of Environment, Safety & Health, Washington, DC (1987).
Dhillon, B. S. and C. Singh. *Engineering Reliability* (New York: John Wiley & Sons, 1981).
Dourson, M. L. and J. F. Stara. "Regulatory History and Experimental Support of Uncertainty (Safety) Factors," *Regul. Toxicol. Pharmacol.* 3:224–238 (1983).
Evans, L. J. "Chemistry of Metal Retention by Soils," *Environ. Sci. Technol.* 23(9):1047–1056.
Freund, J. E. and R. E. Walpole, Eds. *Mathematical Statistics* (Englewood Cliffs, NJ: Prentice-Hall, 1987).
Gilbert, R. O. *Statistical Methods for Environmental Pollution Monitoring* (New York: Van Nostrand-Reinhold, 1987).
Glickman, T. S. and M. Gough, Eds. *Readings in Risk* (Washington, DC: Resources for the Future, 1990).
Hallenbeck, W. H. and Cunningham, K. M. *Quantitative Risk Assessment for Environmental and Occupational Health*, 4th Printing (Chelsea, MI: Lewis Publishers, 1988).
Henley, E. J. and H. Kumamoto. *Reliability Engineering & Risk Assessment* (Englewood Cliffs, NJ: Prentice-Hall, 1981).
Huckle, K. R. *Risk Assessment — Regulatory Need or Nightmare* (Shell Center, London: Shell Publications, 1991).
Klassen, C. D., Amdur, M. O., and Doull, J., Eds. *Casarett and Doull's Toxicology: The Basic Science of Poisons*, 3rd ed. (New York: Macmillan Publishing, 1986).
Larsen, R. J. and M. L. Marx. *An Introduction to Probability and Its Applications* (Englewood Cliffs, NJ: Prentice-Hall, 1985).
Lave, L. B. Ed. *Quantitative Risk Assessment in Regulation* (Washington, DC: Brookings Institute, 1982).
Lave, L. B. and A. C. Upton, Eds. *Toxic Chemicals, Health, and the Environment* (Baltimore: Johns Hopkins University Press, 1987).
Lees, F. B. *Loss Prevention in the Process Industries*, Vol. 1 (Boston: Butterworths, 1980).

Leidel, N. and K. A. Busch. "Statistical Design and Data Analysis Requirements," in *Patty's Industrial Hygiene and Toxicology*, Vol. IIIa, 2nd ed. (New York: John Wiley & Sons, 1985).

Lind, N. C., J. S. Nathwani, and E. Siddall. *Managing Risks in the Public Interest* (University of Waterloo, Ontario: Institute for Risk Research, 1991).

Lyman, W. J., W. F. Reehl and D. H. Rosenblatt. *Handbook of Chemical Property Estimation Methods: Environmental Behavior of Organic Compounds* (Washington, DC: American Chemical Society, 1990).

Mackay, D. and P. J. Leinonen. "Rate of Evaporation of Low-Solubility Contaminants from Water Bodies," *Environ. Sci. Technol.* 9:1178–1180 (1975).

Mackay, D. and A. T. K. Yenn. "Mass Transfer Coefficient Correlations for Volatilization of Organic Solutes from Water," *Environ. Sci. Technol.* 17:211–217(1983).

Meyer, C. R. "Liver Dysfunction in Residents Exposed to Leachate from a Toxic Waste Dump," *Environ. Health Pers.* 48:9–13 (1983).

Miller, I. and J. E. Freund. *Probability and Statistics for Engineers*, 3rd Ed. (Englewood Cliffs, NJ: Prentice-Hall, 1985).

NRC (National Research Council). *Risk Assessment in the Federal Government: Managing the Process* (Washington, DC: NAS Press, 1983).

Paustenbach, D. J., Ed. *The Risk Assessment of Environmental Hazards: A Textbook of Case Studies* (New York: John Wiley & Sons, 1988).

Rappaport, S. M. and J. Selvin. "A Method for Evaluating the Mean Exposure from a LogNormal Distribution," *J. Am. Ind. Hyg. Assoc.* 48:374–379 (1987).

Saccomanno, F. F., J. H. Shortreed, M. Van Averde, and J. Higgs. "Comparison of Risk Measures for the Transport of Dangerous Commodities by Truck and Rail" Presented at the 68th Annual Meeting of the Transportation Research Board, Washington, DC, January 1989.

Santos, S. L. and J. Sullivan. "The Use of Risk Assessment for Establishing Corrective Action Levels at RCRA Sites," in *Hazardous Wastes and Hazardous Materials,* Mary Ann Liebert, Inc. Publishers.

Sedman, R. M. "The Development of Applied Action Levels for Soil Contact: A Scenario for the Exposure of Humans to Soil in a Residential Setting," *Environ. Health Pers.* 79:291–313 (1989).

Sidall, J. N. *Probabilistic Engineering Design: Principles and Applications* (New York: Marcel Dekker, 1983).

Swann, R. L. and A. Eschenroeder, Eds. *Fate of Chemicals in the Environment*, ACS Symp. Ser. 225, (Washington, DC: American Chemical Society, 1983).

Theodore, L., J. P. Reynolds, and F. B. Taylor. *Accident and Emergency Management* (New York: Wiley-Interscience, New York 1989).

Thibodeaux, L. J. and S. T. Hwang. "Landfarming of Petroleum Wastes — Modeling the Air Emission Problem," *Environ. Prog.* 1:42–46 (1982).

USBR (U.S. Bureau of Reclamation). "Guidelines to Decision Analysis," ACER Tech. Memo No. 7, Denver, CO (1986).

U.S. EPA. "Approaches to Risk Assessment for Multiple Chemical Exposures," U.S. Environmental Protection Agency, Environmental Criteria and Assessment Office, Cincinnati, OH, EPA-600/9-84-008 (1984a).

U.S. EPA. "Proposed Guidelines for Carcinogen, Mutagenicity, and Developmental Toxicant Risk Assessment," *Fed. Regist.* 49:46294–46331 (1984b).

U.S. EPA. "Risk Assessment and Management: Framework for Decision Making," EPA 600/9-85-002, Washington, DC (1984c).

U. S. EPA. "Characterization of Hazardous Waste Sites: A Methods Manual, Volume 1, Site Investigations," U.S. Environmental Protection Agency, Environmental Monitoring Systems Laboratory, Las Vegas, EPA-600/4-84-075 (1985a).

U.S. EPA. "Development of Statistical Distribution or Ranges of Standard Factors Used in Exposure Assessments," U.S. Environmental Protection Agency, Office of Health and Environmental Assessment, Washington, DC (1985b).

U.S. EPA. "Rapid Assessment of Exposure to Particulate Emissions From Surface Contamination Sites," EPA/600/8-85/002, NTIS PB85-192219, Office of Health and Environmental Assessment, Washington, DC (1985c).

U.S. EPA. "Guidelines for Carcinogen Risk Assessment," *Fed. Regist.* 51(185):33992–34003, CFR 2984, September 24, 1986 (1986a).

U.S. EPA. "Guidelines for the Health Risk Assessment of Chemical Mixtures," *Fed. Regist.* 51(185):34014–34025, CFR 2984, September 24, 1986 (1986b).

U.S. EPA. "Ecological Risk Assessment," Hazard Evaluation Division Standard Evaluation Procedure, Washington, DC (1986c).

U.S. EPA. "Methods for Assessing Environmental Pathways of Food Contamination: Methods for Assessing Exposure to Chemical Substances," Vol. 8. Exposure Evaluation Division, Office of Toxic Substances. EPA 560/5-85-008, September 1986 (1986d).

U.S. EPA. "Superfund Risk Assessment Information Directory," Office of Emergency and Remedial Response, Washington, DC, EPA/540/1-86/061 (1986e).

U.S. EPA. "Alternate Concentration Limit Guidance," Report No. EPA/530-SW-87-017, OSWER Directive 9481-00-6C, U. S. EPA, Office of Solid Waste, Waste Management Division, Washington, DC (1987a).

U.S. EPA. "RCRA Facility Investigation (RFI) Guidance," EPA/530/SW-87/001, Washington, DC (1987b).

U.S. EPA. "Selection Criteria for Mathematical Models Used in Exposure Assessments: Surface Water Models," EPA-600/8-87/042, Office of Health and Environmental Assessment, Washington, DC (1987c).

U.S. EPA. "Technical Guidance for Hazard Analysis," Washington, DC, December 1987 (1987d).

U.S. EPA. "A Workbook of Screening Techniques for Assessing Impacts of Toxic Air Pollutants," EPA-450/4-88-009, Office of Air Quality Planning and Standards, Research Triangle Park, NC (1988a).

U.S. EPA. "CERCLA Compliance with Other Laws Manual," EPA/540/6-89/006, Office of Solid Waste and Emergency Response, Washington, DC (1988b).

U.S. EPA. "Estimating Toxicity of Industrial Chemicals to Aquatic Organisms Using Structure Activity Relationships," EPA/560/6-88/001, Office of Toxic Substances, Washington, DC (1988c).

U.S. EPA. "GEO-EAS (Geostatistical Environmental Assessment Software) User's Guide," Environmental Monitoring Systems Laboratory, EPA/600/4-88/033a Office of R&D, Las Vegas, NV (1988).

U.S. EPA. "Guidance for Conducting Remedial Investigations and Feasibility Studies Under CERCLA," EPA/540/G-89/004, OSWER Directive 9355. 3–01, Office of Emergency and Remedial Response, Washington, DC (1988d).

U.S. EPA. "Interim Report on Sampling Design Methodology," EPA/600/X-88/408, Environmental Monitoring Support Laboratory, Las Vegas, NV (1988e).

U.S. EPA. "Review of Ecological Risk Assessment Methods," Office of Policy, Planning and Evaluation, Washington, DC (1988f).

U.S. EPA. "Selection Criteria for Mathematical Models Used in Exposure Assessments: Ground-Water Models," EPA-600/8-88/075, Office of Health and Environmental Assessment, Washington, DC (1988g).

U.S. EPA. "Superfund Exposure Assessment Manual," Report No. EPA/540/1-88/001, OSWER Directive 9285. 5–1, U. S. EPA, Office of Remedial Response, Washington, DC (1988h).

U.S. EPA. "Ecological Assessments of Hazardous Waste Sites: A Field and Laboratory Reference Document," EPA/600/3-89/013, Office of Research and Development, Corvallis Environmental Research Laboratory, Corvallis, OR (1989a).

U.S. EPA. "Exposure Factors Handbook," EPA/600/8-89/043, Office of Health and Environmental Assessment, Washington, DC (1989b).

U.S. EPA. "Ground-water Sampling for Metals Analyses," EPA/540/4-89-001, Office of Solid Waste and Emergency Response, Washington, DC (1989c).

U.S. EPA. "Interim Methods for Development of Inhalation Reference Doses," EPA/600/8-88/066F, Office of Health and Environmental Assessment, Washington, DC (1989d).

U.S. EPA. "Application of Air Pathway Analyses for Superfund Activities," Air/Superfund National Technical Guidance Study Series, Procedures for Conducting Air Pathway Analyses for Superfund Applications, Vol. I, EPA-450/1-89-001, Interim Final, Office of Air Quality Planning and Standards, Research Triangle Park, NC (1989e).

U.S. EPA. "Estimation of Air Emissions from Cleanup Activities at Superfund Sites," Air/Superfund National Technical Guidance Study Series, Vol. III, EPA-450/1-89-003, Interim Final, Office of Air Quality Planning and Standards, Research Triangle Park, NC (1989f).

U. S. EPA. "Procedures for Conducting Air Pathway Analyses for Superfund Applications," Vol. IV, Procedures for Dispersion Modeling and Air Monitoring for Superfund Air Pathway Analysis, Air/Superfund National Technical Guidance Study Series, EPA-450/1-89-004, Interim Final, Office of Air Quality Planning and Standards, Research Triangle Park, NC (1989g).

U.S. EPA. "Review and Evaluation of Area Source Dispersion Algorithms for Emission Sources at Superfund Sites," EPA-450/4-89-020, Office of Air Quality Planning and Standards, Research Triangle Park, NC (1989h).

U.S. EPA. "Risk Assessment Guidance for Superfund," Vol. I, Human Health Evaluation Manual (Part A), EPA/540/1-89/002, Office of Emergency and Remedial Response, Washington, DC (1989i).

U.S. EPA. "Risk Assessment Guidance for Superfund," Vol. II, Environmental Evaluation Manual, EPA/540/1-89/001, Office of Emergency and Remedial Response, Washington, DC (1989j).

U.S. EPA. "Soil Sampling Quality Assurance Guide," Experimental Monitoring Support Laboratory, Las Vegas, NV (1989k).

U.S. EPA. "User's Guide to the Contract Laboratory Program," OSWER Dir. 9240. 0–1, Office of Emergency and Remedial Response, Washington, DC (1989l).

U.S. EPA. " Development of Example Procedures for Evaluating the Air Impacts of Soil Excavation Associated with Superfund Remedial Actions," Air/Superfund National Technical Guidance Study Series, EPA-450/4-90-014, Office of Air Quality Planning and Standards, Research Triangle Park, NC (1990a).

U.S. EPA. "Estimation of Baseline Air Emissions at Superfund Sites," Air/Superfund National Technical Guidance Study Series, Procedures for Conducting Air Pathway Analyses for Superfund Applications, Vol. II, EPA-450/1-89-002a, Office of Air Quality Planning and Standards, Research Triangle Park, NC (1990b).

U.S. EPA. "State of the Practice of Ecological Risk Assessment Document," Office of Pesticides and Toxic Substances, U.S. EPA draft report. Washington, DC (1990c).

Wonnacott, T. H. and R. J. Wonnacott. *Introductory Statistics*, 2nd ed. (New York: John Wiley & Sons, 1972).
Zirschy, J. H. and D. J. Harris. "Geostatistical Analysis of Hazardous Waste Site Data," *ASCE J. Environ. Eng.* 112(4) (1986).
Zogg, H. A. "'Zurich' Hazard Analysis," Zurich Insurance Group, Risk Engineering, Zurich, Switzerland (1987).

CHAPTER 6

Hazardous Waste Management Decisions from Risk Assessment

The chief purpose of risk assessment is to aid decisionmaking, and this focus should be maintained throughout a given program. The application of risk assessment can reduce possible ambiguities in the decision-making process. It can also aid in the selection of prudent, technically feasible, and scientifically justifiable corrective actions that will help protect public health and the environment in a cost-effective manner. The results can be used to determine the types of risk management actions necessary for a given hazardous waste problem.

Risk assessment can be used to define the level of risk, which will in turn aid in determining the level of analysis and the type of corrective actions to adopt for a given hazardous waste management problem. It can also be used in the development of performance goals for various response alternatives. The level of risk considered can be depicted by a risk-decision conceptual model, with the risk levels being defined as low, moderate, high, very high, etc. (Figure 6.1) that will help distinguish between imminent health and/or environmental hazards and risks. In general, this can be used as an aid for policy decisions to develop variations in the scope of work necessary for case-specific problems. Indeed, the use of such an approach will help answer the infamous question of "how clean is clean enough?" or "how safe is safe enough?" that must be repeatedly asked when confronted with hazardous waste management issues and decisions.

Risk assessment has several specific applications that could affect the type of decisions to be made in relation to hazardous waste management. Typical decision issues potentially addressed by the use of risk assessment concepts and techniques in hazardous waste management relate to the following (Asante-Duah, 1990):

- Use of specific chemicals in manufacturing processes and industrial activities
- Preliminary screening for potential problems at hazardous waste TSDFs

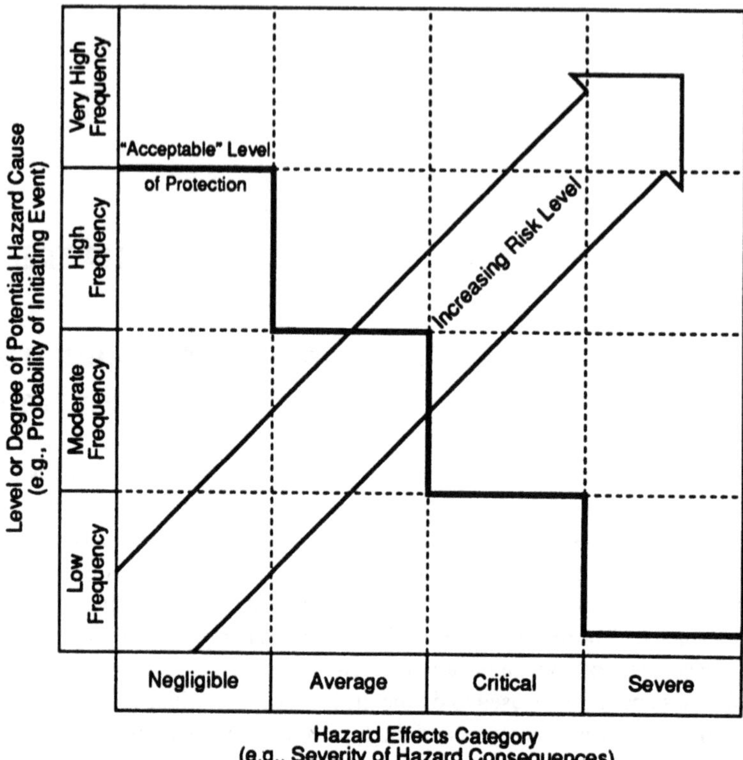

Figure 6.1 A conceptual representation for defining a risk profile in a risk-decision matrix.

- Hazardous waste facility design and operation
- Hazardous waste site selection
- Field sampling design (through identification of data needs)
- Design of monitoring programs (to identify chemicals present in various media and their persistence)
- Prioritization of hazardous waste sites for remedial action
- Development of "acceptable" cleanup criteria and guidelines for contaminated sites
- Corrective measures evaluation and selection of remedial alternatives
- Evaluation and ranking of potential liabilities in hazardous waste management practices
- Ecological/environmental risk assessment for the identification of critical habitats and organisms exposed to environmental contaminants
- Probabilistic risk assessment for the evaluation of transportation risks associated with the movement of hazardous wastes
- Implementation of general risk management and prevention programs for hazardous waste management planning

This is by no means complete and exhaustive, since variations or even completely different and unique problems may be resolved by use of one risk assessment methodology or another.

6.1 RISK ASSESSMENT INPUTS TO RISK MANAGEMENT

Risk management is defined as the process of weighing policy alternatives and selecting the most appropriate action, integrating the results of risk assessment with engineering data and with social, economic, and political concerns to reach a decision (Seip and Heiberg, 1989). It comprises actions evaluated and implemented to help in risk reduction policies. This may include concepts for prioritizing the risks and also an evaluation of the costs and benefits of proposed risk reduction programs. Examples of risk management actions include deciding how much of a chemical a company may discharge into a river; deciding which substances may be stored at a hazardous waste disposal facility; deciding to what extent a hazardous waste site must be cleaned up; setting permit levels for discharge, storage, or transport of hazardous materials; establishing levels for air contaminant emissions; determining allowable levels of contamination in drinking water or food; industry decisions on the use of specific chemicals in manufacturing processes and related industrial activities; and establishing protocols for hazardous waste facility design and operation by the regulated community. All these management decisions are made based on inputs from a prior risk assessment of the applicable case-specific problem.

Risk management combines political, legal, and engineering approaches to manage risks. Risk assessment information is used in the risk management process to help arrive at decisions on how to protect public health and the environment. Essentially, risk assessment provides *information* on the health, environmental, and economic risks, and risk management is the *action* taken based on that information (Figure 6.2). In fact, risk assessment can in principle be carried out objectively, whereas risk management involves preferences and attitudes and therefore is essentially considered a subjective activity (Seip and Heiberg, 1989; NRC, 1983; U.S. EPA, 1984).

6.2 GENERAL RISK MANAGEMENT AND RISK PREVENTION PROGRAMS

Risk management and prevention programs (RMPPs) for hazardous waste facilities are typically implemented to ensure reasonable safety of the equipment and facilities used in the treatment, storage, and/or disposal of acutely hazardous materials (AHMs). Specifically referring to industrial equipment and facilities, OES (1989) defines RMPP as the sum total of programs for minimizing AHM incident risks. The RMPP mechanics can achieve several results, including:

- Improvement in system efficiency and reliability
- Increase in the level of protection of public health and safety
- Reduction in liability

To ensure overall safety, it is important to systematically identify hazards throughout an entire facility, assess the potential consequences due to the hazards, and examine corrective measures for dealing with potential releases associated with the identified hazards. This can be accomplished by use of probablistic risk assessment

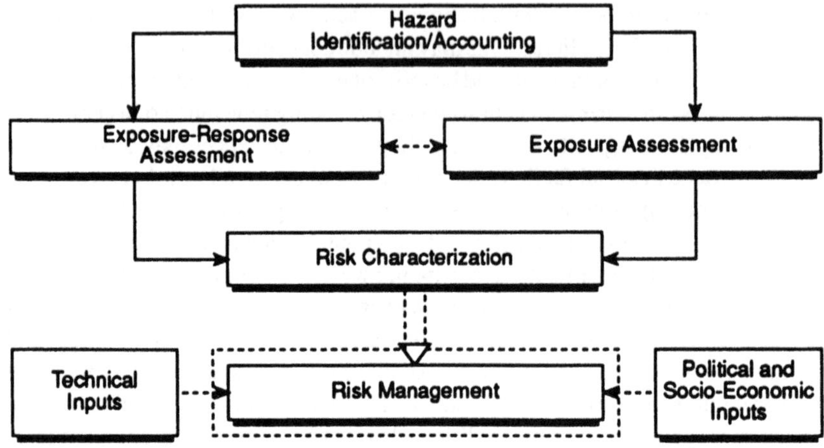

Figure 6.2 The risk assessment process as an input to risk management programs.

(PRA) principles and concepts. The event tree model, used in the context of PRA and/or used for defining potential exposure scenarios and pathways, facilitates the development of good RMPPs. Individual component failure rates for a given system can be used to estimate the potential for adverse consequences for a given hazardous situation.

In general, RMPPs are designed to aid waste and AHM handlers in the conduct of comprehensive evaluations of facilities that handle such materials. This will help to minimize any releases of AHM or similar wastes, as well as help protect public health and safety and also the environment from releases that do occur. The process of developing an RMPP should create a comprehensive facility risk analysis and hazard evaluation that can be used to support implementation of corrective actions for waste and AHM facilities.

6.3 USING SYSTEMATIC TOOLS TO FACILITATE RISK COMMUNICATION

Effective risk communication is important for effective risk management and risk prevention programs. It may therefore be important to give as much attention to risk communication as to risk quantification and/or risk qualification. For instance, knowledge of the so-called NIMBY ("not in my backyard"), NIABY ("not in anyone's backyard"), NIMTO ("not in my term of office"), LULU ("locally unwanted land use"), and related syndromes (e.g., Gregory and Kunreuther, 1990) raining upon several communities gives a signal that risk communication may indeed dictate public perception and therefore public acceptance of risk mitigation alternatives and hazardous waste management decisions. A systematic evaluation using structured decision analysis methods, such as the use of special risk assessment tools as depicted by the event tree approach (Chapter 5), can greatly help in this direction. The event tree illustrates the cause and effect ordering of event scenarios, with each event being shown by a branch of the event tree in the context of a risk assessment. The event tree model structure can aid risk communicators in improving the quality and effectiveness of their performance and presentations.

6.4 REFERENCES

Asante-Duah, D. K. "Quantitative Risk Assessment as a Decision Tool for Hazardous Waste Management," in *Proc. 44th Purdue Industrial Waste Conf. (May 1989),* (Chelsea, MI: Lewis Publishers, 1990), pp. 111–123.

Gregory, R. and H. Kunreuther. "Successful Siting Incentives," *Civil Eng.* 60(4):73–75 (1990).

NRC (National Research Council). *Risk Assessment in the Federal Government: Managing the Process* (Washington, DC: National Academy Press, 1983).

OES (California Office of Emergency Services). "Guidance for the Preparation of a Risk Management and Prevention Program," State of California, California Office of Emergency Services, Hazardous Materials Division, Sacramento (1989).

Seip, H. M. and A. B. Heiberg, Eds. *Risk Management of Chemicals in the Environment,* NATO Committee on the Challenges of Modern Society, Vol. 12 (New York: Plenum Press, 1989).

U.S. EPA. "Risk Assessment and Management: Framework for Decision Making," EPA 600/9-85-002, Washington, DC (1984).

CHAPTER 7

Selected Case Studies and Applications

Hazardous waste sites and facilities are potential sources of atmospheric, groundwater, surface water, soils, and food chain contamination due to possible improper management, design, and/or operation. Numerical examples of case studies and hypothetical problems are presented in the following sections to demonstrate some specific uses and typical applications of risk assessment in hazardous waste management. The use of specific models in evaluating the hypothetical and example case scenarios that follow does not preclude the use of the risk assessment protocols to these models only or vice versa. Indeed, model selection should be carefully done to ascertain the choice of appropriate fate/transport and exposure models for use in any risk assessment.

7.1 CASE 1: CHEMICAL RISK ASSESSMENT FOR AN INACTIVE INDUSTRIAL FACILITY

A numerical example involving the application of health risk assessment methods to a potentially contaminated site problem is discussed here. This consists of a risk characterization for the exposure of nearby residential populations to chemicals in soils at an abandoned industrial facility and the development of target cleanup criteria for the subject site. Risk assessment methodologies previously described in Chapter 5 are followed for investigating this hypothetical contaminated site, ABC, located in a suburban community in California.

The ABC site is fenced with barbed wire to prevent unauthorized entry to the property. An environmental site assessment previously conducted for the site indicated the presence of antimony (Sb), cadmium (Cd), copper (Cu), lead (Pb), nickel (Ni), and zinc (Zn) in soils at the property. It is suspected that several offsite population groups are potentially affected by these metals. A risk assessment is required to develop the risk information pertinent to making appropriate decisions regarding the need for cost-effective corrective actions.

7.1.1 Site-Specific Objectives

The main objectives of the risk assessment for the ABC site is to provide a consistent process for evaluating and documenting potential public health and environmental threats from the site, accomplished through the following steps:

- Determine potential health impacts from the subject site by providing an analysis of baseline ("no action") risks that will help determine the need for corrective actions at this site.
- Develop appropriate health-based cleanup criteria for site remediation, as necessary, and then determine post-remediation risks for potential future scenarios.

Potential receptor exposures to the metals in soils originating from the ABC site and the associated human health risks to potential receptors in the vicinity of the site are evaluated in this study.

7.1.2 Study Design for the ABC Site Risk Assessment

The following specific tasks are considered in order to achieve the objectives of the risk assessment for the ABC site:

- Data evaluation
    - Identify potential chemicals of concern.
    - Analyze relevant site data, including summary statistics.
- Exposure assessment
    - Identify source areas and significant migration pathways, including an evaluation of important fate and transport processes for the chemicals of potential concern.
    - Identify potentially affected populations and determine potential exposure pathways to these potential receptors.
    - Develop site conceptual model(s).
    - Develop exposure scenarios (including the determination of current and future land uses and the analysis of environmental fate and persistence).
    - Identify exposure points and estimate/model exposure point concentrations for the chemicals of potential concern.
    - Compute potential receptor intakes and resultant doses for the chemicals of potential concern, for all potential receptors and significant pathways of concern.
- Toxicity assessment
    - Compile toxicological profiles (including the intrinsic toxicological properties of the chemicals of potential concern, which may include their acute, subchronic, chronic, carcinogenic, and reproductive effects).
    - Determine appropriate toxicity indices (*viz*, the ADIs or RfDs and SFs).

- Risk characterization
    - Estimate carcinogenic risks for carcinogens.
    - Estimate noncarcinogenic hazard quotients and indices for systemic toxicants.
    - Summarize risk information, including an evaluation of uncertainties.
- Development of cleanup criteria
    - Determine "acceptable" level of risks to potential receptors.
    - "Back-model" to obtain target/acceptable cleanup levels for contaminants.

The risks and/or hazards associated with the residual contamination to be left at the case site following remedial action are also evaluated to simulate risk characterization of future land use conditions.

7.1.3 General Site-Specific Data Analyses

For risk assessment purposes, environmental media of concern at a potentially contaminated site include

- Any currently contaminated environmental media to which potential receptors may be exposed or that will serve as a transport pathway for chemicals to reach potential receptors.
- Any currently uncontaminated environmental media, but that which is suspect to contamination at a future date due to contaminant migration from affected media.

For the ABC property, the soils are the main source of contaminants potentially released into other environmental media. Representative soil samples have been collected and analytical results obtained from the soil samples for the ABC site. These are supplemented with modeling information as necessary and appropriate. No air sampling data were available. An air model for fugitive dust emission and dispersion was used to estimate the applicable exposure point concentrations of respirable particulates from this site; the model details (e.g., U.S. EPA, 1985; CDHS, 1986; DOE, 1987; CAPCOA, 1989) are not given in this elaboration. In this model, fugitive dust dispersion concentrations were evaluated as represented by a three-dimensional Gaussian distribution of particulate emissions from the source. Table 7.1 presents summary statistics for the chemicals of potential concern in the various environmental compartments for the ABC site; this includes the respirable concentrations estimated by the air model.

7.1.4 Exposure Assessment

The exposure evaluation for the ABC site considers the migration of the chemicals of potential concern from the contaminated site to potentially exposed populations. The potential exposures considered in this assessment are for current land use conditions, assuming a "no-action" (i.e., baseline) scenario, and also for future land-use scenarios that involve potential onsite exposures to residual contamination that remains after site remediation.

Table 7.1 Summary Statistics for the Chemicals of Potential Concern at the ABC Site

Chemical	Range of Background Levels[a] (ppm)	Onsite Soil Concentration Range (mg/kg)	Offsite DNW[b] Soil Concentration Range (mg/kg)	Mean Exposure Concentrations		
				Onsite Soils (mg/kg)	Offsite (DNW) Soils (mg/kg)	Air Particulates[c] (mg/m^3)
Antimony (Sb)	0.3–5.0	3.0–1,500	0.3–27	2.05E + 02	7.49E + 00	1.10E – 03
Cadmium (Cd)	0.03–3.3	0.4–210	0.03–3.9	1.50E + 01	1.73E + 00	8.16E – 05
Copper (Cu)	8.0–260	19.0–39,600	6.0–310	3.50E + 03	9.85E + 01	1.84E – 02
Lead (Pb)	10.0–1,960	32.0–16,500	42.0–720	4.08E + 03	1.39E + 02	2.14E – 02
Nickel (Ni)	5.0–40.0	8.0–12,600	3.0–150	1.17E + 03	4.65E + 01	6.11E – 03
Zinc (Zn)	32.0–1,070	300.0–299,000	240–8,660	4.49E + 04	1.11E + 03	2.36E – 01

[a] Background levels are represented by offsite locations that are upwind to the ABC site.
[b] DNW = downwind locations.
[c] Refers to respirable concentration of chemicals adsorbed on soil particles and carried as fugitive dust.

Results of the preliminary site assessment indicates that erosional (overland) transport at the ABC site is the major transport mechanism for the site-related chemicals of potential concern. This inference is based on data that indicate that groundwater has apparently not been impacted; this observation is in line with the knowledge that metals generally exhibit relatively low mobilities in soils (Evans, 1989). The following potential routes of exposure are determined to be of primary concern in the risk assessment:

- Particulate inhalation of fugitive dust
- Ingestion of contaminated soils (incidental and pica)
- Dermal exposures and skin adsorption (through skin contact with soils)

This represents the most likely and significant pathways selected for further analyses. In planning for corrective actions, focus will be on these *likely* and *significant pathways* only.

Potentially Exposed Populations

The *critical receptors* potentially exposed to contamination from the ABC site are nearby residential and occupational populations in the vicinity of the site. The difference in sensitivities between adults and children demands that they be treated separately in evaluating their exposure intakes and doses of chemicals. Also, due to the variance in activity and behavior of children at the different ages, child exposure of soils is broken down into two categories for this evaluation, including

- Children aged up to 6 years (to include infants and preschool children)
- Children aged between 6 and 12 years (to include young children of school-going age)

For the purpose of this risk assessment and consistent with EPA guidance (U.S. EPA, 1989b), all population groups aged more than 12 years are included in the adult category.

Adults — In describing the adult population, it is recognized that there is the potential for adults living and/or working in the vicinity of the ABC property to be exposed to chemical contaminants originating from the ABC site. Under the current land use conditions, offsite residents, passers-by, and nearby workers are identified as a target population. Potential exposures for this group is limited to dermal contact with soils, incidental ingestions of soils, and/or inhalation of fugitive dust containing site-related chemicals. Potential exposures of adult populations under future land use conditions will be similar to that for current land use scenarios; but in addition to potential offsite exposures, there will also be the potential for receptor exposure onsite (i.e., within the property boundary).

Children aged 6 to 12 years — It is expected that children aged 6 to 12 years would be exposed via the same pathways described above for adult residents and passers-by under both current and future land use conditions.

Children aged under 6 years — Children aged under 6 years who live within the general vicinity of the ABC site are expected to be exposed via the same

Table 7.2 Receptor Exposure Levels for the Chemicals of Potential Concern at the ABC Site

Exposure Pathway	Chemical of Concern	RME Concentration (mg/m³ or mg/kg)
Inhalation of dust that has been generated from wind erosion (units: mg/m³)	Antimony (Sb)	1.10E − 03
	Cadmium (Cd)	8.16E − 05
	Copper (Cu)	1.84E − 02
	Lead (Pb)	2.14E − 02
	Nickel (Ni)	6.11E − 03
	Zinc (Zn)	2.36E − 01
Incidental Ingestion of soil that has been carried offsite to nearby locations/playgrounds (units: mg/kg)	Antimony (Sb)	7.49E + 00
	Cadmium (Cd)	1.73E + 00
	Copper (Cu)	9.85E + 01
	Lead (Pb)	1.39E + 02
	Nickel (Ni)	4.65E + 01
	Zinc (Zn)	1.11E + 03
Dermal contact with soil deposited along unpaved sidewalk adjoining site and also to nearby playgrounds (units: mg/kg)	Antimony (Sb)	7.49E + 00
	Cadmium (Cd)	1.73E + 00
	Copper (Cu)	9.85E + 01
	Lead (Pb)	1.39E + 02
	Nickel (Ni)	4.65E + 01
	Zinc (Zn)	1.11E + 03

pathways described for children between ages 6 and 12 years. In addition, pica behavior (i.e., the intentional eating/mouthing of large quantities of dirt and other objects) may also be exhibited by children in this age group.

Exposure Scenarios

Realistic sets of exposure scenarios representative of the ABC site and vicinity are developed and evaluated to aid the risk characterization. For the present conditions, the exposures include offsite inhalation of fugitive dust (based on air modeling information), soils ingestion (based on offsite soil samples), and dermal contact with surface soils (based on offsite soil samples). This set of exposures assumes that the fence at the ABC site forms a complete barrier to prevent unauthorized entry. Under future land use conditions, potential exposures will include onsite inhalation of fugitive dust (based on air modeling information), soils ingestion (based on onsite "residual" soil contamination after implementation of corrective actions), and dermal contact with surface soils (based on onsite "residual" soil contamination after implementation of corrective actions).

Table 7.2 lists the exposure point concentrations for the reasonable maximum exposures (RMEs) to be used for intake and dose estimations under the various exposure scenarios for current land use conditions considered in the risk assessment for the ABC site. The RMEs are determined to be the highest exposure judged to be reasonably expected to occur under the site-specific conditions, and these are estimated for the individual pathways; the arithmetic mean concentrations are selected as the RME in this case. The exposure point concentrations for future land use conditions will be represented by the recommended soil cleanup levels to be developed later in this evaluation.

Estimation of Chemical Intakes for Critical Individual Pathways

Chronic intakes are evaluated for the ABC site risk assessment, since it is expected that the potential receptors under the postulated exposure scenarios will experience long-term exposures. The RME concentration values, representing estimates of long-term exposure point concentrations, are used to calculate the CDIs. The equations used for the intake calculations are discussed in Appendix B1.

Parameters which define chemical absorption and the frequency, duration, and amount of exposure are assigned values which more closely represent site-specific exposure conditions associated with the ABC site; these are summarized into the site-specific exposure parameters that is used in the computation of receptor intakes and doses (Table 7.3). Chemical-specific absorption and bioavailability factors are incorporated where appropriate. Where no chemical-specific absorption rate was found in the literature, an absorption value of 100% is used in the calculation of the absorption-adjusted intakes (i.e., the doses). A summary of the estimated intakes and/or doses are presented in Table 7.4 for the current land use conditions.

7.1.5 Toxicity Assessment

The toxicity evaluation of the chemicals of potential concern present at the ABC site is carried out to identify relevant toxicity indices and acceptable daily intakes against which exposure point intakes and doses are compared. The toxicological profiles for the chemicals of potential concern are summarized in Table 7.5. The indices include the cancer SFs for the carcinogens and the RfDs or ADIs for the noncarcinogens. This information has been compiled from the IRIS database, where available, or from other applicable sources referenced in the table.

7.1.6 Risk Characterization Under Current Land Use and No-Action Scenario

Both carcinogenic risks and noncarcinogenic HIs are evaluated for the ABC site under a baseline ("no-action") scenario. This represents the very worst-case scenario that the site could possibly pose to potential receptors in the vicinity. Risk is characterized for all categories of receptor groups and exposure pathways as

- Incremental lifetime risk from identified carcinogens
- Chronic HIs for noncarcinogenic effects of both carcinogens and noncarcinogens

The calculation of potential carcinogenic risks and noncarcinogenic HIs under the worst-case/no-action conditions at the ABC site are performed for the different population groups. The equations used for these calculations are discussed in Appendix B2, and the results are included in Tables 7.6 and 7.7. For the noncarcinogenic effects, it is assumed for simplicity purposes in this evaluation that all the chemicals of concern have the same physiological endpoint.

Table 7.3 Site-Specific Parameters for Exposure Assessment at the ABC Site

Parameter	Children Aged up to 6	Children Aged 6–12	Adult	Reference Sources
Physical characteristics				
Average body weight	16 kg	29 kg	70 kg	a,b,c
Average total skin surface area	6980 cm^2	10,470 cm^2	18,150 cm^2	a,b,e,h
Average lifetime			70 years	a,b,c,e
Average lifetime exposure period	5 years	6 years	58 years	b,e
Activity characteristics				
Inhalation rate	0.25 m^3/h	0.46 m^3/h	0.83 m^3/h	b,e
Retention rate of inhaled air	100%	100%	100%	e
Frequency of fugitive dust inhalation				
Offsite residents, schools, and passers-by	365 days/year	365 days/year	365 days/year	b,e
Offsite workers	—	—	260 days/year	b,e
Duration of fugitive dust inhalation (outside)				
Offsite residents, schools, and passers-by	12 h/day	12 h/day	12 h/day	b,e
Offsite workers	—	—	8 h/day	b,e
Amount of soil ingested incidentally	200 mg/day	100 mg/day	50 mg/day	a,b,c,e,h,i
Frequency of soil contact				
Offsite residents, schools, and passers-by	330 days/year	330 days/year	330 days/year	b,e
Offsite workers	—	—	260 days/year	b,e
Duration of soil contact				
Offsite residents, schools, and passers-by	12 h/day	8 h/day	8 h/day	b,e
Offsite workers	—	—	8 h/day	b,e
Percentage of skin area contacted by soil	20%	20%	10%	b,e,h
Material characteristics				
Soil to skin adherence factor	0.75 mg/cm^2	0.75 mg/cm^2	0.75 mg/cm^2	a,b,e,f,g
Soil matrix attenuation factor	15%	15%	15%	d

Chemical-specific skin absorption factors (soil/dermal)				
Antimony (Sb)	10%	10%	5%	d,e
Cadmium (Cd)	10%	10%	5%	d,e
Copper (Cu)	10%	10%	5%	d,e
Lead (Pb)	10%	10%	5%	d,e
Nickel (Ni)	10%	10%	5%	d,e
Zinc (Zn)	10%	10%	5%	d,e
Chemical-specific bioavailability factors (ingestion)				
Antimony (Sb)	100%	100%	100%	e
Cadmium (Cd)	5%	5%	5%	e
Copper (Cu)	100%	100%	100%	e
Lead (Pb)	53%	10%	10%	e,h,j
Nickel (Ni)	100%	100%	100%	e
Zinc (Zn)	100%	100%	100%	e
Inhalation absorption factors (all chemicals)	100%	100%	100%	e
Chemical availability factor[k] (after site development)	5%	5%	5%	e

[a] U.S. EPA (1989b).
[b] U.S. EPA (1989c).
[c] U.S. EPA (1988a).
[d] Hawley (1985).
[e] Estimate based on site-specific conditions.
[f] Lepow et al. (1975).
[g] Lepow et al. (1974).
[h] Sedman (1989).
[i] Calabrese et al. (1989).
[j] Marcus (1986).
[k] Accounts for the reduction in exposure with potentially contaminated sources, due to change of physical scenario at site.

Table 7.4 Exposure Assessment for Current Land Use Conditions

Population	Exposure Pathway	Chemical of Concern	Absorption Factor (%)	Exposure Concentration[a] (RME)	Chronic Daily Intake (mg/kg/day)	
					Carcinogenic Effects	Noncarcinogenic Effects
Nearby residents (child 1–6 years)	Inhalation of particulates from dust generated by wind erosion (units: mg/m³)	Antimony (Sb)	100	1.10E − 03	—	2.07E − 04
		Cadmium (Cd)	100	8.16E − 05	1.09E − 06	1.53E − 05
		Copper (Cu)	100	1.84E − 02	—	3.46E − 03
		Lead (Pb)	100	2.14E − 02	2.87E − 04	4.02E − 03
		Nickel (Ni)	100	6.11E − 03	—	1.15E − 03
		Zinc (Zn)	100	2.36E − 01	—	4.44E − 02
	Incidental ingestion of soil carried offsite to nearby playgrounds (units: mg/kg)	Antimony (Sb)	100	7.49E + 00	—	8.46E − 05
		Cadmium (Cd)	5	1.73E + 00	—	9.77E − 07
		Copper (Cu)	100	9.85E + 01	—	1.11E − 03
		Lead (Pb)	53	1.39E + 02	5.96E − 05	8.35E − 04
		Nickel (Ni)	100	4.65E + 01	—	5.25E − 04
		Zinc (Zn)	100	1.11E + 03	—	1.25E − 02
	Dermal contact with soil deposited in residential areas and also nearby playgrounds (units: mg/kg)	Antimony (Sb)	10	7.49E + 00	—	6.64E − 06
		Cadmium (Cd)	10	1.73E + 00	—	1.53E − 06
		Copper (Cu)	10	9.85E + 01	—	8.74E − 05
		Lead (Pb)	10	1.39E + 02	8.84E − 06	1.24E − 04
		Nickel (Ni)	10	4.65E + 01	—	4.12E − 05
		Zinc (Zn)	10	1.11E + 03	—	9.80E − 04
Nearby residents (child 6–12 years)	Inhalation of particulates from dust generated by wind erosion (units: mg/m³)	Antimony (Sb)	100	1.10E − 03	—	2.09E − 04
		Cadmium (Cd)	100	8.16E − 05	1.33E − 06	1.55E − 05
		Copper (Cu)	100	1.84E − 02	—	3.50E − 03
		Lead (Pb)	100	2.14E − 02	3.49E − 04	4.07E − 03
		Nickel (Ni)	100	6.11E − 03	—	1.16E − 03
		Zinc (Zn)	100	2.36E − 01	—	4.48E − 02

	Incidental ingestion of soil carried offsite to nearby playgrounds (units: mg/kg)	Antimony (Sb)	100	7.49E + 00	—	2.34E – 05
		Cadmium (Cd)	5	1.73E + 00	—	2.70E – 07
		Copper (Cu)	100	9.85E + 01	—	3.07E – 04
		Lead (Pb)	10	1.39E + 02	3.72E – 06	4.35E – 05
		Nickel (Ni)	100	4.65E + 01	—	1.45E – 04
		Zinc (Zn)	100	1.11E + 03	—	3.45E – 03
	Dermal contact with soil deposited in residential areas and also nearby playgrounds (units: mg/kg)	Antimony (Sb)	10	7.49E + 00	—	5.50E – 06
		Cadmium (Cd)	10	1.73E + 00	—	1.27E – 06
		Copper (Cu)	10	9.85E + 01	—	7.23E – 05
		Lead (Pb)	10	1.39E + 02	8.78E – 06	1.02E – 04
		Nickel (Ni)	10	4.65E + 01	—	3.41E – 05
		Zinc (Zn)	10	1.11E + 03	—	8.11E – 04
Nearby residents (adults)	Inhalation of particulates from dust generated by wind erosion (units: mg/m^3)	Antimony (Sb)	100	1.10E – 03	—	1.56E – 04
		Cadmium (Cd)	100	8.16E – 05	9.63E – 06	1.16E – 05
		Copper (Cu)	100	1.84E – 02	—	2.61E – 03
		Lead (Pb)	100	2.14E – 02	2.53E – 04	3.04E – 03
		Nickel (Ni)	100	6.11E – 03	—	8.68E – 04
		Zinc (Zn)	100	2.36E – 01	—	3.35E – 02
	Incidental ingestion of soil carried offsite to nearby playgrounds (units: mg/kg)	Antimony (Sb)	100	7.49E + 00	—	4.84E – 06
		Cadmium (Cd)	5	1.73E + 00	—	5.59E – 08
		Copper (Cu)	100	9.85E + 01	—	6.36E – 05
		Lead (Pb)	10	1.39E + 02	7.46E – 06	9.00E – 06
		Nickel (Ni)	100	4.65E + 01	—	3.00E – 05
		Zinc (Zn)	100	1.11E + 03	—	7.14E – 04
	Dermal contact with soil deposited in residential areas and along unpaved sidewalks (units: mg/kg)	Antimony (Sb)	5	7.49E + 00	—	9.89E – 07
		Cadmium (Cd)	5	1.73E + 00	—	2.28E – 07
		Copper (Cu)	5	9.85E + 01	—	1.30E – 05
		Lead (Pb)	5	1.39E + 02	1.53E – 05	1.84E – 05
		Nickel (Ni)	5	4.65E + 01	—	6.14E – 06
		Zinc (Zn)	5	1.11E + 03	—	1.46E – 04

[a] Units are mg/m^3 for air respirable concentrations and mg/kg for concentrations in soil.

Table 7.5 Toxicity Values for the Chemicals of Concern at the ABC Site

Chemical of Concern	Oral Exposure Route[a]					Inhalation Exposure Route[a]				
	Carcinogen Class	SF 1/(mg/kg/day)	SF Source	Chronic RfD (mg/kg/day)	RfD Source	Carcinogen Class	SF 1/(mg/kg/day)	SF Source	RfD (mg/kg/day)	RfD Source
Antimony (Sb)	NC	—	—	4.00E – 04	IRIS	NC	—	—	NA	IRIS
Cadmium (Cd)	NC	—	—	5.00E – 04	IRIS	B1	6.10E + 00	IRIS	NA	IRIS
Copper (Cu)	NC	—	—	3.70E – 02	HEA	NC	—	—	1.00E – 02	HEA
Lead (Pb)	B2	NA	IRIS	b	HEA[c]	B2	NA	IRIS	b	HEA[c]
Nickel (Ni)	NC	—	—	2.00E – 02	IRIS	NC	—	—	NA	IRIS
Zinc (Zn)	NC	—	—	2.00E – 01	HEAST	NC	—	—	1.00E – 02	HEA

[a] SF = slope factor; RfD = Reference dose; — = Not applicable; NA = Not available; IRIS = Integrated Risk Information System (1990); HEAST = Health Effects Assessment Summary Tables (1990); and HEA = Health Effects Assessment document (Environmental Criteria and Assessment Office, U.S. EPA, 1986).

[b]

	Oral Exposure	Inhalation Exposure
AIC for children under 6 years	1.19E – 03	5.63E – 04
AIC for children 6–12 years	6.55E – 04	5.71E – 04
AIC for adults	6.86E – 04	4.29E – 04

[c] For the quantification of the toxicological effects of lead, ADIs are used to calculate the AICs for adults and children, which serve as a surrogate for the chronic RfDs for lead.

It is determined that Cd is the only site-related chemical likely to pose any cancer risks to the potential receptors. However, cancer risks of 7.0×10^{-6}, 8.0×10^{-6}, and 6.0×10^{-5} estimated for children under 6, children aged 6 to 12, and adult populations, respectively, are all within the acceptable range of 1.0×10^{-7} to 1.0×10^{-4}. Thus, within the limits of uncertainty, carcinogenic risks to human health are not indicated, even under the baseline conditions at the ABC site. On the other hand, potential noncarcinogenic hazards are indicated for some receptors under the baseline conditions. For children aged up to 6 years, the total exposure HI of 13.1 is well above the acceptable level of 1. Exposure through dermal contact is an insignificant contributor to the HI for this population group; however, ingestion and inhalation exposures are significant to the potential health threats posed by the contaminants of concern — with the single greatest contributor to the high hazard index (HI) for this receptor group coming from Pb. For children aged between 6 and 12 years, it is apparent that the total exposure HI of 12.3 is well above the acceptable level of 1. Exposures through soil ingestion and dermal contact do not appear to be significant contributors to the HI for this population group; however, inhalation exposure is critical to the potential health threats. For adult populations, the total exposure HI of 10.7 is well above the acceptable level of 1. Exposures through soil ingestion and dermal contact do not appear to be a problem for this population group; however, inhalation exposure is critical to the potential health threats posed by the contaminants of concern.

Pictorial presentations (e.g., Figures 7.1 and 7.2) of the risk assessment results will generally aid in visualizing several impacts and pathways that can aid in preliminary management decisions and risk communication programs. For instance, with inhalation so obviously dominating the overall site risks, an interim corrective measure for the ABC site involving the application of dust suppressants to the surface soil (so as to minimize fugitive dust generation) will generally be better received by the impacted community (who will appreciate the decision that is tied in to the graphical representations). After an application of dust suppressants to the site, the effects of airborne contaminants will be minimized. This leads to a situation where the inhalation pathway in this risk assessment is practically eliminated.

7.1.7 Risk Characterization Under Future Land Use: Light Industrial/Office Developments Scenario

Both carcinogenic risks and noncarcinogenic HIs are evaluated for a scenario that will prevail following development of the ABC site into light industrial, office, and/or retail complex(es), prior to remediation. Under such a development program, it is assumed that 85 to 90% of the site will be paved and/or covered with concrete slabs, and that there will be about 10% landscaping done using imported fill materials. Under these conditions, it is expected that the maximally exposed potential receptors will receive only about 5% of what will prevail at the undeveloped site with respect to the inhalation, soil ingestion, and/or dermal exposure pathways for as long as the site is controlled in this manner.

Table 7.6A Cancer Risk Estimates for Children Aged up to 6 Years (Offsite) — Current Land Use Conditions (No-Action Scenario) for the ABC Site

Exposure Pathway	Chemical of Concern	CDI[a] (mg/kg/day)	CDI Adjusted for Absorption	SF $(mg/kg/day)^{-1}$	Weight-of-Evidence Classification	Type of Cancer	SF Basis (Vehicle)	SF Source	Chemical Specific Risk	Total Pathway Risk	Total Exposure Risk
Inhalation of fugitive dust	Antimony (Sb)	—	—	—	NC	—	—	—	—		
	Cadmium (Cd)	1.09E – 06	No	6.10E + 00	B1	Lung; trachea; bronchus	Air	IRIS	7E – 06		
	Copper (Cu)	—	—	—	NC	—	—	—	—		
	Lead (Pb)	2.87E – 04	No	NA	B2	NA	—	—	—		
	Nickel (Ni)	—	—	—	NC	—	—	—	—		
	Zinc (Zn)	—	—	—	NC	—	—	—	—	7E – 06	
Ingestion of soil	Antimony (Sb)	—	—	—	NC	—	—	—	—		
	Cadmium (Cd)	—	—	—	NC	—	—	—	—		
	Copper (Cu)	—	—	—	NC	—	—	—	—		
	Lead (Pb)	5.96E – 05	Yes	NA	B2	NA	—	—	—		
	Nickel (Ni)	—	—	—	NC	—	—	—	—		
	Zinc (Zn)	—	—	—	NC	—	—	—	—	0E + 00	
Dermal contact	Antimony (Sb)	—	—	—	NC	—	—	—	—		
	Cadmium (Cd)	—	—	—	NC	—	—	—	—		
	Copper (Cu)	—	—	—	NC	—	—	—	—		
	Lead (Pb)	8.84E – 06	Yes	NA	B2	NA	—	—	—		
	Nickel (Ni)	—	—	—	NC	—	—	—	—		
	Zinc (Zn)	—	—	—	NC	—	—	—	—	0E + 00	
Nearby residential population potentially affected by the ABC site — total cancer risk for children aged under 6 years (weight-of-evidence predominantly B1)											7E – 06

[a] CDI = Chronic daily intake (carcinogenic effects); NA = Not available; — = Not applicable.

Table 7.6B Cancer Risk Estimates for Children Aged 6–12 Years (Offsite) — Current Land Use Conditions (No-Action Scenario) for the ABC Site

Exposure Pathway	Chemical of Concern	CDI[a] (mg/kg/day)	CDI Adjusted for Absorption	SF $(mg/kg/day)^{-1}$	Weight-of-Evidence Classification	Type of Cancer	SF Basis (Vehicle)	SF Source	Chemical Specific Risk	Total Pathway Risk	Total Exposure Risk
Inhalation of fugitive dust	Antimony (Sb)	—	—	—	NC	—	—	—	—		
	Cadmium (Cd)	1.33E – 06	No	6.10E + 00	B1	Lung; trachea; bronchus	Air	IRIS	8E – 06		
	Copper (Cu)	—	—	—	NC	—	—	—	—		
	Lead (Pb)	3.49E – 04	No	NA	B2	NA	—	—	—		
	Nickel (Ni)	—	—	—	NC	—	—	—	—		
	Zinc (Zn)	—	—	—	NC	—	—	—	—	8E – 06	
Ingestion of soil	Antimony (Sb)	—	—	—	NC	—	—	—	—		
	Cadmium (Cd)	—	—	—	NC	—	—	—	—		
	Copper (Cu)	—	—	—	NC	—	—	—	—		
	Lead (Pb)	3.72E – 06	Yes	NA	B2	NA	—	—	—		
	Nickel (Ni)	—	—	—	NC	—	—	—	—		
	Zinc (Zn)	—	—	—	NC	—	—	—	—	0E + 00	
Dermal contact	Antimony (Sb)	—	—	—	NC	—	—	—	—		
	Cadmium (Cd)	—	—	—	NC	—	—	—	—		
	Copper (Cu)	—	—	—	NC	—	—	—	—		
	Lead (Pb)	8.78E – 06	Yes	NA	B2	NA	—	—	—		
	Nickel (Ni)	—	—	—	NC	—	—	—	—		
	Zinc (Zn)	—	—	—	NC	—	—	—	—	0E + 00	
Nearby residential population potentially affected by the ABC site — total cancer risk for children aged 6–12 years (weight-of-evidence predominantly B1)											8E – 06

[a] CDI = Chronic daily intake (carcinogenic effects); NA = Not available; — = Not applicable.

Table 7.6C Cancer Risk Estimates for Adults in the Site Vicinity — Current Land Use Conditions (No-Action Scenario) for the ABC Site

Exposure Pathway	Chemical of Concern	CDI[a] (mg/kg/day)	CDI Adjusted for Absorption	SF $(mg/kg/day)^{-1}$	Weight-of-Evidence Classification	Type of Cancer	SF Basis (Vehicle)	SF Source	Chemical Specific Risk	Total Pathway Risk	Total Exposure Risk
Inhalation of fugitive dust	Antimony (Sb)	—	—	—	NC	—	—	—	—		
	Cadmium (Cd)	9.63E – 06	No	6.10E + 00	B1	Lung; trachea; bronchus	Air	IRIS	6E – 05		
	Copper (Cu)	—	—	—	NC	—	—	—	—		
	Lead (Pb)	2.53E – 04	No	NA	B2	NA	—	—	—		
	Nickel (Ni)	—	—	—	NC	—	—	—	—		
	Zinc (Zn)	—	—	—	NC	—	—	—	—		
										6E – 05	
Ingestion of soil	Antimony (Sb)	—	—	—	NC	—	—	—	—		
	Cadmium (Cd)	—	—	—	NC	—	—	—	—		
	Copper (Cu)	—	—	—	NC	—	—	—	—		
	Lead (Pb)	7.46E – 06	Yes	NA	B2	NA	—	—	—		
	Nickel (Ni)	—	—	—	NC	—	—	—	—		
	Zinc (Zn)	—	—	—	NC	—	—	—	—		
										0E + 00	
Dermal contact	Antimony (Sb)	—	—	—	NC	—	—	—	—		
	Cadmium (Cd)	—	—	—	NC	—	—	—	—		
	Copper (Cu)	—	—	—	NC	—	—	—	—		
	Lead (Pb)	1.53E – 05	Yes	NA	B2	NA	—	—	—		
	Nickel (Ni)	—	—	—	NC	—	—	—	—		
	Zinc (Zn)	—	—	—	NC	—	—	—	—		
										0E + 00	
Nearby residential population potentially affected by the ABC site — total cancer risk for adult residents in site vicinity (weight-of-evidence predominantly B1)											6E – 05

[a] CDI = Chronic daily intake (carcinogenic effects); NA = Not available; — = Not applicable.

Table 7.7A Chronic HI Estimates for Children Aged up to 6 years (Offsite) — Current Land Use Conditions (No-Action Scenario) for the ABC Site

Exposure Pathway	Chemical of Concern	CDI[a] (mg/kg/day)	CDI Adjusted for Absorption	RfD[b] (mg/kg/day)	RfD Source	Hazard Quotient	Pathway HI	Total Exposure HI
Inhalation of fugitive dust	Antimony (Sb)	2.07E – 04	No	NA[c]	—[d]	—		
	Cadmium (Cd)	1.53E – 05	No	NA	—	—		
	Copper (Cu)	3.46E – 03	No	1.00E – 02	IRIS	0.35		
	Lead (Pb)	4.02E – 03	No	5.60E – 04	IRIS	7.18		
	Nickel (Ni)	1.15E – 03	No	NA		—		
	Zinc (Zn)	4.44E – 02	No	1.00E – 02	IRIS	4.44		
							11.97	
Ingestion of soil	Antimony (Sb)	8.46E – 05	No	4.00E – 04	IRIS	0.21		
	Cadmium (Cd)	9.77E – 07	Yes	5.00E – 04	IRIS	0.00		
	Copper (Cu)	1.11E – 03	No	3.70E – 02	IRIS	0.03		
	Lead (Pb)	8.35E – 04	Yes	1.20E – 03	IRIS	0.70		
	Nickel (Ni)	5.25E – 04	No	2.00E – 02	IRIS	0.03		
	Zinc (Zn)	1.25E – 02	No	2.00E – 01	IRIS	0.06		
							1.03	
Dermal contact	Antimony (Sb)	6.64E – 06	Yes	4.00E – 04	IRIS	0.02		
	Cadmium (Cd)	1.53E – 06	Yes	5.00E – 04	IRIS	0.00		
	Copper (Cu)	8.74E – 05	Yes	3.70E – 02	IRIS	0.00		
	Lead (Pb)	1.24E – 04	Yes	1.20E – 03	IRIS	0.10		
	Nickel (Ni)	4.12E – 05	Yes	2.00E – 02	IRIS	0.00		
	Zinc (Zn)	9.80E – 04	Yes	2.00E – 01	IRIS	0.00		
							0.13	
Nearby residential population potentially affected by the ABC site — total chronic HI for children aged under 6 years								13.1

[a] CDI = Chronic daily intake (noncarcinogenic effects).
[b] RfD = Reference dose (or equivalent ADI where no RfD exists).
[c] NA = Not available.
[d] — = Not applicable.

Table 7.7B Chronic HI Estimates for Children Aged 6–12 Years (Offsite) — Current Land Use Conditions (No-Action Scenario) for the ABC Site

Exposure Pathway	Chemical of Concern	CDI[a] (mg/kg/day)	CDI Adjusted for Absorption	RfD[b] (mg/kg/day)	RfD Source	Hazard Quotient	Pathway HI	Total Exposure HI
Inhalation of fugitive dust	Antimony (Sb)	2.09E – 04	No	NA[c]	—[d]	—		
	Cadmium (Cd)	1.55E – 05	No	NA	—	—		
	Copper (Cu)	3.50E – 03	No	1.00E – 02	IRIS	0.35		
	Lead (Pb)	4.07E – 03	No	5.70E – 04	IRIS	7.13		
	Nickel (Ni)	1.16E – 03	No	NA		—		
	Zinc (Zn)	4.48E – 02	No	1.00E – 02	IRIS	4.48	11.97	
Ingestion of soil	Antimony (Sb)	2.34E – 05	No	4.00E – 04	IRIS	0.06		
	Cadmium (Cd)	2.70E – 07	Yes	5.00E – 04	IRIS	0.00		
	Copper (Cu)	3.07E – 03	No	3.70E – 02	IRIS	0.01		
	Lead (Pb)	4.35E – 04	Yes	6.60E – 04	IRIS	0.07		
	Nickel (Ni)	1.45E – 04	No	2.00E – 02	IRIS	0.01		
	Zinc (Zn)	3.45E – 03	No	2.00E – 01	IRIS	0.02	0.16	
Dermal contact	Antimony (Sb)	5.50E – 06	Yes	4.00E – 04	IRIS	0.01		
	Cadmium (Cd)	1.27E – 06	Yes	5.00E – 04	IRIS	0.00		
	Copper (Cu)	7.23E – 05	Yes	3.70E – 02	IRIS	0.00		
	Lead (Pb)	1.02E – 04	Yes	6.60E – 04	IRIS	0.16		
	Nickel (Ni)	3.41E – 05	Yes	2.00E – 02	IRIS	0.00		
	Zinc (Zn)	8.11E – 04	Yes	2.00E – 01	IRIS	0.00	0.18	
Nearby residential population potentially affected by the ABC site — total chronic HI for children aged 6–12 years								12.3

[a] CDI = Chronic daily intake (noncarcinogenic effects).
[b] RfD = Reference dose (or equivalent ADI where no RfD exists).
[c] NA = Not available.
[d] — = Not applicable.

Table 7.7C Chronic HI Estimates for Adult Residents in Vicinity of Site — Current Land Use Conditions (No-Action Scenario) for the ABC Site

Exposure Pathway	Chemical of Concern	CDI[a] (mg/kg/day)	CDI Adjusted for Absorption	RfD[b] (mg/kg/day)	RfD Source	Hazard Quotient	Pathway HI	Total Exposure HI
Inhalation of fugitive dust	Antimony (Sb)	1.56E – 04	No	NA[c]	—[d]	—		
	Cadmium (Cd)	1.16E – 05	No	NA	—	—		
	Copper (Cu)	2.61E – 03	No	1.00E – 02	IRIS	0.26		
	Lead (Pb)	3.04E – 03	No	4.30E – 04	IRIS	7.07		
	Nickel (Ni)	8.68E – 04	No	NA	—	—		
	Zinc (Zn)	3.35E – 02	No	1.00E – 02	IRIS	3.35		
							10.68	
Ingestion of soil	Antimony (Sb)	4.84E – 06	No	4.00E – 04	IRIS	0.01		
	Cadmium (Cd)	5.59E – 08	Yes	5.00E – 04	IRIS	0.00		
	Copper (Cu)	6.36E – 05	No	3.70E – 02	IRIS	0.00		
	Lead (Pb)	9.00E – 06	Yes	6.90E – 04	IRIS	0.01		
	Nickel (Ni)	3.00E – 05	No	2.00E – 02	IRIS	0.00		
	Zinc (Zn)	7.14E – 04	No	2.00E – 01	IRIS	0.00		
							0.03	
Dermal contact	Antimony (Sb)	9.89E – 07	Yes	4.00E – 04	IRIS	0.00		
	Cadmium (Cd)	2.28E – 07	Yes	5.00E – 04	IRIS	0.00		
	Copper (Cu)	1.30E – 05	Yes	3.70E – 02	IRIS	0.00		
	Lead (Pb)	1.84E – 05	Yes	6.90E – 04	IRIS	0.03		
	Nickel (Ni)	6.14E – 06	Yes	2.00E – 02	IRIS	0.00		
	Zinc (Zn)	1.46E – 04	Yes	2.00E – 01	IRIS	0.00		
							0.03	
Nearby residential population potentially affected by the ABC site — total chronic HI for adult residents								10.7

[a] CDI = Chronic daily intake (noncarcinogenic effects).
[b] RfD = Reference dose (or equivalent ADI where no RfD exists).
[c] NA = Not available.
[d] — = Not applicable.

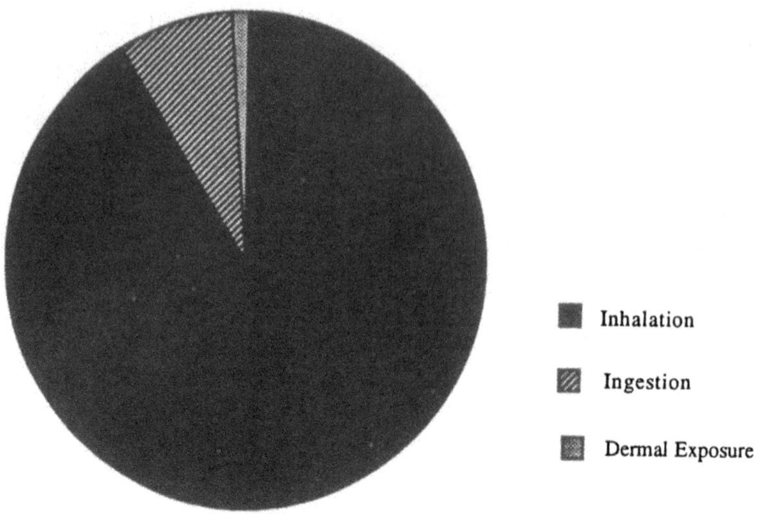

Figure 7.1A Pie chart schematic for HI contributions (%) for child 0 to 6 years.

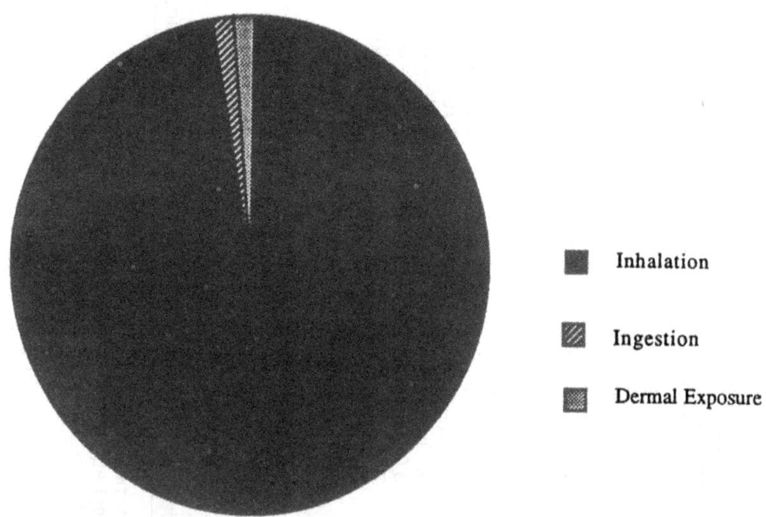

Figure 7.1B Pie chart schematic for HI contributions (%) for child 6 to 12 years.

The potential risks to human health associated with exposure to the various chemicals of concern under future land use conditions are evaluated for the site; two different receptor groups are evaluated under this scenario for the same exposure pathways previously analyzed:

- Children aged between 6 and 12 years visiting the site
- Adult workers onsite

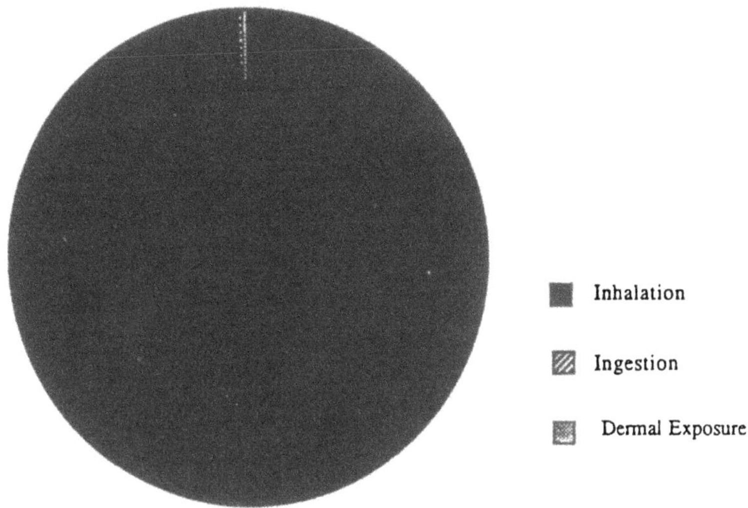

Figure 7.1C Pie chart schematic for HI contributions (%) for adult residents.

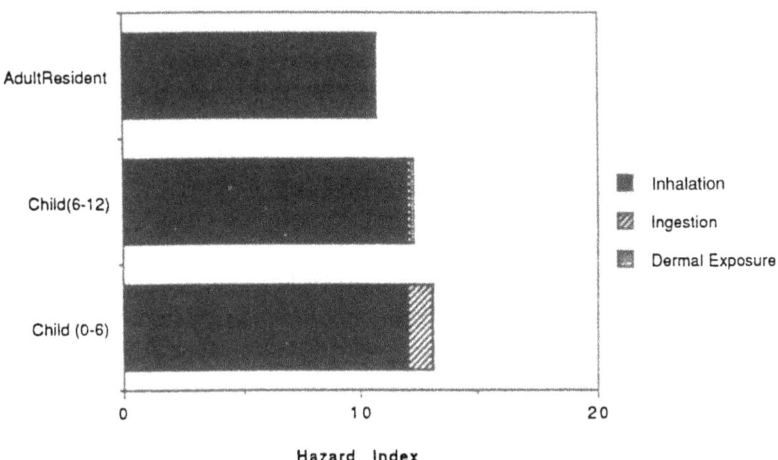

Figure 7.2A Schematic of HI representation (arithmetic scale).

The potential chemical intakes and doses under the new conditions (that will be approximately 5% of previous exposures) are used to characterize risks and hazards for the different population groups and exposure pathways; the results are included in Tables 7.8 and 7.9. Again, Cd is the only site-related chemical evaluated as having the potential to pose cancer risks to the potential receptors in the vicinity of the ABC site. However, cancer risks of 4×10^{-7} and 1×10^{-6} estimated for children aged 6 to 12 and adult workers, respectively, are all within the acceptable range of 1×10^{-7} to 1×10^{-4}. Thus, carcinogenic risks to human health are not indicated under the conditions that will prevail after the site is developed into industrial, office, and/or

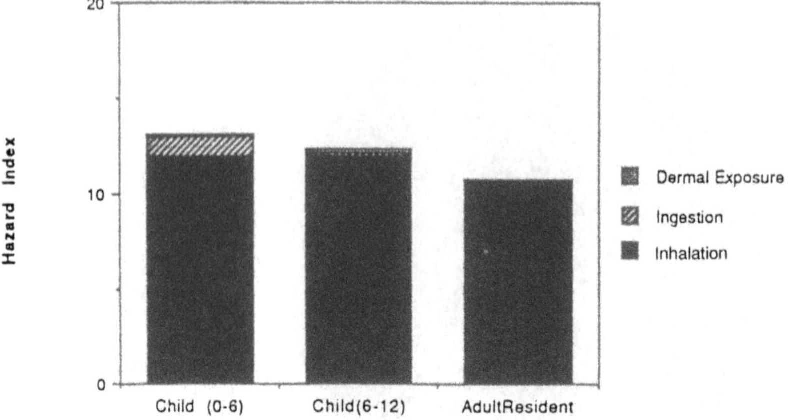

Figure 7.2B Schematic of HI representation (arithmetic scale).

retail complexes. Also, no significant noncarcinogenic hazards are indicated either. For children aged between 6 and 12 years, the total exposure HI of 1.1 may be considered acceptable (compared with the acceptable level of 1). For adult workers, the total exposure hazard index of 0.3 is considered acceptable (compared with the acceptable level of 1) for the conservative assumptions used in the evaluation. Thus, after development of the site into light industrial and/or office complexes, the ABC site would indicate practically no health threats to adult workers and children present at the site.

7.1.8 Postremediation (to Soil ALs) Risk Characterization — Future Onsite Land Uses

The future land use conditions are determined to be comprised of potential onsite receptor exposures. The soil ALs are developed for the ABC site according to the relationships discussed in Appendix B3. For the ABC site, noncarcinogenic effects are the significant contributors to any health risks indicated. Consequently, in determining the health-based cleanup criteria for this site, the relationship for developing soil ALs for systemic toxicants is applied to the critical receptors. The soil ALs developed for the site and the exposure point concentrations are presented in Table 7.10; the most stringent values are conservatively selected as the health-based soil ALs for the chemicals of concern.

For onsite exposure to chemicals present at the soil ALs, the exposure point concentrations under the RMEs are taken to be the site-specific ALs. Residual risks that remain after the implementation of a remedial action to bring chemical concentration levels at the site to the soil ALs are assessed for the critical receptors that are assumed to be onsite; this represents potential future land use for light industrial, office, and/or retail facilities. Under the future land use conditions, up to about 90% of the site is assumed to be paved or covered with concrete slabs and about 10% covered with imported fill and landscaped. It is therefore reasonable to assume that only approximately 5% of chemicals present onsite would actually be available for possible intake by potential receptors at the developed site. From these exposures, the carcinogenic risks (Table 7.11) and the chronic HIs (Table 7.12) for the critical receptor groups are computed. These indicate significantly reduced health threats if the site is cleaned up to the estimated soil ALs. It may therefore not be necessary to clean up the ABC site to levels lower than the ALs, even if direct onsite exposure to children is anticipated under the future land use conditions.

7.1.9 The Recommended Soil Cleanup Limit

In situations when the soil ALs result in unacceptable levels from the postremediation risk characterization, a recommended soil cleanup limit (RSCL) will be determined. However, since the soil ALs would present no significant health threats to potential receptors, the AL values become the actual RSCLs for remedial action planning (Table 7.13). In the implementation of corrective actions, it may become apparent that some chemicals can be more easily removed than others. In that case, it still may be possible to clean up to lower levels than the RSCL for some constituents and to higher levels than the RSCL for others and still attain an acceptable HI for residual contamination remaining at the site after remediation.

Residual risks that remain after the implementation of a remedial action to bring contaminant concentrations at the ABC site to the RSCL do not indicate significant health threats if the site is cleaned up to the RSCLs (which are numerically equivalent to the ALs in this case). Tables 7.14 and 7.15 summarize the carcinogenic risk and HI, respectively, for the changing trends in health threats posed under the different conditions and land uses for the ABC site.

7.1.10 Identification and Evaluation of Uncertainties

The main sources of uncertainties would be from the the conservative assumptions used in developing the exposure scenarios for this risk assessment, including, for instance, the assumption of 100% chemical absorption in situations where no justifiable lower level was obtainable from the literature. The conservative assumptions will obviously lead to overestimates of potential risks and hazards. This will tend to protect human health to an even greater degree if remedial alternatives are developed on this basis.

It is noteworthy that although the single greatest contributor to the high hazard level for some receptor groups is attributable to Pb, these may not all be site related. That is, the overall contribution of the site to any health threats from Pb in the vicinity of the site may not be fully attributed to the site, but also to other possible sources such as result of automoblile emissions and industrial activity in the area. The observation that other industrial activities and vehicular traffic may have caused high background levels for the chemicals of concern at the site can be demonstrated by collecting background samples from upwind locations to the site.

Other principal uncertainties surrounding the assessment are the absence of SFs and RfDs for some of the chemicals of potential concern originating from the site. This means that the contributions from such chemicals to potential risks are not completely accounted for in this risk assessment. However, conservative assump-

Table 7.8A Cancer Risk Estimates for Children Aged 6–12 Years — Future Site Conditions (Onsite Exposure for Developed Site Scenario)

Exposure Pathway	Chemical of Concern	CDI[a] (mg/kg/day)	CDI Adjusted for Absorption	SF $(mg/kg/day)^{-1}$	Weight-of-Evidence Classification	Type of Cancer	SF Basis (Vehicle)	SF Source	Chemical Specific Risk	Total Pathway Risk	Total Exposure Risk
Inhalation of fugitive dust	Antimony (Sb)	—[b]	—	—	NC	—	—	—	—		
	Cadmium (Cd)	6.65E – 08	No	6.10E + 00	B1	Lung; trachea; bronchus	Air	IRIS	4E – 07		
	Copper (Cu)	—	—	—	NC	—	—	—	—		
	Lead (Pb)	1.74E – 05	No	NA[c]	B2	NA	—	—	—		
	Nickel (Ni)	—	—	—	NC	—	—	—	—		
	Zinc (Zn)	—	—	—	NC	—	—	—	—		
										4E – 07	
Ingestion of soil	Antimony (Sb)	—	—	—	NC	—	—	—	—		
	Cadmium (Cd)	—	—	—	NC	—	—	—	—		
	Copper (Cu)	—	—	—	NC	—	—	—	—		
	Lead (Pb)	5.44E – 06	Yes	NA	B2	NA	—	—	—		
	Nickel (Ni)	—	—	—	NC	—	—	—	—		
	Zinc (Zn)	—	—	—	NC	—	—	—	—		
										0E + 00	
Dermal contact	Antimony (Sb)	—	—	—	NC	—	—	—	—		
	Cadmium (Cd)	—	—	—	NC	—	—	—	—		
	Copper (Cu)	—	—	—	NC	—	—	—	—		
	Lead (Pb)	1.28E – 05	Yes	NA	B2	NA	—	—	—		
	Nickel (Ni)	—	—	—	NC	—	—	—	—		
	Zinc (Zn)	—	—	—	NC	—	—	—	—		
										0E + 00	
Nearby residential population potentially affected by the ABC site — total cancer risk for children aged 6–12 years (weight-of-evidence predominantly B1)											4E – 07

[a] CDI = Chronic daily intake (carcinogenic effects). Onsite mean chemical concentrations are used to estimate the CDIs in this case.
[b] — = Not applicable.
[c] NA = Not available.

Table 7.8B Cancer Risk Estimates for Adult Workers Onsite — Future Site Conditions (Onsite Exposure for Developed Site Scenario)

Exposure Pathway	Chemical of Concern	CDI[a] (mg/kg/day)	CDI Adjusted for Absorption	SF $(mg/kg/day)^{-1}$	Weight-of-Evidence Classification	Type of Cancer	SF Basis (Vehicle)	SF Source	Chemical Specific Risk	Total Pathway Risk	Total Exposure Risk
Inhalation of fugitive dust	Antimony (Sb)	—[b]	—	—	NC	—	—	—	—		
	Cadmium (Cd)	2.28E – 07	No	6.10E + 00	B1	Lung; trachea; bronchus	Air	IRIS	1E – 06		
	Copper (Cu)	—	—	—	NC	—	—	—	—		
	Lead (Pb)	5.99E – 05	No	NA[c]	B2	NA	—	—	—		
	Nickel (Ni)	—	—	—	NC	—	—	—	—		
	Zinc (Zn)	—	—	—	NC	—	—	—	—		
										1E – 06	
Ingestion of soil	Antimony (Sb)	—	—	—	NC	—	—	—	—		
	Cadmium (Cd)	—	—	—	NC	—	—	—	—		
	Copper (Cu)	—	—	—	NC	—	—	—	—		
	Lead (Pb)	8.60E – 06	Yes	NA	B2	NA	—	—	—		
	Nickel (Ni)	—	—	—	NC	—	—	—	—		
	Zinc (Zn)	—	—	—	NC	—	—	—	—		
										0E + 00	
Dermal contact	Antimony (Sb)	—	—	—	NC	—	—	—	—		
	Cadmium (Cd)	—	—	—	NC	—	—	—	—		
	Copper (Cu)	—	—	—	NC	—	—	—	—		
	Lead (Pb)	1.75E – 05	Yes	NA	B2	NA	—	—	—		
	Nickel (Ni)	—	—	—	NC	—	—	—	—		
	Zinc (Zn)	—	—	—	NC	—	—	—	—		
										0E + 00	
Nearby worker population potentially affected by the ABC site — total cancer risk for onsite adult workers (weight-of-evidence predominantly B1)											1E – 06

[a] CDI = Chronic daily intake (carcinogenic effects). Onsite mean chemical concentrations are used to estimate the CDIs in this case.
[b] — = Not applicable.
[c] NA = Not available.

Table 7.9A Chronic HI Estimates for Children Aged 6–12 Years — Future Site Conditions (Onsite Exposure for Developed Site Scenario)

Exposure Pathway	Chemical of Concern	CDI[a] (mg/kg/day)	CDI Adjusted for Absorption	RfD[b] (mg/kg/day)	RfD Source	Hazard Quotient	Pathway HI	Total Exposure HI
Inhalation of fugitive dust	Antimony (Sb)	1.05E – 05	No	NA[c]	—[d]	—[d]		
	Cadmium (Cd)	7.75E – 07	No	NA	—	—		
	Copper (Cu)	1.75E – 04	No	1.00E – 02	IRIS	0.02		
	Lead (Pb)	2.03E – 04	No	5.70E – 04	IRIS	0.36		
	Nickel (Ni)	5.80E – 05	No	NA	—	—		
	Zinc (Zn)	2.24E – 03	No	1.00E – 02	IRIS	0.02		
							0.60	
Ingestion of soil	Antimony (Sb)	3.20E – 05	No	4.00E – 04	IRIS	0.08		
	Cadmium (Cd)	1.17E – 07	Yes	5.00E – 04	IRIS	0.00		
	Copper (Cu)	5.46E – 04	No	3.70E – 02	IRIS	0.01		
	Lead (Pb)	6.36E – 05	Yes	6.60E – 04	IRIS	0.10		
	Nickel (Ni)	1.82E – 04	No	2.00E – 02	IRIS	0.01		
	Zinc (Zn)	7.01E – 03	No	2.00E – 01	IRIS	0.04		
							0.24	
Dermal contact	Antimony (Sb)	7.52E – 06	Yes	4.00E – 04	IRIS	0.02		
	Cadmium (Cd)	5.51E – 07	Yes	5.00E – 04	IRIS	0.00		
	Copper (Cu)	1.28E – 04	Yes	3.70E – 02	IRIS	0.00		
	Lead (Pb)	1.50E – 04	Yes	6.60E – 04	IRIS	0.23		
	Nickel (Ni)	4.28E – 05	Yes	2.00E – 02	IRIS	0.00		
	Zinc (Zn)	1.65E – 03	Yes	2.00E – 01	IRIS	0.01		
							0.26	
Nearby residential population potentially affected by the ABC site — total chronic HI for children aged 6–12 years								1.1

[a] CDI = Chronic daily intake (noncarcinogenic effects). On-site mean chemical concentrations are used to estimate the CDIs in this case.
[b] RfD = Reference dose (or equivalent ADI where no RfD exists).
[c] NA = Not available.
[d] — = Not applicable.

Table 7.9B Chronic HI Estimates for Adult Workers (Onsite) — Future Site Conditions (Onsite Exposure for Developed Site Scenario)

Exposure Pathway	Chemical of Concern	CDI[a] (mg/kg/day)	CDI Adjusted for Absorption	RfD[b] (mg/kg/day)	RfD Source	Hazard Quotient	Pathway HI	Total Exposure HI
Inhalation of fugitive dust	Antimony (Sb)	3.72E – 06	No	NA[d]	—[d]	—		
	Cadmium (Cd)	2.76E – 07	No	NA	—	—		
	Copper (Cu)	6.22E – 05	No	1.00E – 02	IRIS	0.01		
	Lead (Pb)	7.23E – 05	No	4.30E – 04	IRIS	0.17		
	Nickel (Ni)	2.07E – 05	No	NA	—	—		
	Zinc (Zn)	7.98E – 04	No	1.00E – 02	IRIS	0.08		
							0.25	
Ingestion of soil	Antimony (Sb)	5.22E – 06	No	4.00E – 04	IRIS	0.01		
	Cadmium (Cd)	1.91E – 08	Yes	5.00E – 04	IRIS	0.00		
	Copper (Cu)	8.91E – 05	No	3.70E – 02	IRIS	0.00		
	Lead (Pb)	1.04E – 05	Yes	6.90E – 04	IRIS	0.02		
	Nickel (Ni)	2.97E – 05	No	2.00E – 02	IRIS	0.00		
	Zinc (Zn)	1.14E – 03	No	2.00E – 01	IRIS	0.01		
							0.04	
Dermal contact	Antimony (Sb)	1.07E – 06	Yes	4.00E – 04	IRIS	0.00		
	Cadmium (Cd)	7.80E – 08	Yes	5.00E – 04	IRIS	0.00		
	Copper (Cu)	1.82E – 05	Yes	3.70E – 02	IRIS	0.00		
	Lead (Pb)	2.12E – 05	Yes	6.90E – 04	IRIS	0.03		
	Nickel (Ni)	6.07E – 06	Yes	2.00E – 02	IRIS	0.00		
	Zinc (Zn)	2.34E – 04	Yes	2.00E – 01	IRIS	0.00		
							0.04	
Nearby worker population potentially affected by the ABC site — total chronic HI for onsite adult workers								0.3

[a] CDI = Chronic daily intake (noncarcinogenic effects). Onsite mean chemical concentrations are used to estimate the CDIs in this case.
[b] RfD = Reference dose (or equivalent ADI where no RfD exists).
[c] NA = Not available.
[d] — = Not applicable.

Table 7.10A Health-Based Soil AL — Future Land Use Conditions (Onsite Scenario)

Chemical of Concern	Carcinogen Class (Oral Route)	Oral RfD (mg/kg/day)		Absorption Factor (%)	Receptor-Specific Soil AL[a]		Maximum Soil Concentration (mg/kg)	Health-Based Soil AL[b] (mg/kg)
		Child (6–12)	Adult		@ Child (6–12) (mg/kg)	@ Adults (mg/kg)		
Antimony (Sb)	NC	4.00E – 04	4.00E – 04	100	116	560	1,500	116
Cadmium (Cd)	NC	5.00E – 04	5.00E – 04	5	2,900	14,000	210	2,900
Copper (Cu)	NC	3.70E – 02	3.70E – 02	100	10,730	51,800	39,600	10,730
Lead (Pb)	B2	6.55E – 04	6.86E – 04	10	1,900	9,604	16,500	1,900
Nickel (Ni)	NC	2.00E – 02	2.00E – 02	100	5,800	28,000	12,600	5,800
Zinc (Zn)	NC	2.00E – 01	2.00E – 01	100	58,000	280,000	299,000	58,000

[a] The soil AL is calculated according to the relationship in Appendix B3. AL = (RfD × Body weight)/(soil intake × chemical absorption factor).

[b] Health-based soil AL is the minimum value calculated for critical receptors (for RME) identified for site.

Table 7.10B Exposure Point Concentrations for Intake and Dose Estimation (Reasonable Maximum Exposure Under Future Land Use Conditions)

Exposure Pathway	Chemical of Concern	Onsite RME	
		Exposure Point Concentration[a]	Comments
Inhalation of fugitive dust	Antimony (Sb)	1.73E − 04	Closest receptor concentrations obtained from air modeling of fugitive dust generation and dispersion for site vicinity
	Cadmium (Cd)	2.51E − 03	
	Copper (Cu)	1.60E − 02	
	Lead (Pb)	6.06E − 04	
	Nickel (Ni)	4.33E − 03	
	Zinc (Zn)	9.09E − 02	
Ingestion of soil	Antimony (Sb)	1.16E + 02	Maximum concentrations for the closest potential receptors at site and its vicinity @ ALs
	Cadmium (Cd)	2.90E + 03	
	Copper (Cu)	1.07E + 04	
	Lead (Pb)	1.90E + 03	
	Nickel (Ni)	5.80E + 03	
	Zinc (Zn)	5.80E + 04	
Dermal contact	Antimony (Sb)	1.16E + 02	Maximum concentrations for the closest potential receptors at site and its vicinity @ ALs
	Cadmium (Cd)	2.90E + 03	
	Copper (Cu)	1.07E + 04	
	Lead (Pb)	1.90E + 03	
	Nickel (Ni)	5.80E + 03	
	Zinc (Zn)	5.80E + 04	

[a] Represents exposure point concentrations to potential receptors judged to be reasonably expected under the site-specific conditions; units are mg/m^3 for air respirable concentrations and mg/kg for concentrations in soil.

tions have been used throughout the assessment process, including the assumption that all the chemicals present have common toxicological endpoints and would therefore affect the same physiological organs. It is expected that the use of conservative assumptions will lead to overestimates of potential risks that will compensate for possible underestimations in the risk characterizations.

7.1.11 Conclusions

The goal of the assessment was to evaluate the risk information pertinent to selecting appropriate corrective actions for the ABC site. A risk characterization for the site under the "baseline" conditions showed no significant cancer risks to the populations in the vicinity of the site; long-term noncancer risks, however, were indicated. Under such conditions, therefore, corrective actions to reduce risks for the site would be necessary to prevent potential long-term health impacts.

Under future land use conditions, it was determined that development (that incorporates asphalting and landscaping) on the site will drastically reduce the inhalation pathway which dominates the risk levels posed by the ABC site. The risk characterization for the conditions after site development (but with no site remediation performed) showed no significant cancer health risks to the population in the vicinity of the site, and neither was long-term noncancer risks indicated under this site scenario. Under these conditions, therefore, further corrective actions may not be required for the ABC site.

Table 7.11A Cancer Risk Estimates for Children Aged 6–12 Years — Future Land Use Conditions (Onsite Scenario)

Exposure Pathway	Chemical of Concern	CDI[a] (mg/kg/day)	CDI Adjusted for Absorption	SF $(mg/kg/day)^{-1}$	Weight-of-Evidence Classification	Type of Cancer	SF Basis (Vehicle)	SF Source	Chemical Specific Risk	Total Pathway Risk	Total Exposure Risk
Inhalation of fugitive dust	Antimony (Sb)	—[b]	—	—	NC	—	—	—	—		
	Cadmium (Cd)	2.05E – 06	No	6.10E + 00	B1	Lung; trachea; bronchus	Air	IRIS	1E – 05		
	Copper (Cu)	—	—	—	NC	—	—	—	—		
	Lead (Pb)	4.94E – 07	No	NA[c]	B2	NA	—	—	—		
	Nickel (Ni)	—	—	—	NC	—	—	—	—		
	Zinc (Zn)	—	—	—	NC	—	—	—	—	1E – 05	
Ingestion of soil	Antimony (Sb)	—	—	—	NC	—	—	—	—		
	Cadmium (Cd)	—	—	—	NC	—	—	—	—		
	Copper (Cu)	—	—	—	NC	—	—	—	—		
	Lead (Pb)	2.54E – 06	Yes	NA	B2	NA	—	—	—		
	Nickel (Ni)	—	—	—	NC	—	—	—	—		
	Zinc (Zn)	—	—	—	NC	—	—	—	—	0E + 00	
Dermal contact	Antimony (Sb)	—	—	—	NC	—	—	—	—		
	Cadmium (Cd)	—	—	—	NC	—	—	—	—		
	Copper (Cu)	—	—	—	NC	—	—	—	—		
	Lead (Pb)	5.98E – 06	Yes	NA	B2	NA	—	—	—		
	Nickel (Ni)	—	—	—	NC	—	—	—	—		
	Zinc (Zn)	—	—	—	NC	—	—	—	—	0E + 00	
Nearby residential population potentially affected by the ABC site — total cancer risk for children aged 6–12 years (weight-of-evidence predominantly B1)											1E – 05

[a] CDI = Chronic daily intake (carcinogenic effects).
[b] — = Not applicable.
[c] NA = Not available.

Table 7.11B Cancer Risk Estimates for Adults Onsite — Future Land Use Conditions (Onsite Scenario)

Exposure Pathway	Chemical of Concern	CDI[a] (mg/kg/day)	CDI Adjusted for Absorption	SF $(mg/kg/day)^{-1}$	Weight-of-Evidence Classification	Type of Cancer	SF Basis (Vehicle)	SF Source	Chemical Specific Risk	Total Pathway Risk	Total Exposure Risk
Inhalation of fugitive dust	Antimony (Sb)	—[b]	—	—	NC	—	—	—	—		
	Cadmium (Cd)	7.03E – 06	No	6.10E + 00	B1	Lung; trachea; bronchus	Air	IRIS	4E – 05		
	Copper (Cu)	—	—	—	NC	—	—	—	—		
	Lead (Pb)	1.70E – 06	No	NA[c]	B2	NA	—	—	—		
	Nickel (Ni)	—	—	—	NC	—	—	—	—		
	Zinc (Zn)	—	—	—	NC	—	—	—	—	14 – 05	
Ingestion of soil	Antimony (Sb)	—	—	—	NC	—	—	—	—		
	Cadmium (Cd)	—	—	—	NC	—	—	—	—		
	Copper (Cu)	—	—	—	NC	—	—	—	—		
	Lead (Pb)	4.01E – 06	Yes	NA	B2	NA	—	—	—		
	Nickel (Ni)	—	—	—	NC	—	—	—	—		
	Zinc (Zn)	—	—	—	NC	—	—	—	—	0E + 00	
Dermal contact	Antimony (Sb)	—	—	—	NC	—	—	—	—		
	Cadmium (Cd)	—	—	—	NC	—	—	—	—		
	Copper (Cu)	—	—	—	NC	—	—	—	—		
	Lead (Pb)	8.17E – 06	Yes	NA	B2	NA	—	—	—		
	Nickel (Ni)	—	—	—	NC	—	—	—	—		
	Zinc (Zn)	—	—	—	NC	—	—	—	—	0E + 00	
Nearby residential population potentially affected by the ABC site — total cancer risk for onsite adult workers (weight-of-evidence predominantly B1)											4E – 05

[a] CDI = Chronic daily intake (carcinogenic effects).
[b] — = Not applicable.
[c] NA = Not available.

Table 7.12A Chronic HI Estimates for Children Aged 6–12 Years — Future Land Use Conditions (Onsite Scenario)

Exposure Pathway	Chemical of Concern	CDI[a] (mg/kg/day)	CDI Adjusted for Absorption	RfD[b] (mg/kg/day)	RfD Source	Hazard Quotient	Pathway HI	Total Exposure HI
Inhalation of fugitive dust	Antimony (Sb)	1.64E – 06	No	NA[c]	—[d]		—	
	Cadmium (Cd)	2.38E – 05	No	NA	—	—		
	Copper (Cu)	1.52E – 04	No	1.00E – 02	IRIS	0.02		
	Lead (Pb)	5.76E – 06	No	5.70E – 04	IRIS	0.01		
	Nickel (Ni)	4.11E – 05	No	NA	—	—		
	Zinc (Zn)	8.64E – 04	No	1.00E – 02	IRIS	0.09	0.11	
Ingestion of soil	Antimony (Sb)	1.81E – 05	No	4.00E – 04	IRIS	0.05		
	Cadmium (Cd)	2.26E – 05	Yes	5.00E – 04	IRIS	0.05		
	Copper (Cu)	1.67E – 03	No	3.70E – 02	IRIS	0.05		
	Lead (Pb)	2.96E – 05	Yes	6.60E – 04	IRIS	0.04		
	Nickel (Ni)	9.05E – 04	No	2.00E – 02	IRIS	0.05		
	Zinc (Zn)	9.05E – 03	No	2.00E – 01	IRIS	0.05	0.27	
Dermal contact	Antimony (Sb)	4.26E – 06	Yes	4.00E – 04	IRIS	0.01		
	Cadmium (Cd)	1.06E – 04	Yes	5.00E – 04	IRIS	0.21		
	Copper (Cu)	3.94E – 04	Yes	3.70E – 02	IRIS	0.01		
	Lead (Pb)	6.97E – 05	Yes	6.60E – 04	IRIS	0.11		
	Nickel (Ni)	2.13E – 04	Yes	2.00E – 02	IRIS	0.01		
	Zinc (Zn)	2.13E – 03	Yes	2.00E – 01	IRIS	0.01	0.36	
Nearby residential population potentially affected by the ABC site — total chronic HI for children aged 6–12 years								0.7

[a] CDI = Chronic daily intake (noncarcinogenic effects).
[b] RfD = Reference dose (or equivalent ADI where no RfD exists).
[c] NA = Not available.
[d] — = Not applicable.

Table 7.12B Chronic HI Estimates for Adults Onite — Future Land Use Conditions (Onsite Scenario)

Exposure Pathway	Chemical of Concern	CDI[a] (mg/kg/day)	CDI Adjusted for Absorption	RfD[b] (mg/kg/day)	RfD Source	Hazard Quotient	Pathway HI	Total Exposure HI
Inhalation of fugitive dust	Antimony (Sb)	5.85E – 07	No	NA[c]	—[d]	—		
	Cadmium (Cd)	8.48E – 06	No	NA	—	—		
	Copper (Cu)	5.41E – 05	No	1.00E – 02	IRIS	0.01		
	Lead (Pb)	2.05E – 06	No	4.30E – 04	IRIS	0.00		
	Nickel (Ni)	1.46E – 05	No	NA	—	—		
	Zinc (Zn)	3.07E – 04	No	1.00E – 02	IRIS	0.03		
							0.04	
Ingestion of soil	Antimony (Sb)	2.95E – 06	No	4.00E – 04	IRIS	0.01		
	Cadmium (Cd)	3.69E – 06	Yes	5.00E – 04	IRIS	0.06		
	Copper (Cu)	2.73E – 04	No	3.70E – 02	IRIS	0.04		
	Lead (Pb)	4.83E – 06	Yes	6.90E – 04	IRIS	0.01		
	Nickel (Ni)	1.48E – 04	No	2.00E – 02	IRIS	0.04		
	Zinc (Zn)	1.48E – 03	No	2.00E – 01	IRIS	0.03		
							0.04	
Dermal contact	Antimony (Sb)	6.03E – 07	Yes	4.00E – 04	IRIS	0.00		
	Cadmium (Cd)	1.51E – 05	Yes	5.00E – 04	IRIS	0.03		
	Copper (Cu)	5.58E – 05	Yes	3.70E – 02	IRIS	0.00		
	Lead (Pb)	9.88E – 06	Yes	6.90E – 04	IRIS	0.01		
	Nickel (Ni)	3.02E – 05	Yes	2.00E – 02	IRIS	0.00		
	Zinc (Zn)	3.02E – 04	Yes	2.00E – 01	IRIS	0.00		
							0.05	
Nearby worker population potentially affected by the ABC site — total chronic HI for onsite adult workers								0.1

[a] CDI = Chronic daily intake (noncarcinogenic effects).
[b] RfD = Reference dose (or equivalent ADI where no RfD exists).
[c] NA = Not available.
[d] — = Not applicable.

Table 7.13 Recommended Soil Cleanup Limits for the ABC Site Based on Future Land Use Conditions (Onsite Scenario)

Chemical of Concern	Carcinogen Class (Oral Route)	Maximum Soil Chemical Levels (mg/kg)	Estimated Soil AL (mg/kg)	Trace Chemical Content of Natural Soils[a] (mg/kg)		RSCL (mg/kg)	Degree of Cleanup[b] (%)
Antimony (Sb)	NC	1,500	116	2–10[c]	(–)[c]	110	7
Cadmium (Cd)	NC	210	2,900	0.01–0.7	(0.06)	—[d]	—
Copper (Cu)	NC	39,600	10,730	2–100	(30)	10,730	27
Lead (Pb)	B2	16,500	1,900	2–200	(10)	1,900	12
Nickel (Ni)	NC	12,600	5,800	5–500	(40)	5,800	46
Zinc (Zn)	NC	299,000	58,000	2–10	(50)	58,000	19

[a] Shields (1985).
[b] Recommended cleanup fraction relative to maximum onsite chemical concentrations in soil.
[c] 2–10 is common range; (–) is average.
[d] — = Not applicable , i.e., no cleanup necessary.

Table 7.14 Risk Characterization Summary — Estimated Incremental Lifetime Risks Associated with Different Site Scenarios

Group/Scenario[a]	Inhalation	Ingestion	Dermal	Total
		I. Age 0–6 Years		
Baseline	7.00E − 06	—	—	7.00E − 06
Developed (I)[b]	—	—	—	—
Developed (II)[c]	—	—	—	—
		II. Age 6–12 Years		
Baseline	8.00E − 06	—	—	8.00E − 06
Developed (I)	4.00E − 07	—	—	4.00E − 07
Developed (II)	1.00E − 05	—	—	1.00E − 05[d]
		III. Adults		
Baseline	6.00E − 05	—	—	6.00E − 05
Developed (I)	—	—	—	—
Developed (II)	—	—	—	—

[a] Baseline estimates are for potential offsite impacts. Developed scenarios are for potential onsite exposures. Risks calculated for developed site (II) > (I), because cadmium levels at site is less than the soil ALs used for condition (II).
[b] Refers to developed onsite landuse without site remediation.
[c] Refers to developed onsite landuse after site remediation.
[d] Level of risk represented here is higher because the AL for Cd computed and used for risk characterization under the site conditions here was far greater than the maximum existing Cd levels at the site.

Table 7.15 Risk Characterization Summary — Chronic HIs Associated with Different Site Scenarios

Group/Scenario[a]	Inhalation	Ingestion	Dermal	Total
		I. Age 0–6 Years		
Baseline	12.0	1.0	0.1	13.1
Developed (I)[b]	—	—	—	—
Developed (II)[c]	—	—	—	—
		II. Age 6–12 Years		
Baseline	12.0	0.2	0.2	12.4
Developed (I)	0.6	0.2	0.3	1.1
Developed (II)	0.1	0.3	0.4	0.8
		III. Adults		
Baseline	10.7	0.0	0.0	10.7
Developed (I)	—	—	—	—
Developed (II)	—	—	—	—

[a] Baseline estimates are for potential offsite impacts. Developed scenarios are for potential onsite exposures.
[b] Refers to developed onsite landuse without site remediation.
[c] Refers to developed onsite landuse after site remediation.

A second evaluation under future land use scenario involves a post remediation risk characterization associated with receptor exposure to the chemicals of concern present at the ABC site; this represents residual risks that remains after implementation of remedial actions to bring contamination at the site to the ALs. This showed significantly reduced risks and practically no health threats from the ABC site.

The soil cleanup levels recommended for the chemicals of concern present at the ABC site are shown in Table 7.13. This includes the RSCL as well as the percent cleanup necessary, in relation to the maximum onsite chemical concentrations. Selective remediation of the hot spots to bring chemical concentrations to the stipulated levels may be adequate for future onsite use of this property for light industrial, office, and/or retail developments.

7.2 CASE 2: NUMERICAL EVALUATION FOR MULTIMEDIA RISK ASSESSMENT

The use of site remediation risk assessment to define corrective action plans for a case involving multiple media is demonstrated in this section. The health risk assessment methodology previously described in Chapter 5 is applied to a hypothetical site contaminated by petroleum products. Leakage of petroleum products and other chemicals from underground tanks and pipelines is a frequent occurrence within commercial, industrial, and even private individual uses. Such leakages subsequently can lead to contamination of several environmental media, in particular, soils and groundwater. When petroleum products enter the soil, gravitational forces act to draw the fluid in a downward direction. Other forces act to retain it, which is either adsorbed to soil particles or trapped in soil pores. The amount of product retained in the soil is of importance, because this could determine both the degree of contamination and the likelihood of subsequent contaminant transport to other environmental compartments.

7.2.1 Overview of the Hypothetical Case

A hypothetical case is presented that involves an abandoned process facility used for handling heavy/waste oils at the XYZ site located in eastern U.S. Petroleum products are believed to have leaked from this facility, leading to surface soil contamination. An environmental site assessment conducted for the XYZ site indicated that the major contaminants of concern present at this site are the organic constituents listed in Table 7.16. It is apparent from the site setting that contaminated soils may be carried offsite from the XYZ property via wind erosion, surface runoff, and by leaching into an underlying aquifer.

It is required to protect all potential receptors from possible health impacts. It is also required by environmental laws/regulations to protect the surface water quality (from the migration of contaminated soil that may be washed into a nearby creek) and also the groundwater resources (from contaminated leachate migrating into an underlying aquifer). However, current levels of chemicals in soil at the site would likely lead to unacceptable health impacts; it will further lead to the exceedance of acceptable contaminant levels in the adjoining creek and also the aquifer following loadings from erosion runoff and leachate, respectively. The PRPs are required to clean up the site to such levels that will not significantly impact human health, nor impact the creek after receiving runoff carrying eroded soil, and also not impact the aquifer following percolation of leachate from this site.

7.2.2 Exposure Assessment

A contaminant pathway analysis is conducted to help assess the routes of migration and areas where hydrocarbon phases may impact human health and/or the environment. Several factors affect the fate and transport of the mobile liquid, vapor, and dissolved hydrocarbon phases; these include the volume released, the adsorptive capacity of the earth materials, the rates and directions of groundwater movement, and indeed all processes which attenuate concentrations and/or limits the area of the contaminated zones. A consideration of the exposure setting for the XYZ site indicates that the chemicals of potential concern at the XYZ site are likely to migrate primarily by two major transport mechanisms:

- Erosional (overland) transport by water and wind
- Leaching (by infiltration and groundwater movement)

The following *potential* exposure scenarios are considered to be representative of the XYZ site and vicinity, given the present level of knowledge about this site and vicinity:

- Human residential exposure offsite via inhalation of fugitive dust with site-related contaminants, via ingestion of soils with site-related contaminants, and through dermal contact with soils containing site-related contaminants
- Surface water contamination through erosion of contaminated soils into creek adjoining the contaminated site
- Groundwater contamination from leachate migration into an underlying aquifer

Table 7.17 presents a summary of statistics for the chemicals of potential concern in the various environmental compartments for the XYZ site. An air model for fugitive dust emission and dispersion was used to estimate the applicable exposure point concentrations of respirable particulates from this site (e.g., U.S. EPA, 1985; CDHS, 1986; DOE, 1987; CAPCOA, 1989). In this model, fugitive dust dispersion concentrations were evaluated as represented by a three-dimensional Gaussian distribution of particulate emissions from the source. Site-specific exposure parameters used in the computation of receptor intakes and doses for the XYZ site are given in Table 7.18. The applicable equations for completing the exposure estimates are discussed in Appendix B1.

7.2.3 The Risk Characterization for Potential Receptors

The calculation of potential carcinogenic risks and noncarcinogenic hazards under the existing conditions at the XYZ site were performed for the different population groups, using the applicable equations presented in Appendix B2. The

Table 7.16 Chemicals of Potential Concern Present at the XYZ Site

Chemical	Potential Migration Pathways
Benzene	Multiple exposure and migration pathways
Benz(a)anthracene	Preferentially adsorbed onto soil particles
Benzo(a)pyrene	Preferentially adsorbed onto soil particles
Ethylbenzene	Multiple exposure and migration pathways
Naphthalene	Multiple exposure and migration pathways
Phenanthrene	Preferentially adsorbed onto soil particles
Phenol	Preferentially solubilized
Toluene	Multiple exposure and migration pathways

pathway-specific and total risks/hazards for the receptor groups are summarized in Table 7.19. Carcinogenic risks to human health are indicated under the present conditions at the XYZ site. Noncarcinogenic impacts are, however, not indicated even under the present conditions at the site.

Development of Soil ALs

Soil ALs are developed for the XYZ site according to the relations discussed in Appendix B3. For the XYZ site, carcinogenic effects are the significant contributors to health risks indicated by this site. Consequently, in determining the health-based cleanup criteria for this site, the relationship for developing soil ALs for carcinogens is used, where SFs exists, and that for noncarcinogens used otherwise. Table 7.20 gives the applicable soil ALs for the chemicals of potential concern.

7.2.4 Cleanup Criteria for Groundwater Protection

Several methods exist for representing the leachate generation and contaminant flux estimation processes necessary for establishing cleanup limits with respect to groundwater contamination from leachates (e.g., U.S. EPA, 1988). A vertical and horizontal spreading (VHS) model is used here for simulating and estimating contaminant concentrations migrating into an aquifer. The VHS model [*Federal Register*, Vol. 50, No. 229, Section 48897]—that is used in this analysis—has been developed based on an earlier model proposed by Domenico and Palciauskas (1982) for the predictions of maximum concentration of contaminants in groundwater. The VHS model simulates soluble toxic constituents dissolving into percolating precipitation and moving with the groundwater. The overall approach aids in the prediction of reasonable worst-case contaminant levels in groundwater in nearby receptor/compliance points. The model basically estimates the capacity of an aquifer to dilute the contaminants from a specific volume of waste. To be on the conservative side, retardation and attenuation factors are not incorporated in the model. The model thus provides a conservative estimate of whether or not minimum performance standards will be achieved at a compliance boundary.

The maximum concentration, C_y, as measured at the receptor location, is described by the following analytical relationship:

Table 7.17 Summary Data for the Chemicals of Potential Concern Present at the XYZ Site

			Mean Concentrations		
Chemical	Onsite Soil Concentration Range (mg/kg)	Offsite Soil Concentration Range (mg/kg)	Onsite Soils (mg/kg)	Offsite Soils (mg/kg)	Air[a] (μ/m^3)
Benzene	3.0–1,500	0.3–27	5.39E + 01	4.70E + 00	1.10E – 03
Benz(a)antharacene	0.4–210	0.3–39	8.42E + 01	1.36E + 00	8.16E – 05
Benzo(a)pyrene	19.0–39,600	6.0–310	1.08E + 02	7.55E + 01	1.84E – 02
Ethylbenzene	32.0–16,500	20.0–1,110	6.52E + 02	3.43E + 02	2.14E – 02
Naphthalene	8.0–12,600	5.0–150	4.80E + 02	2.96E + 01	6.11E – 03
Phenanthrene	5.0–2,000	0.8–280	2.68E + 02	1.20E + 02	2.50E – 02
Phenol	22.0–9,800	15.0–980	7.90E + 02	4.30E + 02	3.20E – 02
Toluene	300.0–299,000	57.00–8,660	3.38E + 03	1.15E + 03	2.36E – 01

[a] Refers to respirable concentration of chemicals adsorbed on soil particles.

Table 7.18 Case-Specific Parameters for Doses and Intakes Calculation for the XYZ Site

Parameter	Children Aged up to 6	Children Aged 6–12	Adult	Reference Sources
Physical characteristics				
Average body weight	16 kg	29 kg	70 kg	a,b,c
Average total skin surface area	6980 cm^2	10,470 cm^2	18,150 cm^2	a,b,e,h
Average lifetime			70 years	a,b,c,e
Average lifetime exposure period	5 years	6 years	58 years	b,e
Activity characteristics				
Inhalation rate	0.25 m^3/h	0.46 m^3/h	0.83 m^3/h	b,e
Retention rate of inhaled air	100%	100%	100%	e
Frequency of fugitive dust inhalation				
Offsite residents, schools, and passers-by	365 days/year	365 days/year	365 days/year	b,e
Offsite workers	—	—	260 days/year	b,e
Duration of Fugitive Dust Inhalation (Outside)				
Offsite residents, schools, and passers-by	12 h/day	12 h/day	12 h/day	b,e
Offsite workers	—	—	8 h/day	b,e
Amount of soil ingested incidentally	200 mg	100 mg	50 mg	a,b,c,e,h,i
Frequency of soil contact				
Offsite residents, schools, and passers-by	330 days/year	330 days/year	330 days/year	b,e
Offsite workers	—	—	260 days/year	b,e
Duration of soil contact				
Offsite residents, schools, and passers-by	12 h/day	8 h/day	8 h/day	b,e
Offsite workers	—	—	8 h/day	b,e
Percentage of skin area contacted by soil	20%	20%	10%	b,e,h

Material characteristics				
Soil to skin adherence factor	0.75 mg/cm^2	0.75 mg/cm^2	0.75 mg/cm^2	a,b,e,f,g
Soil matrix attenuation factor	15%	15%	15%	d

[a] U.S. EPA (1989b).
[b] U.S. EPA (1989c).
[c] U.S. EPA (1988a).
[d] Hawley (1985).
[e] Estimate based on site-specific conditions.
[f] Lepow et al. (1975).
[g] Lepow et al. (1974).
[h] Sedman (1989).
[i] Calabrese et al. (1989).

$$C_y = C_0 \times erf\left\{\frac{Z}{\left[2(\alpha_t Y)^{0.5}\right]}\right\} \times erf\left\{\frac{X}{\left[4(\alpha_t Y)^{0.5}\right]}\right\}$$

where

C_y = contaminant concentration in aquifer at receptor/compliance location (mg/L)
C_0 = initial contaminant concentration in leachate (mg/L)
erf = error function (dimensionless)
Z = aquifer penetration depth (mixing zone) (i.e., depth to which plume can reach at distance Y from source) (ft). The aquifer penetration depth, Z, is computed via a mass balance approach (e.g., Asante-Duah et al., 1989)
α_t = transverse dispersivity (ft)
Y = distance from disposal source to receptor/compliance location (ft)
X = length of disposal site (amount of waste) (= site width perpendicular to flow direction [ft])

For a situation where the permeable unit transmitting the contaminant is thin and is underlain by a continuous low permeability unit, the above model is modified into the following:

$$C_y = C_0 \times \left\{\frac{Z}{H}\right\} \times erf\left\{\frac{X}{\left[4\left((\alpha_t Y)^{0.5}\right)\right]}\right\}$$

where H is the thickness of the available vertical spreading zone (ft), and all other parameters are as defined before. This evaluation is based on the assumption of a thin aquifer, utilizing this modification to the VHS model.

Recommended Cleanup Limits for Groundwater Protection

By performing back-calculations, based on contaminant concentrations in aquifer as a result of the current constituents loading from the site, a conservative estimate is made as to what the maximum acceptable concentration could be on the site that will not adversely impact the aquifer. The back-calculation is carried out according to the following relation:

$$C_{max} = \left(\frac{C_{std}}{C_{gw}}\right) \times C_s$$

Table 7.19 Risk Characterization Summary (Scenario Under Current Land Use Conditions)

Population	Exposure Type/Pathway	Pathway Carcinogenic Risk[a]	Pathway HI[b]
Child, up to 6 years	Inhalation (of particulates)	2E – 03	0.33
	Ingestion (of soil — incidentai and pica)	2E – 04	0.17
	Dermal (contact with soil)	2E – 04	0.14
Total exposure carcinogenic risk to child aged 1–6 years		2E – 03	
Total exposure chronic HI to child aged 1–6 years			0.6
Child, 6–12 years	Inhalation (of particulates)	2E – 03	0.33
	Ingestion (of soil — incidental and pica)	6E – 05	0.05
	Dermal (contact with soil)	2E – 04	0.11
Total exposure carcinogenic risk to child aged 6–12 years		2E – 03	
Total exposure chronic HI to child aged 6–12 years			0.5
Adult populations	Inhalation (of particulates)	1E – 02	0.25
	Ingestion (of soil — incidentai and pica)	1E – 04	0.01
	Dermal (contact with soil)	5E – 04	0.04
Total exposure carcinogenic risk to adults		1E – 02	
Total exposure chronic HI to adults			0.3

[a] Acceptable level is in 1E – 07 to 1E – 05 range.
[b] Acceptable level ≤ 1.

Table 7.20 Health-Based Soil ALs for the XYZ Site — Future Land Use Conditions (Onsite Scenario)

Chemical of Concern	Oral RfD	Oral SF	Absorption Factor (%)	Receptor-Specific Soil AL			Health-Based Soil AL (mg/kg)
				@ Adults (mg/kg)	@ Child (6–12) (mg/kg)	@ Child (1–6) (mg/kg)	
Benzene	7.00E − 04	2.90E − 02	100	5.83E + 01	1.17E + 02	3.86E + 01	3.86E + 01
Benz(a)anthracene	4.00E − 01	4.66E − 01	100	3.63E + 00	7.26E + 00	2.40E + 00	2.40E + 00
Benzo(a)pyrene	4.00E − 01	3.22E + 00	100	5.25E − 01	1.05E + 00	3.48E − 01	3.48E − 01
Ethylbenzene	1.00E − 01	—	100	1.40E + 05	2.90E + 04	8.00E + 03	8.00E + 03
Naphthalene	4.00E − 01	—	100	5.60E + 05	1.16E + 05	3.20E + 04	3.20E + 04
Phenanthrene	4.00E − 01	—	100	5.60E + 05	1.16E + 05	3.20E + 04	3.20E + 04
Phenol	6.00E − 01	—	100	8.40E + 05	1.74E + 05	4.80E + 04	4.80E + 04
Toluene	3.00E − 01	—	100	4.20E + 05	8.70E + 04	2.40E + 04	2.40E + 04

where

C_{max} = maximum acceptable soil concentration on site (mg/kg)
C_{std} = applicable groundwater quality criteria (mg/L)
C_{gw} = constituent concentration in aquifer (mg/L)
C_s = soil chemical concentration prior to cleanup (mg/kg)

Based on the appropriate maximum acceptable soil concentration value, C_{max}, the site may be cleaned up to such levels as *not* to impact the groundwater quality. The overall computational process for the analysis of groundwater contamination is presented in Table 7.21.

7.2.5 Development of Soil Cleanup Levels for Surface Water Protection

Contaminated runoff and overland flow of toxic contaminants constitutes one source of concern for surface water contamination at uncontrolled hazardous waste sites. Runoff release estimation procedures (e.g., U.S. EPA, 1988) can be applied to such uncontrolled sites. Also, surface waters may be contaminated by inflow of groundwater through bank seepage and springs. The rates of such inflows may be estimated through modeling of the groundwater-surface water linkages. Only the former, involving surface runoff releases of chemicals to surface, is discussed here.

Modeling the Migration of Contaminated Soil to Surface Runoff: Sediment Loss and Contaminant Flux Calculations

Assuming that all contamination is from adsorbed waste oil contaminants on the XYZ site, surface runoff release of chemicals can be estimated by means of the modified universal soil loss equation (MUSLE) and sorption partition coefficients derived from the compounds octanol-water partition coefficient. The MUSLE allows estimation of the amount of surface soil eroded in a storm event of given intensity, while sorption coefficients allow the projection of the amount of contaminant carried along with the soil and the amount carried in dissolved form; the procedures used here are fully described in the literature (e.g., DOE, 1987; U.S. EPA, 1988). "Back modeling" can then be performed to estimate allowable contaminant levels that could be left in the soil environment after remediation without adversely impacting aquatic life and/or the environment due to contaminated runoff entering the adjoining creek.

Soil Loss Calculation

The soil loss caused by a storm event is given by the MUSLE (Williams, 1975; DOE, 1987):

$$Y(S)_E = a\left(V_r q_p\right)^{0.56} KLSCP$$

where

$Y(S)_E$ = sediment yield (tons per event)
a = conversion constant
V_r = volume of runoff (acre-feet)
q_p = peak flow rate (cubic feet per second)
K = soil erodibility factor (commonly expressed in tons/acre/dimensionless rainfall erodibility unit)
L = the slope-length factor (dimensionless ratio)
S = the slope-steepness factor (dimensionless ratio)
C = the cover factor (dimensionless ratio)
P = the erosion control practice factor (dimensionless ratio)

Computation of Vr — The storm runoff volume generated at the case site is calculated by the following equation (Mills et al., 1982):

$$V_r = 0.083 \ A \ Q_r$$

where

A = contaminated area ≈ 6 acres (estimated during remedial investigation)
Q_r = depth of runoff from site (inches)

The depth of runoff, Q_r, is determined by (Mockus, 1972)

$$Q_r = \frac{\left(R_t - 0.2S_w\right)^2}{\left(R_t + 0.8S_w\right)}$$

where

Q_r = the depth of runoff from the site (inches)
R_t = the total storm rainfall (inches)
S_w = water retention factor (inches)

In deciding on a value for the total storm rainfall, R_t, to use, the Soil Conservation District may be consulted; in this case, a suggested typical annual storm is one represented by the 2-year, 24-h event, giving an R_t value of about 3.5 in. Also, an average annual storm of between 44 and 48 in. can be expected for this area.

On this basis and for a 2-year, 24-h storm event, R_t = 3.5 in. and average annual rainfall ≈ 46 in. This means that the average number of average rainfall events per year ≈ 46/3.5 ≈ 13 events, which also compounds to 920 storm events over an assumed 70-year lifetime period.

The value of the water retention factor (S_w) is obtained as follows (Mockus, 1972):

Table 7.21 Development of Soil Cleanup Levels — Modeling Leachate Migration into an Aquifer

Constituent	Soil Chemical Concentration Cs (mg/L)	α_t [m]	$AA = \frac{X}{4(\alpha_t Y)^{0.5}}$	erf(AA)	$\frac{Z}{H}$	Concentration in Aquifer (mg/L)	Chronic Water Quality Standards (mg/L)	Attenuation Factor	Maximum Acceptable Soil Concentration (mg/kg)
Benzene	5.39E + 01	4.00E + 00	4.61E – 01	4.85E – 01	3.02E – 01	7.90E + 00	5.10E + 00	1.46E – 01	3.48E + 01
Benz(a)anthracene	8.42E + 01	4.00E + 00	4.61E – 01	4.85E – 01	3.02E – 01	1.23E + 01	2.89E + 01	1.46E – 01	1.97E + 02
Benzo(a)pyrene	1.08E + 02	4.00E + 00	4.61E – 01	4.85E – 01	3.02E – 01	1.58E + 01	1.13E + 02	1.46E – 01	7.71E + 02
Ethylbenzene	6.52E + 02	4.00E + 00	4.61E – 01	4.85E – 01	3.02E – 01	9.55E + 01	2.24E + 02	1.46E – 01	1.53E + 03
Naphthalene	4.80E + 02	4.00E + 00	4.61E – 01	4.85E – 01	3.02E – 01	7.03E + 01	1.69E + 01	1.46E – 01	1.15E + 02
Phenanthrene	2.68E + 02	4.00E + 00	4.61E – 01	4.85E – 01	3.02E – 01	3.93E + 01	9.02E + 00	1.46E – 01	6.16E + 01
Phenol	7.90E + 02	4.00E + 00	4.61E – 01	4.85E – 01	3.02E – 01	1.16E + 02	5.00E + 00	1.46E – 01	3.41E + 01
Toluene	3.38E + 03	4.00E + 00	4.61E – 01	4.85E – 01	3.02E – 01	4.95E + 02	2.00E + 00	1.46E – 01	1.37E + 01

$$S_w = \left\{ \frac{1000}{CN} - 10 \right\}$$

where

S_w = water retention factor (in.)
CN = the Soil Conservation Services runoff curve number (dimensionless)

The CN factor for the site is determined by the soil type at the XYZ site, its condition, and other parameters that establish a value indicative of the tendency of the soil to absorb and hold precipitation or to allow precipitation to run off the site. CN values of uncontrolled hazardous waste sites are estimated from tables in the literature (e.g., U.S. EPA, 1988; Schwab et al., 1966); charts have also been developed for determining CN values (e.g., USBR, 1977), and these may be used for estimating CN. Based on existing site conditions of the fill material that is very heterogeneous, higher infiltration and relatively low to moderate runoff is anticipated at the site. An estimate of CN = 74 for group B soil type is taken as conservative enough for our purposes (U.S. EPA, 1988). This represents a moderately low runoff potential and an above average infiltartion rate of 4 to 8 mm/h seems reasonable for the site. For this, S_w, Q_r, and V_r are computed to be

$$S_w = \left\{ \frac{1000}{74} - 10 \right\} = 3.51 in.$$

$$Q_r = \frac{\left[3.5 - (0.2)(S_w)\right]^2}{\left[3.5 + (0.8)S_w\right]} = 1.24 in.$$

$$V_r = (0.083)(5.9)(Q_r) = 0.61\, acre\text{-}ft.$$

Computation of q_p —The peak runoff rate, q_p, is determined by (Haith, 1980)

$$Q_p = \frac{1.01\, A\, R_t Q_r}{T_r\left(R_t - 0.2 S_w\right)}$$

where

q_p = the peak runoff rate (ft^3/sec)
A = contaminated area (acres)
R_t = the total storm rainfall (in.)
Q_r = the depth of runoff from the watershed area (in.)
T_r = storm duration (h)
S_w = water retention factor (in.)

For the typical storm represented by the 2-year, 24-h rainfall event suggested for this scenario, and given the following parameters:

$A = 5.9$ acres
$R_t = 3.5$ in.
$Q_r = 1.24$ in.
$R_t = 24$ h
$S_w = 3.51$ in.

$$q_p = 0.38 cfs$$

Estimation of K — The soil erodibility factors are indicators of the erosion potential of given soil types and are therefore site specific. The value of the soil erodibility factor, K, as obtained from the Soil Conservation Service is 0.32. This compares reasonably well with estimates given by charts and nomographs found in the literature (e.g., DOE, 1987; Erickson, 1977; Goldman et al., 1986). That is,

$$K = 0.32 tons / acre / rainfall\ erodibility\ unit$$

Estimation of LS — The product of the slope-length and slope-steepness factors, LS, is determined from charts/nomographs given in the literature (e.g., U.S. EPA, 1988; DOE, 1987; Mitchell and Bubenzer, 1980). This is based on a slope length of 570 ft and a slope of 0.5% obtained from information from the remedial investigations, yielding

$$LS = 0.14$$

C and P Factors — A C value of 1.0 is used, assuming no vegetative cover exists at this site. This will help simulate a worst-case scenario. Similarly a worst-case (conservative) P value of 1 for uncontrolled sites is used.

Summary of estimated parameters for soil loss computation —

$a = 95$
$V_r = 0.61$ acre-ft
$q_p = 0.38$ cfs
$K = 0.32$ tons/acre/rainfall erodibility unit
$LS = 0.14$
$C = 1.0$
$P = 1.0$

Substitution of these estimated parameters into the MUSLE yields

$$Y(S)_E = 1.89\ \ tons / event$$

for a typical rainstorm event of 2-year, 24-h magnitude.

Dissolved Sorbed Contaminant Loading

After computing the soil loss during a storm event, the amounts of adsorbed and dissolved substance loadings on the creek are calculated. The amounts of adsorbed and dissolved substances are determined by the following equations (Haith, 1980; DOE, 1987):

$$S_s = \left[\frac{1}{\left(1 + q_c / K_d \beta\right)}\right] C_i A$$

$$D_s = \left[\frac{1}{\left(1 + K_d \beta / q_c\right)}\right] C_i A$$

where

S_s = sorbed substance quantity (kg)
q_c = available water capacity of the top centimeter of soil (dimensionless)
K_d = sorption partition coefficient (cm^3/g)
= f_{oc} x K_{oc}, where foc is the organic carbon content/fraction of the soil and K_{oc} is the soil/water distribution coefficient, normalized for organic content
β = soil bulk density (g/cm^3)
C_i = total substance concentration (kg/ha-cm)
= $C_s \times \beta \times \phi$, where C_s is the chemical concentration in soil (mg/kg), β is the soil bulk density (g/cm^3), and ϕ is a conversion factor
A = contaminated area (ha-cm)
D_s = dissolved substance quantity (kg)

The model assumes that only the contaminant in the top 1 cm of soil is available for release via runoff.

Estimating K_d—The soil sorption partition coefficient for a given chemical can be determined from known values of certain other physical/chemical parameters, primarily the chemical's octanol-water partition coefficient, solubility in water, or bioconcentration factor. The sorption partition coefficient, K_d, is given by the following relationship:

$$K_d = f_{oc} K_{oc}$$

where

f_{oc} = the organic carbon content/fraction of the soil
K_{oc} = the soil/water distribution coefficient, normalized for organic content

Organic carbon fraction/content of 0.1% is assumed for this site.

Estimating β, θ_c — The contaminated site area, A, is 6 acres. The soil bulk density, β, is estimated to be about 1.4 g/cm^3 (Walton, 1984). The available water capacity of the top 1 cm of soil is estimated at about $\theta_c \approx 120$ mm/m, i.e., $\theta_c \approx 0.12$ (Walton, 1984).

Estimation of C_i — The total substance concentration, C_i (in kg/ha-cm), is obtainable by multiplying the chemical concentration in soil, C_s (mg/kg) by the soil bulk density, β (g/cm^3) and an appropriate conversion factor, ø:

$$C_i[kg/ha\text{-}cm] = C_s[mg/kg]\beta[g/cm^3]\phi$$

where ø is a conversion factor, equal to 0.1. This results in

$$C_i[kg/ha-cm] = 0.1\beta[g/cm^3]C_s[mg/kg] \ or \ C_i = 0.14C_s$$

for the estimated value of $\beta = 1.40$ g/cm^3.

It must be noted that although in deriving the above model it was assumed that only contaminants in the top 1 cm of soil is available for release via runoff, maximum chemical concentrations found within the whole sampled soil media at the XYZ site is utilized for subsequent calculations. That is, the highest constituent concentrations for soil samples from the remedial investigations are used for the C_s values. Maximum concentrations found within the XYZ site unit are used in this computation to help simulate the possible worst-case scenario.

Parameters for Computation of Dissolved/Sorbed Contaminant Loading—

A = 5.9 acres
$\theta = 0.12$
$\beta = 1.4$ g/cm^3
$f_{oc} = 0.1$ %

The equation for sorbed substance quantity, S_s (kg) is simplified into the following:

$$S_s = \left\{\frac{1}{\left(\left(1+\frac{0.12}{[1.4K_d]}\right)\right)}\right\}(0.14C_s)(2.388) = \left\{\frac{1.4K_d}{1.4K_d+0.12}\right\}(0.334C_s)$$

$$= \frac{0.468K_dC}{((0.12+0.028K_{oc}))} = \frac{0.009K_{oc}C_s}{(0.12+1.4K_d)}$$

Similarly, the equation for dissolved substance quantity, D_s (kg) is simplified into:

$$D_s = \left\{ \frac{1}{1+\left[K_a\beta/\theta_c\right]} \right\}\left(0.334C_s\right) = \left\{ \frac{1}{1+\left[1.4K_d/0.12\right]} \right\}\left(0.334C_s\right)$$

$$= \left(\frac{0.12}{1+\left\{1.4K_d/\left[0.12+1.4K_d\right]\right\}} \right)\left(0.334C_s\right) = \frac{0.04C_s}{\left[0.12+1.4K_d\right]}$$

$$= \frac{0.04C_s}{\left[0.12+0.028K_{oc}\right]}$$

Computation of Total Contaminant Loading on Creek

After calculating the amount of sorbed and dissolved contaminant, the total loading to the receiving water body is calculated as follows (Haith, 1980; DOE, 1987):

$$PX_i = \left[\frac{Y(S)_E}{100\beta} \right] S_s$$

$$PQ_i = \left[\frac{Q_r}{R_t} \right] D_s$$

where

PX_i = sorbed substance loss per event (kg)
$Y(S)_E$ = sediment yield (tons per event, metric tons)
β = soil bulk density (g/cm^3)
S_s = sorbed substance quantity (kg)
PQ_i = dissolved substance loss per event (kg)
Q_r = total storm runoff depth (cm)
R_t = total storm rainfall (cm)
D_s = dissolved substance quantity (kg)

PX_i and PQ_i can be converted to mass per volume terms for use in estimating contaminant concentration in the receiving water body by dividing by the site storm runoff volume, V_r (where V_r = a A Q_r). The contaminant concentrations in the surface runoff, C_{sr}, are then given by

$$C_{sr} = \frac{\left(PS_i + PQ_i\right)}{V_r}$$

Next, the contaminant concentrations in the creek are computed by a mass balance analysis according to the following relationship:

$$C_{cr} = \frac{\left(C_{sr} \times q_p\right)}{\left[q_p + Q_{cr}\right]}$$

where

C_{cr} = concentration of contaminant in creek (mg/L)
C_{sr} = concentration of contaminant in surface runoff (mg/L)
q_p = peak runoff rate (cfs)
Q_{cr} = volumetric flow rate of creek (cfs)

Maximum Acceptable Contaminant Concentration for Soil Cleanup

By performing back-calculations, based on contaminant concentrations in creek as a result of the current constituents loading from the site, a conservative estimate is made as to what the maximum acceptable concentration should be on the site so as not to adversely impact the creek. The back-calculation is carried out as follows:

$$C_{\max} = \left(\frac{C_{std}}{C_{cr}}\right) \times C_s$$

where

C_{max} = maximum acceptable soil concentration on site (mg/kg)
C_{std} = applicable surface water quality criteria (mg/L)
C_{cr} = constituent concentration in creek (mg/L)
C_s = soil chemical concentration prior to cleanup (mg/kg)

Based on the appropriate maximum acceptable soil concentration value, C_{max}, the site may be cleaned up to such levels as *not* to impact the surface water quality. The overall computational process for the analysis of surface water contamination is presented in Table 7.22.

7.2.6 Summary of Recommended Soil Cleanup Levels

Cleanup levels established for the scenarios defined previously are compiled in Table 7.23. The stricter of the cleanup levels from these scenarios will generally be selected as the RSCL in the implementation of corrective actions. Remediation to residual contamination levels corresponding to the RSCLs will be protective of human health, due to future onsite land uses, as well as protect surface and groundwater resources. No significant potential health or environmental impacts will be expected if remediation is carried out to the recommended levels estimated for the XYZ site.

7.3 CASE 3: RISK COMPARISON OF DISPOSAL ALTERNATIVES FOR HOUSEHOLD HAZARDOUS WASTES

The general public has become increasingly concerned about chemical contamination of the environment. This public concern about the effects of potentially toxic chemicals entering the environment has resulted in an increasing demand for more stringent controls over chemical waste disposal practices. One area of concern is the disposal of household hazardous materials (including used dry cell batteries) in landfills. Used dry cell batteries are generally discarded in mixed municipal solid wastes (MSW) from domestic and business garbage. Eventually, such wastes are incinerated or become landfilled. Recycling is becoming an important option for the management of used household batteries. In any one of the applicable waste management options, metals from the household batteries may pose different health and environmental concerns. However, environmental impact studies reviewed on the disposal of dry cell batteries in municipal waste management systems do not indicate that such disposal practices pose significant threat to the environment (IRR, 1992). In fact, most battery manufacturers have taken steps to significantly reduce (up to about 98% reduction over the past decade alone) the mercury content of household batteries, mercury being one of the most toxic metals used in household batteries (IRR, 1992).

While the public tends to support recycling in principle, the separate collection of dry cell batteries may introduce such hazardous situations as fire, explosion, and human exposures to chemicals causing skin irritation. In addition, separate collection of dry cell batteries for the purpose of recycling is not seen to be a viable option currently due to the large energy requirements to recycle the products, the poor quality of recoverable materials, and the lack of markets for recovered materials (IRR, 1992). The recycling process may also result in the production of new and different kinds of waste streams of potential concern.

The discussion in this section is excerpted and adapted in part from a study conducted in 1991–92 by the Institute for Risk Research (IRR), Waterloo, Ontario, Canada, concerning the implications of dry cell battery disposal practices. This concerns a comparison of incineration vs. landfilling of used dry cell batteries. The case region for this investigation was Canada, with specific evaluations conducted for case sites in the province of Ontario.

7.3.1 Study Objectives and Scope

The overall agenda of this presentation is to investigate the potential health impacts from the disposal practices applied to used household batteries. The study considers only AAA, AA, C, D, and 9-V batteries and which belong to the alkaline, zinc-carbon/zinc-chloride, and nickel-cadmium families. Recommendations are then made on preferred disposal methods that minimizes overall environmental and health hazards. Several types of information are reviewed and analyzed, including the following (IRR, 1992):

- Information on the composition of household batteries
- Average domestic population consumption of household batteries
- Contribution of the various metals of concern to MSW streams from household batteries
- Research studies pertaining to health and environmental impacts of management of used dry cell batteries

The study bases the assessment on the known chemical constituents of the household batteries under review. Pertinent properties of the chemical composition of household batteries, including the toxicity, fate and transport, persistence, and attenuation effects are determined. Information obtained on pertinent properties of proprietary organic surfactants used as substitute for mercury (Hg) in the household batteries are extrapolated for completing the environmental impact and risk assessment. Environmental impacts and potential risks posed by household batteries are evaluated and compared for the various disposal options based on such factors and properties.

7.3.2 Dry Cell Battery Composition

For all the different types, household batteries contain specific metals that may potentially impact the environment. The specific metals of concern for the dry cells identified for this study include cadmium (Cd), manganese (Mn), mercury (Hg), nickel (Ni), and zinc (Zn). Of all the heavy metals used in the manufacture of primary and secondary cells, Cd and Hg are of most concern. Alkaline and Ni-Cad batteries account, respectively, for greater proportions of the mercury and cadmium used in dry cell batteries. Mercury in particular has become widely recognized as one of the most hazardous elements to human health. The potential for mercury contamination exists where disposal practices create conditions conducive for conversion of mercury to toxic forms, such as methyl-mercury and other organic mercury compounds. Mercury moves very slowly through soils under field conditions.

7.3.3 The Changing Composition of Dry Cell Batteries

One major contributor of mercury from household batteries to MSW is from alkaline cells. Thus, reducing the amount of mercury in alkaline batteries could have a significant impact on the total amount of mercury that may enter MSW due to used battery disposal. Due to its high toxicity and its relatively high cost, manufacturers have been encouraged to reduce the amount of mercury contained in dry cell batteries. During the past decade, the total amount of mercury used in dry cell batteries manufactured in North America has decreased by about 95%, and this continues to decrease. It is believed that the present concentration of mercury in alkaline batteries manufactured in North America is at most 0.8%, and less than 0.05% in the case of zinc-carbon/zinc-chloride batteries (Environment Canada, 1991). In fact, most zinc-chloride batteries contain either no mercury or less than

Table 7.22 Development of Soil Clean-up Levels — Modeling the Migration of Contaminated Soil in Surface Runoff into Creek

Chemical Constituent	$K_{(oc)}$ (m^3/kg)	$K_{(oc)}$ (cm^3/g)	K_d (cm^3/g)	Soil Chemical Concentration C_s(mg/kg)	Sorbed Quantity S_s (kg)	Dissolved Quantity D_s (kg)	Sorbed Loss PX_i (kg)	Dissolved Loss PQ_i (kg)	Total Loss (kg)
Benzene	8.30E – 02	8.30E + 01	8.30E – 02	5.39E + 01	9.33E + 00	4.99E – 03	1.21E – 01	1.77E – 03	1.23E – 01
Benz(a)Anthracene	1.38E + 03	1.38E + 06	1.38E + 03	8.42E + 01	2.92E + 01	1.00E – 02	3.50E + 02	3.55E – 03	3.50E + 02
Benzo(a)Pyrene	5.50E + 03	5.50E + 06	5.50E + 03	1.08E + 02	3.74E + 01	1.00E – 02	4.48E + 02	3.55E – 03	4.49E + 02
Ethylbenzene	1.10E + 00	1.10E + 03	1.10E + 00	6.52E + 02	2.10E + 02	9.31E – 03	2.74E + 00	3.30E – 03	2.74E + 00
Naphthalene	1.45E + 00	1.45E + 03	1.45E + 00	4.80E + 02	1.57E + 02	9.47E – 03	2.05E + 00	3.36E – 03	2.05E + 00
Phenanthrene	1.40E + 01	1.40E + 04	1.40E + 01	2.68E + 02	9.24E + 01	9.96E – 03	1.20E + 00	3.53E – 03	1.21E + 00
Phenol	1.42E – 02	1.42E + 01	1.42E – 02	7.90E + 02	3.99E + 01	1.46E – 03	5.19E – 01	5.17E – 04	5.20E – 01
Toluene	3.00E – 01	3.00E + 02	3.00E – 01	3.38E + 03	9.17E + 02	7.84E – 03	1.19E + 01	2.78E – 03	1.19E + 01

K = 3.15E – 01
L = 5.75E + 2
S = 5.00E – 03
LS = 1.40E – 01
C = 1.00E + 00
P = 1.00E + 00
A = 5.95E + 00
R_t = 3.50E + 00

T_r = 2.40E + 01
Average Annual Storm = 4.60E + 01
River Flow Rate, QV_r = 8.45E – 01
CN = 7.40E + 01
Bulk Density = 1.44E + 00
Available Water Capacity 1.20E – 01
Organic Fraction, f_{oc} = 1.00E – 03
Storm Events/year = 1.31E + 01

Storm Events in 70 years = 9.20E + 02
S_w = 3.51E + 00
Q_r = 1.24E + 00
V_r = 6.12E – 01
Peak Flow Q_p = 3.88E – 01
$Y(S)_e$ = 1.87E + 00

Concentration in Runoff (mg/L)	Concentration in Surface Water (mg/L)	Chronic Water Quality Standards (mg/L)	Attenuation Factor	Maximum Acceptable Soil Concentration (mg/kg)
1.63E – 01	5.13E – 02	5.10E + 00	9.52E – 04	5.35E + 03
4.63E + 02	1.46E + 02	3.00E – 01	1.73E + 00	1.73E – 01
5.94E + 02	1.87E + 02	3.00E – 01	1.73E + 00	1.73E – 01
3.63E + 00	1.14E + 00	4.30E – 01	1.75E – 03	2.46E + 02
2.72E + 00	1.25E + 00	3.00E – 01	2.60E – 03	1.15E + 02
1.60E + 00	7.34E – 01	3.00E – 01	2.74E – 03	1.10E + 02
6.88E – 01	2.17E – 01	5.80E + 00	2.74E – 04	2.12E + 04
1.58E + 01	4.98E + 00	5.00E + 00	1.47E – 03	3.40E + 03

[a] A soil organic carbon fraction of 0.1% is assumed; $K_d = f_{oc} \times K_{(oc)}$.

Table 7.23 Recommended Soil Cleanup Limits — Future Land Use Conditions

Chemical of Concern	Soil Maximum Chemical Levels (mg/kg)	Estimated Soil AL (mg/kg)	ALs for SW[a] Protection (mg/kg)	ALs for GW[b] Protection (mg/kg)	RSCL (mg/kg)
Benzene	5.39E + 01	3.86E + 01	5.35E + 03	3.48E + 01	3.48E + 01
Benz(a)anthracene	8.42E + 01	2.40E + 00	1.73E – 01	1.97E + 02	1.73E – 01
Benzo(a)pyrene	1.08E + 02	3.48E – 01	1.73E – 01	7.71E + 02	1.73E – 01
Ethylbenzene	6.52E + 02	8.00E + 03	2.46E + 02	1.53E + 03	2.46E + 02
Naphthalene	4.80E + 02	3.20E + 04	1.15E + 02	1.15E + 02	1.15E + 02
Phenanthrene	2.68E + 02	3.20E + 04	1.10E + 02	6.16E + 01	6.16E + 01
Phenol	7.90E + 02	4.80E + 04	2.12E + 04	3.41E + 01	3.41E + 01
Toluene	3.38E + 03	2.40E + 04	3.40E + 03	1.37E + 01	1.37E + 01

[a] SW = Surface water.
[b] GW = Groundwater.

than 0.01% of total cell weight when present. Most of the reduction in mercury usage has occurred in alkaline batteries. Indeed, current mercury content in alkaline batteries manufactured in North America does not exceed 0.025% by weight in most cases. To achieve this, a number of patented organic substitutes are being used in several makes of household batteries. On the other hand, cadmium is used as the negative electrode in rechargeable Ni-Cad batteries, and as such its use cannot be reduced without proportionally reducing the energy content of the batteries (Balfour, 1990).

7.3.4 Material Used as Substitute for Mercury

Proprietary organic surfactants are used as a substitute for mercury in the anode production process to help minimize the amount of mercury used in the production of alkaline batteries. This organic substitute is used at a level of approximately 100 ppm. Typical material safety data sheet (MSDS) for the proprietary complex organic phosphate ester have been reviewed and determined to present far less toxicity effects than mercury; in addition, it is not volatile and is generally a more stable compound (IRR, 1992). For instance, the organic substitute has an LD_{50} of about 5 to 15 times higher than that reported for mercury. However, this material is known to be soluble in water and is also corrosive. This could therefore possibly facilitate the degradation of batteries under landfill conditions.

7.3.5 Feasible Disposal Options for Used Household Batteries

Traditionally, landfilling has been the method of choice for disposal of MSW. With the growing stringent regulations and the lack of suitable landfill sites, disposal practices are undergoing several changes, especially with respect to the management of hazardous wastes. Incineration and recycling are becoming prominent alternatives to landfilling. These alternative waste management processes, however, are not without their own problems. Although incineration tends to reduce the volume of waste, the residual ash which has to be disposed of is usually more toxic due to the concentration of toxic remains and by-products; in addition, there is the potential for release of toxic contaminants into the air. Much as recycling seems a more acceptable option, its application to management of used household batteries is yet to be proven as efficient. In fact, the economic feasibility of recycling household batteries is a major issue — apart from potential risks due to collection of large quantities of such batteries, with residual voltage which could cause them to short together and generate enough heat to cause fires. The residual power also makes batteries more susceptible to corrosion; this factor should be considered when discussing the condition of batteries in landfills over long periods of time.

7.3.6 Landfilling with MSW

Since the past decade, the general public has shown increased concern over ground- and surface water contamination from landfill waste disposal (Greghian et al., 1981). The potential environmental impacts from landfilling of MSW are a major reason for many regulations enforced, the technical innovations, and the public opposition to siting of such facilities. The actual impact which a landfill has on its surrounding environment, however, is highly dependent on the practice and operation of the facility (Christensen et al., 1989). The contamination potential of a landfill depends on three main factors (Miller and Mishra 1989):

- The migration of leachate out of the landfill
- The potential harm of the leachate constituents
- The corresponding concentrations of each contaminant in the leachate

In addition, there should be completeness of exposure pathways due to the presence of potential human or ecological receptors that are potentially at risk.

Metals in batteries will generally not be released rapidly from landfills, but the overall load of metals into the soil and the specific soil's ability to adsorb metals should be viewed as critical factors to consider in deciding whether or not to landfill household batteries. The mobility of any of the metals of concern within the soil system will be controlled by the extent of fixation, adsorption, exclusion, complex formation, reaction kinetics, as well as the soils overall physical and chemical properties (MPCA, 1991). Indeed, physical and chemical properties of soils may control leachate migration. The combined physical and chemical properties of texture, particle size, hydrous oxides and organic matter, cation exchange capacity (CEC), and pH are the most significant soil properties that may determine leachate migration (MPCA, 1991). Thus, the nature of soils surrounding a landfill greatly affect the rate of heavy metal migration; for instance, clay material appears to greatly retard, if not stop, heavy metal migration, whereas sand or gravel soils are less adsorptive and also offer less resistance to water flow (Fochtman and Haas, 1975). Cd, Ni, and Zn are believed to have low mobility in most clay soils, but mercury may have moderate mobility in the same group of soils.

When batteries are disposed of in a landfill they usually are buried together with MSW with variable moisture content. Capped landfills are expected to have a lower moisture content than uncapped landfills. Because of the compressibility of MSW, it is likely that the loads on battery cases are too small to damage them. Maintaining these conditions will limit the release of heavy metals from batteries. Batteries with broken cases present a different problem scenario. Additionally, batteries may be subjected to varying stresses between disposal in landfills, which may lead to rupture of the casing. A study of batteries in landfills has shown that some batteries are broken, corroded, or otherwise damaged in a manner that would allow their contents to leach out (Little, 1989). Batteries that are landfilled go through various physical changes after they are buried. Landfill conditions may promote corrosion and deterioration of battery cases, which consequently may release metals to leachate generated within the landfill. Leachate is produced in landfills as water and other liquids within the waste and any moisture that collects on top of the landfill flow through the waste. As this liquid moves down through the wastes, it flushes out and collects pieces of other materials. If a battery is cracked or corroded, the liquids in the landfill will pick up the metals contained in it. Nonetheless, laboratory studies show that neither MSW leachate nor heavy metal leachate significantly affect polymeric or admix liners (CMU, 1989). Furthermore, because metals tend to be adsorbed on clay materials, where natural clay liners serve as landfill liners, the possibility of metals escaping from the landfill into groundwater below is limited.

7.3.7 Incineration with MSW

Incineration of batteries does not destroy the heavy metals contained in them. Thermal processes transfer heavy metals either to stack gas as fine particles or to bottom ash. An important characteristic of the ash is its toxicity and its status as a waste. In an incineration, the solid waste is burned, with bottom and fly ash remaining as end products; the bottom ash falls through the grates of the incinerator, while the fly ash rises through the smokestack and may be trapped by a collection device. Incineration reduces the volume of waste by about 80 to 95%, and the weight by about 50 to 75%.

Most of the metals in incinerator emissions may be captured by the pollution control systems on an incinerator. It is believed that lime scrubbing systems using filter fabric typically remove over 99.9% of the fly ash; over 99% of cadmium and zinc and 90 to 95% of mercury are removed by the filters (CMU, 1989). Conservatively estimated, U.S. EPA (1989) also indicates air pollution control devices (APCDs) efficiencies for controlling cadmium of up to 99% and up to 98% for mercury. MPCA (1991) agrees that existing air pollution control equipment are very effective at removing particulates and gases containing metals from household batteries; current dry and semidry air pollution control systems are believed to be able to collect 95% or more of the metals found in MSW, except for mercury. A wet scrubber or a wet/dry scrubber can achieve higher collection efficiencies for mercury. The control of mercury emission to the atmosphere from incineration of household batteries in domestic waste is an important pollution control issue. Different APCDs vary in their effectiveness in capturing mercury. A spray dryer and baghouse system can achieve 75 to 85% mercury removal, thus capturing the mercury in the ash; a spray dryer plus electrostatic precipitator can achieve about 35 to 45% removal (Ellison, 1986; MPCA, 1991). Other metals from the incineration of household batteries, including cadmium, are largely collected in the ash from the incineration process; about 92.4% of the cadmium in incinerated waste stream is captured in the total ash (Vogg et al., 1986; MPCA, 1991). Mercury creates the most concern in the incineration process; because of its low boiling point, it may not always condense on the fly ash and therefore might be released into the atmosphere. If it is determined that mercury collection is a problem, removing high-content mercury batteries from the MSW stream could abate the problem.

Fly ash consists of the airborne particles that are captured by filters in the incinerator stack. Bottom ash is the heavy residue formed at the bottom of the incinerator after the MSW is burned. In the process of burning wastes, some of the heavy metals present may volatize and then condense on particulate matter as it rises through the smokestack. Consequently, the fly ash will have a high concentration of heavy metals. Part of the metals do not volatize, though, and these contribute to the bottom ash generated in the process. Of the ash created through the combustion process, about 10% is fly ash and 90% is bottom ash (CMU, 1989; Environment Canada, 1991). The partitioning of metals between bottom and fly ash varies between incinerators and may be influenced by such factors as the operating temperature of the incinerator and whether the material being burned is suspended in the air or on the floor of the incinerator. In fact, the partitioning of metals between the bottom and fly ash is a poorly understood aspect of incineration. Since most incinerator facilities mix the bottom and captured fly ash for disposal, the total amount of metals contributed by batteries to the total ash will approximately be the same, regardless of how it is partitioned between the bottom and fly ash, assuming none or only minimum amounts of the metals escape through the smokestack. Laboratory studies show that cadmium, one of the toxic substances of concern associated with household batteries, is one of two metals (together with lead) that tends to be more concentrated in the fly ash than in the bottom ash (Ujihara and Gough, 1989). Since cadmium is carcinogenic by the inhalation pathway, it is crucial that adequate scrubbers are used in MSW incinerators, that will capture as much of the fly ash as possible and to minimize the amounts that could eventually reach potential human receptors. In the absence of that, Ni-Cads which may be great contributors of cadmium to MSW, may have to be removed from the wastes to be incinerated.

7.3.8 Recycling as an Option for the Management of Used Batteries

IRR (1992) gives a detailed review of the logistics and viability of recycling household batteries; battery collection systems; current status of battery recycling technology for the various battery systems; and the costs of battery recycling. Such knowledge will aid in evaluating the feasibility of implementing battery recycling programs. In fact, it is believed that the collection and recycling of used household batteries poses several unanswered technical and economic problems that need to be resolved before any widespread implementation of such a program.

7.3.9 Potential Health and Environmental Impacts from Disposal Practices

The metals of potential concern present in the household batteries studied are cadmium, manganese, mercury, nickel, and zinc. In whatever disposal or management practice that is adopted for the used household batteries, there is the potential for the release of one or more of these metals which might affect human health directly or indirectly or which might impact the environment. Currently, used household batteries are almost exclusively disposed of in domestic garbage, which is eventually incinerated or landfilled. Lately, the idea of used battery collection, separation, and possible recycling is becoming another focus of attention.

Used household batteries will undergo degradation under landfill conditions, with the rate and degree of decay depending on the battery types, state of charge in battery, and the physical conditions at the landfill site. Beyond the degradation process, it is important to determine if metals from the batteries will leach from a landfill into an underlying aquifer. Several variables, including landfill management practices, will determine this. Indeed, under ideal landfill conditions, metals will not leach rapidly through landfills and soils into groundwater. On the other hand, metals do not decompose or degrade and thus have the potential of leaching into aquifers over long periods of time.

Metals are of critical concern in an incineration process since they are not combustible. Thus, the protection of the environment in an incineration management option depends on the ability of the incinerator to capture and remove metals from air emissions. Although technologies exist for most metals removal, which makes such process practical, albeit expensive, it is not completely effective in abating mercury emissions due to the low vapor pressure of mercury. The presence of other metals, including cadmium in the incinerator ash due to the incineration of household batteries, renders such ash potentially highly toxic; such concentrated ash may not be disposed of at MSW landfills.

In general, incineration of used household batteries results in cadmium being deposited in the ash, while the mercury will likely be in both the ash and the air emissions. Landfilling of the used batteries, on the other hand, has the potential of impacting the groundwater resources, in which case groundwater becomes the primary potential exposure pathway of concern. Regarding recycling, the separate collection of dry cell batteries may contribute to various hazardous situations including fire, explosion, and direct human exposures to chemicals.

7.3.10 Household Battery Consumption in the Case Region

There has been a rapid increase in the number of battery-operated devices in use in North America in recent years. Table 7.24 gives an indication of the Canadian battery market estimate for household dry cell batteries. These estimates are based on industry sales primarily through retail outlets. This may exclude relatively minor sales of batteries to medical, military, industrial, and commercial users and also of batteries included in consumer devices imported/sold with batteries.

7.3.11 Potential Metal Contributions to MSW Streams

Table 7.25 shows the potential metal contributions from the various household battery types to the MSW. The calculations are made by multiplying the average percent of metals found in each battery type by the average weight of the battery and also by the number of the particular battery types sold to consumers. An overly conservative assumption has been made that all battery sales each year equals the amount of discards, a situation that is definitely not applicable to the rechargeable nickel-cadmium batteries. However, assuming batteries used from previous years are added to the waste stream, then the assumption of sales equals discards is fair, though still conservative. In fact, mercury and cadmium from discarded household batteries may be contributing up to 20 and 33%, respectively, of the total metal compositions appearing in MSW streams in much of North America. Cadmium is used primarily in nickel-cadmium batteries, which make up the largest portion of the dry cell secondary battery market. Ni-Cad batteries present the potential problem of cadmium entering MSW streams and subsequently the various environmental compartments or media following disposal; if landfilled, cadmium

Table 7.24 Canadian Household Battery Market Estimates (1990/1991)

		Battery Size						
Battery Type	Total in Millions	AAA	AA	C	D	9 V	Other	% of Totals
Alkaline	100	9	60	9	8	9	5	65.4
Zinc-carbon	25 }	1	21	7	9	6	1	{ 16.3
Zinc-chloride	20							13.1
Nickel-cadmium	8	0.5	5	1	1	0.5	—	5.2
Totals	153	10.5	86	17	18	15.5	6	
% of totals		6.9	56.2	11.1	11.8	10.1	3.9	100

Source: CBMA, Canada.

may impact groundwater resources, and if incinerated, potential receptors may be exposed via inhalation of chemicals escaping into the air.

7.3.12 Risk Assessment of Battery Disposal Alternatives

Heavy metals can have detrimental effects on ecosystems. It is apparent that ecosystems have naturally occurring levels of heavy metals; however, additional amounts of such metals can stress the delicate balance necessary for the efficient performance and/or survival of the ecological community. In a similar manner, human populations can tolerate only limited levels of exposure to heavy metals. However, the method of choice for disposal of used dry cells could potentially expose various ecological receptors to unacceptable levels of heavy metals.

This section examines the risks associated with alternative disposal options for used dry cell batteries that becomes part of household waste streams. A conservative assumption is made in this evaluation, that all batteries purchased in a given year enter MSW in that year. Since lesser quantities than this will in practice be disposed of, estimated quantities of heavy metals purported to enter the MSW and which is used in characterizing risks for used battery disposal options will represent overestimates that will be more protective of human health and the environment. This helps to build a factor of safety that can counteract uncertainties potentially present in the analytical processes.

The concepts of risk assessment are employed in evaluating and comparing the feasible disposal options identified for the management of used household batteries. Data on total sales (conservatively assumed to equal total amount of disposal) of dry cell batteries in Canada are used to estimate the potential risks of the battery disposal alternatives. The risk assessment model utilized in this evaluation incorporates methods for estimating the amounts of the metals of concern potentially entering the various environmental compartments due to the alternative disposal practices.

The metals of concern in the battery types being investigated in this study may be sources of potential risks to humans and the environment. The potential risks are the results of potential releases of the metals into the environment following disposal of used batteries in the MSW stream. Cadmium and mercury are recognized as the critical metals of concern. Different exposures may result due to battery disposal practices. The generic equation for calculating chemical intakes previously

Table 7.25 Estimates of Metals Contributions to MSW Streams from Household Batteries[a]

Battery Type	Battery Size	Average Weight (g)	Cd Content		Mn Content		Hg Content		Ni Content		Zn Content		Total Battery Sales (millions)
			% Weight	per Cell (g)	% Weight	per Cell (g)	% Weight	per Cell (g)	% Weight	per Cell (g)	% Weight	per Cell (g)	
Alkaline	AAA	11	0.4	0.04	30	3.30	0.025	0.00	0	0.00	12	1.32	9
	AA	23	0.4	0.09	34	7.82	0.025	0.01	0	0.00	14	3.22	60
	C	66	0.4	0.26	32	21.12	0.025	0.02	0	0.00	14	9.24	9
	D	133	0.4	0.53	32	42.56	0.025	0.03	0	0.00	16	21.28	8
	9 V	46	0.4	0.18	30	13.80	0.025	0.01	0	0.00	10	4.60	9
													95
Zinc-carbon/ Zinc-chloride	AAA	8	0.01	0.00	38	3.04	0.01	0.00	0	0.00	38	3.04	1
	AA	17	0.01	0.00	28	4.76	0.01	0.00	0	0.00	22	3.74	21
	C	44	0.01	0.00	28	12.32	0.01	0.00	0	0.00	18	7.92	7
	D	87	0.01	0.01	28	24.36	0.01	0.01	0	0.00	18	15.66	9
	9 V	37	0.01	0.00	22	8.14	0.01	0.00	0	0.00	14	5.18	6
													44
Nickel-cadmium	AAA	11	13	1.43	0	0.00	0	0.00	20	2.20	0	0.00	0.5
	AA	23	13	2.99	0	0.00	0	0.00	20	4.60	0	0.00	5
	C	52	13	6.76	0	0.00	0	0.00	20	10.40	0	0.00	1
	D	62	13	8.06	0	0.00	0	0.00	20	12.40	0	0.00	1
	9 V	38	13	4.94	0	0.00	0	0.00	20	7.60	0	0.00	0.5
													8
Total Metal (million grams)			47.33		1610.98		1.06		50.70		808.92		

[a] Estimates of metals entering MSW streams from household batteries are based on annual sales figures, percent metal compositions, and average weights for the various battery types.

discussed in Chapter 5 is adapted to estimate actual exposure intakes by potential receptors as a result of releases of metals into various environmental media from the battery disposal practices. A conservative mass balance approach is employed in this evaluation, according to the following relationship:

$$I = \{mass / year\} \times \{1 / BW\} \times \{CF\}$$

where

I = intake, adjusted for absorption (mg/kg/day)
CF = conversion factor (to cater for consistency in units used)
BW = body weight (kg)

and mass/year refers to the amount of chemical available per year to potential receptors. This equation is used to estimate potential intakes by receptors potentially impacted from the method of choice for the disposal of used household batteries. The carcinogenic and noncarcinogenic effects of the chemicals of concern are subsequently calculated according to the relationship presented in Appendix B.

Case simulations are used to demonstrate the potential effects of potential receptor exposures to metals of concern contributed by dry cells to MSW. Three different case scenarios are evaluated, including the following:

- Landfilling disposal only for MSW
- Incineration disposal only for MSW
- Combined landfilling and incineration disposal for MSW

The model utilizes specific information pertaining to the following:

- Total annual amount of batteries entering MSW streams
- Total annual amount from batteries going to landfill and/or incinerator used for MSW management

Additional information necessary for a complete evaluation are identified subsequently.

Risk Characterization for the Landfill Disposal Alternative

Landfilling of used household batteries together with MSW is a popular disposal option for dry cell batteries. The landfilling of used batteries leads to consequential decomposition of the batteries. Subsequently, the metals of concern present in the batteries are released into leachates generated at the landfills. The migration of such leachates into underlying aquifers can result in the contamination of potential water supply sources.

In order to characterize risks associated with the landfilling option, the following information on exposure modeling parameters will be necessary for a complete evaluation :

- For leachate generated, it is assumed that ≈ 0.05% of metals from batteries entering a landfill will escape in form of leachate to contaminate groundwater; this represents a conservative estimate, compared with the transfer coefficients for leachate in MSW landfills.
- It is assumed that there is ingestion intake from potentially contaminated groundwater, following the migration of leachate into a water supply aquifer.

A complete listing of the exposure modeling assumptions are included in Table 7.26, together with the estimates of potential receptor exposures. A summary of the risk characterization computations is given in Table 7.27. It is apparent from the risk characterization results that landfilling even concentrated volumes of the used household batteries will present no significant health and environmental impacts, since the HIs are <1 for all potential receptors.

Disposal Practices in the Regional Municipality of Waterloo as an Approximate Example. To support results obtained from the case model, field information is used to confirm the conclusion reached. A landfill site is chosen in the Regional Municipality of Waterloo. The Waterloo landfill site is located in the extreme southwest corner of the City of Waterloo, Regional Municipality of Waterloo, Ontario. The area is underlain by a major sand and gravel aquifer which supplies about 45 million gallons per day (mgpd) of groundwater to the Kitchener-Waterloo area. In fact, this is one of the most significant groundwater resources in Ontario. Over 100 monitoring wells have been installed at and in the vicinity of the landfill since 1971. Table 7.28 summarizes information obtained by monitoring the Waterloo landfill site area. None of the data for the upper unit aquifer or aquifer unit No. 1 exceed the Ministry of the Environment (MOE) objective. Nonetheless, the higher of these contaminant levels (worst-case scenario) is used to estimate the level of potential risks posed by aquifer contamination from *all* sources of the metals of concern. It should be noted that the dry cell batteries of concern are contributing only a fraction of this total amount of metals monitored as present in the aquifer; for instance, mercury and cadmium from dry cell batteries contribute only about 20% and 33%, respectively, to the quantities appearing in MSW.

Tables 7.29 and 7.30, respectively, show the potential receptor exposures and the risk characterization for the impacts of the Waterloo landfill on underlying aquifers. The HI for the adult population is 0.7 (i.e., <1), indicating no problem. On the other hand, child exposure shows an HI >1. However, this value of 1.6 is not very much above the acceptable index of unity. Considering the conservative assumptions used in the evaluation (including the implicit assumption that all the metals of concern have the same physiologic effect, or toxicological endpoint), it may be concluded that the landfill disposal of household batteries are not posing any risks due to the presence of any of the metals of concern identified for these batteries. No carcinogenic risks are anticipated, since none of the chemicals is a carcinogen by the oral exposure route. Thus, it is apparent from the risk characterization results that landfilling MSW of which used household batteries may be a part is presenting no significant health and environmental impacts. It is also worth mentioning that these

Table 7.26 Exposure Modeling and Evaluation Results: Case Model I — Landfilling Disposal only for MSW

Modeling Assumptions

1. All MSW generated in region are disposed of at municipal landfills.
2. Conservatively assumed that up to 0.05% of all landfill waste will become leachate annually.
3. Populations potentially impacted chosen to be the size of residential population in case region of landfill location; population of case region is taken at 100,000 and that for the Province of Ontario approximated to 10 million.
4. Child average weight is 16 kg and adult average weight is 70 kg; these potential receptors may be impacted via ingestion of contaminated groundwater.
5. Contribution of metals (in household batteries) to MSW in case region is proportional to the population.

Battery Constituent	Tota Annual Amount from Batteries into MSW Streams (National — million g)	Total Annual Amount from Batteries into MSW Streams (Ontario — g)	Total Annual Amount from Batteries into MSW Streams (Case Region1-g)	Total Annual Amount from Batteries into Landfill (Case Region1-mg)	Leachate	Potential Ingestion Intake (mg/person/year)	Potential Receptor Exposure Via Ingestion of Leachate Releases into Groundwater (mg/kg/day)	
							Child	Adult
Cadmium (Cd)	4.73E + 01	1.70E + 07	1.70E + 05	1.70E + 08	8.52E + 04	8.52E − 01	1.46E − 04	3.33E − 05
Manganese (Mn)	1.61E + 03	5.80E + 08	5.80E + 06	5.80E + 09	2.90E + 06	2.90E + 01	4.97E − 03	1.13E − 03
Mercury (Hg)	1.06E + 00	3.80E + 05	3.80E + 03	3.80E + 06	1.90E + 03	1.90E − 02	3.25E − 06	7.44E − 07
Nickel (Ni)	5.07E + 01	1.83E + 07	1.83E + 05	1.83E + 08	9.13E + 04	9.13E − 01	1.56E − 04	3.57E − 05
Zinc (Zn)	8.09E + 02	2.91E + 08	2.91E + 06	2.91E + 09	1.46E + 06	1.46E + 01	2.49E − 03	5.70E − 04

Table 7.27 Risk Characterization: Case Model 1 — Landfilling Disposal Only for MSW

Battery Constitutent	Potential Ingestion Intake (mg/person/year)	Potential Receptor Exposure Via Ingestion of Leachate Releases into Groundwater (mg/kg/day)		Oral RfD (mg/kg/day)	Oral SF (1/[mg/kg/day])	HI for Child Exposure	HI for Adult Exposure	Carcinogenic Risks for Child Exposure	Carcinogenic Risks for Adult Exposure
		Child	Adult						
Cadmium (Cd)[a]	8.52E – 01	1.46E – 04	3.33E – 05	5.00E – 04	—	0.29	0.07	0.00E + 00	0.00E + 00
Manganese (Mn)	2.90E + 01	4.97E – 03	1.13E – 03	1.00E – 01	—	0.05	0.01	0.00E + 00	0.00E + 00
Mercury (Hg)	1.90E – 02	3.25E – 06	7.44E – 07	3.00E – 04	—	0.01	0.00	0.00E + 00	0.00E + 00
Nickel (Ni)	9.13E – 01	1.56E – 04	3.57E – 05	2.00E – 02	—	0.01	0.00	0.00E + 00	0.00E + 00
Zinc (Zn)	1.46E + 01	2.49E – 03	5.70E – 04	2.00E – 01	—	0.01	0.00	0.00E + 00	0.00E + 00
Total						0.4	0.1	—	—

[a] Cadmium is a B1 carcinogen (i.e., a probable human carcinogen) by the inhalation pathway only.

Table 7.28 Representative Maximum Concentrations (mg/L) of the Chemicals of Concern in Leachate and Groundwater Near the Waterloo Landfill Site in the Regional Municipality of Waterloo

Parameter	Upper Fine-Grained Unit Aquifer	Aquifer Unit No. 1	Typical Leachate Quality	Worst-Case Leachate Quality[a]	Collector-Case Leachate Quality[b]	MOE Objective[c]
Cadmium (Cd)	<0.01	<0.01	0.024	0.065	0.013	0.005
Manganese (Mn)	0.01	0.18	12	39	—	0.05
Mercury (Hg)	<0.00001	0.00003	—	—	—	0.001
Nickel (Ni)	<0.01	<0.01	0.23	7.0	0.43	—
Zinc (Zn)	0.35	0.55	6.5	50	2.60	5.0

Source: CRA, 1990a, b.

[a] Worst-case leachate quality represents the MCL from leachate generated from landfill.
[b] Collector-case characterization of leachate quality is based on average value from leachate pumping station.
[c] MOE objective is maximum acceptable contaminant levels.

levels of metals present in the aquifer would have been contributed from all sources other than dry cell batteries only.

Risk Characterization for the Incineration Disposal Alternative

Incineration of used household batteries generally will result in the release of metals as fumes and particulates. Exposure to such contaminants can consequently occur via direct inhalation by humans, or contamination of other environmental media may occur with subsequent impacts on humans (such as through consumption of contaminated foods and water). The incineration process yields fly and bottom ash, which both may contain some metals. Based on an assumption of reported figures that 10% of incinerated wastes become fly ash and 90% become bottom ash (Goldstein, 1989), and that fly ash has a smaller unit weight than bottom ash, it is estimated that about 70% by weight of the metals are in the bottom ash and 30% by weight in the fly ash (CMU, 1989). It is further assumed that 99% of the fly ash is captured by the incinerator's APCD, and this fly ash is added to the bottom ash for disposal at a hazardous waste facility.

A complete listing of the exposure modeling assumptions are indicated in Table 7.31 together with the estimates for potential receptor exposures. A summary of the risk characterization calculations for the incineration option is given in Table 7.32. It is apparent that incineration of concentrated amounts of the dry cell batteries may present both carcinogenic and noncarcinogenic risks to potential receptors. Much of this risk is contributed by cadmium, known to be a probable human carcinogen by the inhalation pathway. This cadmium will be from concentrated sources of Ni-Cads. Removal of large amounts of Ni-Cads from MSW streams to be incinerated will therefore remove the potential risks posed by the incineration alternative for managing the used dry cell batteries.

Disposal Practices in the Regional Municipality of Hamilton-Wentworth as an Approximate Example. To support the results obtained for the case model

Table 7.29 Exposure Assessment: CDIs from Landfill Releases (Waterloo Landfill Facility, Waterloo, Ontario)

Modeling Assumptions

1. Potential receptors are potentially exposed via ingestion of contaminated groundwater.
2. Child average weight is 16 kg and adult average weight is 70 kg; these potential receptors may be impacted via ingestion of groundwater potentially impacted by landfill leachate.
3. Ingestion rates assumed to be 2 L/day, and 1 L/day for child and adult, respectively; chemical bioavailability factors for ingested water is conservatively taken to be 100%, and water is taken for 365 days/year.

Chemical of Concern	Absorption Factor (%)	Exposure Concentration (mg/L)	CDI — Child (mg/kg/day)		CDI — Adult (mg/kg/day)	
			Carcinogenic Effects	Noncarcinogenic Effects	Carcinogenic Effects	Noncarcinogenic Effects
Cadmium (Cd)	100	1.00E − 02	—	6.23E − 04	—	2.86E − 04
Manganese (Mn)	100	1.80E − 01	—	1.13E − 02	—	5.15E − 03
Mercury (Hg)	100	3.00E − 05	—	1.88E − 06	—	8.58E − 07
Nickel (Ni)	100	1.00E − 02	—	1.88E − 04	—	2.86E − 04
Zinc (Zn)	100	5.50E − 01	—	3.44E − 02	—	1.57E − 02

Table 7.30 Risk Characterization For Landfill Leachate Released into Aquifers (Waterloo Landfill Facility, Waterloo, Ontario)

Battery Constitutent	Potential Receptor Exposure via Ingestion of Landfill Releases (Carcinogenic Effects) (mg/kg/day)		Potential Receptor Exposure Via Ingestion of Landfill Releases (Noncarcinogenic Effects) (mg/kg/day)		Oral RfD (mg/kg/day)	Oral SF (1/[mg/kg/day])	HI for Child Exposure	HI for Adult Exposure	Carcinogenic Risks for Child Exposure	Carcinogenic Risks for Adult Exposure
	Child	Adult	Child	Adult						
Cadmium (Cd)[a]	—	—	6.25E – 04	2.86E – 04	5.00E – 04	—	1.25	0.57	0.00E + 00	0.00E + 00
Manganese (Mn)	—	—	1.13E – 02	5.15E – 03	1.00E – 01	—	0.11	0.05	0.00E + 00	0.00E + 00
Mercury (Hg)	—	—	1.18E – 06	8.58E – 07	3.00E – 04	—	0.01	0.00	0.00E + 00	0.00E + 00
Nickel (Ni)	—	—	6.25E – 04	2.86E – 04	2.00E – 02	—	0.03	0.01	0.00E + 00	0.00E + 00
Zinc (Zn)	—	—	3.44E – 02	1.57E – 02	2.00E – 01	—	0.17	0.08	0.00E + 00	0.00E + 00
Total							1.6	0.7	—	—

[a] Cadmium is a B1 carcinogen (i.e., a probable human carcinogen) by the inhalation pathway only.

Table 7.31 Exposure Modeling and Evaluation Results: Case Model II — Incineration Disposal Only for MSW

Modeling Assumptions

1. All MSW generated in region are disposed of at municipal incinerators.
2. It is assumed that 30% of incinerator wastes go to fly ash and 70% into bottom ash.
3. Assumed that pollution control efficiency of up to 99% attained, so that only 1% of the available fly ash is released through the stack into air.
4, All captured fly ash and bottom ash are sent to hazardous waste landfill elsewhere.
5. Populations potentially impacted chosen to be the size of residential population in case region of incinerator location; population of case region is taken at 100,000 and that for the Province of Ontario approximated to 10 million.
6. Child average weight is 16 kg and adult average weight is 70 kg; these potential receptors may be impacted via inhalation of air emissions.
7. Contribution of metals (in household batteries) to MSW in case region is proportional to the population.

Battery Constituent	Total Annual Amount from Batteries into MSW Streams (National — million g)	Total Annual Amount from Batteries into MSW Streams (Ontario — g)	Total Annual Amount from Batteries into MSW Streams (Case Region 2 — g)	Total Annual Amount from Batteries into Incinerator (Case Region 2 — mg)	Bottom Ash (mg)	Incinerator Fly Ash Component (mg)
Cadmium (Cd)	4.73E + 01	1.70E + 07	1.70E + 05	1.70E + 08	1.19E + 08	5.11E + 07
Manganese (Mn)	1.61E + 03	5.80E + 08	5.80E + 06	5.80E + 09	4.06E + 09	1.74E + 09
Mercury (Hg)	1.06E + 00	3.80E + 05	3.80E + 03	3.80E + 06	2.66E + 06	1.14E + 06
Nickel (Ni)	5.07E + 01	1.83E + 07	1.83E + 05	1.83E + 08	1.28E + 08	5.48E + 07
Zinc (Zn)	8.09E + 02	2.91E + 08	2.91E + 06	2.91E + 09	2.04E + 09	8.74E + 08

Table 7.31 (continued)

Total Fly Ash Captured by Pollution Control Devices (mg/year)	Total Fly Ash Released Into Air (mg/year)	Potential Inhalation Intake (mg/person/year)	Potential Receptor Exposure Via Inhalation of Incinerator Releases into Air (mg/kg/day)	
			Child	Adult
5.06E + 07	5.11E + 05	5.11E + 00	8.75E − 04	2.00E − 04
1.72E + 09	1.74E + 07	1.74E + 02	2.98E − 02	6.81E − 03
1.13E + 06	1.14E + 04	1.14E − 01	1.95E − 05	4.46E − 06
5.42E + 07	5.48E + 05	5.48E + 00	9.38E − 04	2.14E − 04
8.65E + 08	8.74E + 06	8.74E + 01	1.50E − 02	3.42E − 03

Table 7.32 Risk Characterization: Case Model II — Incineration Disposal Only for MSW

Battery Constitutent	Potential Inhalation Intake (mg/person/year)	Potential Receptor Exposure Via Inhalation of Incinerator Releases into Air (mg/kg/day)		Inhalation RfD (mg/kg/day)	Inhalation SF (1/[mg/kg/day])	HI for Child Exposure	HI for Adult Exposure	Carcinogenic Risks for Child Exposure	Carcinogenic Risks for Adult Exposure
		Child	Adult						
Cadmium (Cd)[a]	5.11E + 00	8.75E − 04	2.00E − 04	5.00E − 04	6.10E + 00	1.75	0.40	5.34E − 03	1.22E − 03
Manganese (Mn)	1.74E + 02	2.98E − 02	6.81E − 03	2.00E − 01	—	0.15	0.03	0.00E + 00	0.00E + 00
Mercury (Hg)	1.14E − 01	1.95E − 05	4.46E − 06	3.00E − 04	—	0.07	0.01	0.00E + 00	0.00E + 00
Nickel (Ni)	5.48E + 00	9.38E − 04	2.14E − 04	2.00E − 02	—	0.05	0.01	0.00E + 00	0.00E + 00
Zinc (Zn)	8.74E + 01	1.50E − 02	3.42E − 03	1.00E − 02	—	1.50	0.34	0.00E + 00	0.00E + 00
Total						3.5	0.8	5.34E − 03	1.22E − 03

[a] Cadmium is a B1 carcinogen (i.e., a probable human carcinogen) by the inhalation pathway only.

presented, field information is used to confirm the conclusion reached for the incineration management of used dry cell batteries. A solid waste incinerator facility located in Hamilton, with a population of about 430,000, is chosen for this purpose. The Solid Waste Reduction Unit (SWARU) in Hamilton, Ontario, processes and incinerates MSW. The facilities are owned by the Region of Hamilton-Wentworth and operated by Laidlaw Waste Systems. Table 7.33 shows a summary of typical monitoring/modeling data associated with the operation of the SWARU incinerator facility in Hamilton. It should be noted that the dry cell batteries of concern are contributing only a fraction of the total amount of metals present in wastes incinerated at the facility.

Tables 7.34 and 7.35, respectively, show the potential receptor exposures and the risk characterization for the potential impacts of the Hamilton SWARU due to atmospheric emissions. There is no significantly measurable noncarcinogenic risks experienced by potential human receptors, since HIs approaching zero values were estimated for both child and adult population. On the other hand, carcinogenic risks of about 1.6×10^{-7} and 1.4×10^{-6} were estimated for the child and adult groups, respectively. However, these numbers fall within the acceptable carcinogenic risk range of 1.0×10^{-4} to 1.0×10^{-7}. Hence, it may be concluded that metal emissions from the Hamilton-Wentworth facility present no risks of concern.

Risk Characterization of Combined Landfilling and Incineration Option

Part of dry cell batteries present in MSW may be landfilled, while the remainder is incinerated. Tables 7.36 and 7.37 respectively, show the exposure assessment and risk characterization of a management option that involves both the incineration and landfilling of used household batteries. HI values of 1.9 and 0.4 were estimated for the child and adult populations, respectively. Also, carcinogenic risks of 2.7×10^{-3} and 6.1×10^{-4} were estimated for the child and adult groups, respectively. By comparing these numbers with an acceptable HI of 1 and an acceptable carcinogenic risk range of 1.0×10^{-4} to 1.0×10^{-7}, it is noted that risks of concern are presented due to cadmium (from Ni-Cads) being incinerated. In reality, these metals will appear in diluted forms due to mixing with large volumes of MSW. Concentrated amounts of Ni-Cads are less likely to be found. This means that the levels of risks estimated here will actually be lower. Nonetheless, it is recommended that Ni-Cads be landfilled or recycled if in concentrated forms.

Risk Comparisons for Used Household Battery Disposal Alternatives

It is necessary to determine the extent to which risks can be reduced, as well as the pathway for implementation of risk reduction policies. Table 7.38 summarizes the relative level of risks associated with the preferred disposal alternatives for used household batteries. Theoretically, incineration of the batteries of concern in this study will present the greatest risks; in practice, mixed with MSW, these batteries may safely be incinerated with MSW without any significant risks. Landfilling of the dry cell batteries with MSW will generally present no significant risks of concern. Although the recycling of the household batteries has not been quantified, the qualitative indicators are that it is not the best disposal option for the alkaline

Table 7.33 Average Stack Gas Metal Emission Data and Dispersion Modeling Results for the Chemicals of Concern

	Average Stack Gas Metals Emission Data			Dispersion Modeling Results			
Chemical of Concern	Actual Concentration (mg/m^3)	Emission Rate (mg/s)	Average Baghouse Efficiency for Metal Removal (%)[a]	Emission Rate (mg/s)	Maximum Ground-Level Impingement Concentration (mg/m^3)	Allowable Concentration (mg/m^3)	Percent of Allowable Concentration (%)
Cadmium	0.011	0.42	98.3	0.84	0.0019	5	0.04
Manganese	0.048	1.92	98.5	3.84	0.0087	30	0.03
Mercury	0.026	1.05	91.6	2.10	0.0048	5	0.10
Nickel	0.028	1.11	88./4	30.44	0.0691	5	1.38
Zinc	0.312	12.43	98.7	24.86	0.0564	100	0.06

Source: "Review of Emission Studies at the SWARU Facilities, Hamilton," A Report to Tricil, Ltd., Hamilton, Prepared by ORTECH International, Mississauga (August 1990).

[a] Average baghouse efficiency for particulate removal estimated at about 98%.

Table 7.34 Exposure Assessment: CDIs from Incinerator Releases (Hamilton SWARU Facility, Hamilton, Ontario)

Modeling Assumptions

1. Potential receptors are potentially exposed via inhalation of particulates from stack emissions.
2. Child average weight is 16 kg and adult average weight is 70 kg; these potential receptors may be impacted via inhalation of air emissions.
3. Inhalation rates assumed to be 0.25 m^3/h and 0.83 m^3/h for child and adult, respectively; retention rate of inhaled air is conservatively taken to be 100% for 365 days/year.

Chemical of Concern	Absorption Factor (%)	Exposure Concentration (mg/m^3)	CDI — Child (mg/kg/day)		CDI — Adult (mg/kg/day)	
			Carcinogenic Effects	Noncarcinogenic Effects	Carcinogenic Effects	Noncarcinogenic Effects
Cadmium (Cd)	100	1.90E – 06	2.55E – 08	3.57E – 07	2.24E – 07	2.70E – 07
Manganese (Mn)	100	8.70E – 06	1.17E – 07	1.64E – 06	1.03E – 06	1.24E – 06
Mercury (Hg)	100	4.80E – 06	6.43E – 08	9.02E – 07	5.66E – 07	6.82E – 07
Nickel (Ni)	100	6.91E – 05	9.26E – 07	1.30E – 05	8.15E – 06	9.81E – 06
Zinc (Zn)	100	5.64E – 05	7.56E – 07	1.06E – 05	6.66E – 06	8.01E – 06

Table 7.35 Risk Characterization For Incinerator Releases (Hamilton SWARU Facility, Hamilton, Ontario)

Battery Constitutent	Potential Receptor Exposure via Inhalation of Incinerator Releases (Carcinogenic Effects) (mg/kg/day)		Potential Receptor Exposure Via Inhalation of Incinerator Releases (Noncarcinogenic Effects) (mg/kg/day)		Inhalation RfD (mg/kg/day)	Inhalation SF (1/[mg/kg/day])	HI for Child Exposure	HI for Adult Exposure	Carcinogenic Risks for Child Exposure	Carcinogenic Risks for Adult Exposure
	Child	Adult	Child	Adult						
Cadmium (Cd)[a]	2.55E − 08	2.24E − 07	3.57E − 07	2.70E − 07	5.00E − 04	6.10E + 00	0.00	0.00	1.55E − 07	1.37E − 06
Manganese (Mn)	1.17E − 07	1.03E − 06	1.64E − 06	1.24E − 06	2.00E − 01	—	0.00	0.00	0.00E + 00	0.00E + 00
Mercury (Hg)	6.43E − 08	5.66E − 07	9.02E − 07	6.82E − 07	3.00E − 04	—	0.00	0.00	0.00E + 00	0.00E + 00
Nickel (Ni)	9.26E − 07	8.15E − 06	1.30E − 05	9.81E − 06	2.00E − 02	—	0.00	0.00	0.00E + 00	0.00E + 00
Zinc (Zn)	7.56E − 07	6.66E − 06	1.06E − 05	8.01E − 06	1.00E − 02	—	0.00	0.00	0.00E + 00	0.00E + 00
Total							0.0	0.0	1.55E − 07	1.37E − 06

[a] Cadmium is a B1 carcinogen (i.e., a probable human carcinogen) by the inhalation pathway only.

(manganese) and the zinc-carbon/zinc-chloride cells; Ni-Cad recycling programs may, however, be a worthwhile effort.

Conclusions and Recommendations

Waste disposal and management practices are shaped in part by federal, state, regional, and provincial regulation and legislation. Regulations governing waste disposal practices attempt to distinguish between hazardous and nonhazardous materials. Materials are determined to be hazardous based on a set of tests that examine their toxicity, flammability, explosivity, corrosivity, and/or infectiousness. Despite the toxicity of some of the composition of household batteries, dry cell batteries are themselves not affected by hazardous waste regulations, since all household wastes entering the MSW stream are generally classified as nonhazardous. Concerns about battery disposal practices stem from the possibility of hazardous materials/chemicals leaching from landfills or entering the atmosphere through incineration of MSW. On the one hand, the amount of household battery usage seems to be going up, which augments the concern about its impact when disposed together with MSW. On the other hand, some of the amounts of more toxic chemicals used in some of the batteries are going down and/or being substituted with potentially less toxic ones, thus minimizing potential impacts of the presence of dry cell batteries in MSW. Table 7.39 presents the recommended methods for the management of spent dry cell batteries. The preferred management option refers to the best available method of disposal that is recommended for use, whereas the alternative option is what can be called the second best method to adopt when necessary.

Most used household batteries become an integral part of the MSW stream. In practice, all solid waste is either landfilled or incinerated, with recycling becoming an integral component. Several conclusions and recommendations are drawn based on this investigation:

- The dry cell batteries investigated (i.e. the alkaline, zinc-carbon/zinc-chloride, and Ni-Cads) do not generally represent a concentrated source of heavy metals in MSW.
- There is no clear evidence to suggest that the codisposal of dry cell batteries with MSW via incineration or landfilling presents environmental or health problems.
- Risks to the environment from battery disposal by landfilling and incineration are not likely to be significant. Thus, with most household batteries may be safely disposed of in municipal landfills or municipal incinerators; Ni-Cads are better landfilled.
- At present, recycling is more likely to present significant risks. There appear to be significant health-related problems associated with the separate collection, storage, and disposal of most household batteries. Thus, with the current reduced levels of mercury in most primary cells (especially the alkaline and zinc-carbon/zinc chloride batteries), recycling of alkaline and zinc-carbon/zinc chloride cells is not necessary or needed. However, recycling for Ni-Cads may be a more viable and desirable measure to adopt.
- There should be a policy implemented that requires all municipal incinerators to be equipped with wet gas scrubbers. In that case, mercury emitted during combustion of household waste can then be removed. Also, since cadmium is carcinogenic by the inhalation pathway, it is crucial that adequate scrubbers are used in MSW incinerators that will capture as much of the fly ash as possible and to minimize the amounts that could eventually reach potential human receptors. In the absence of that, Ni-Cads which may be a great contributor of cadmium to MSW, may have to be removed from the wastes to be incinerated.

Table 7.36 Exposure Modeling and Evaluation Results Case Model III — Combined Landfilling and Incineration Disposal for MSW

Modeling Assumptions

1. All MSW generated in region are disposed of at municipal landfills and incinerators — 50% to each.
2. Conservatively assumed that up to 0.05% of all landfill waste will become leachate annually.
2. It is assumed that 30% of incinerator wastes go to fly ash, and 70% into bottom ash.
3. Assumed that pollution control efficiency of up to 99% attained, so that only 1% of the available fly ash is released through the stack into air.
4. All captured fly ash and bottom ash are sent to a hazardous waste landfill elsewhere.
5. Populations potentially impacted chosen to be the size of residential population in case region of incinerator location; population of case region is taken at 100,000 and that for the Province of Ontario approximated to 10 million.
6. Child average weight is 16 kg and adult average weight is 70 kg; these potential receptors may be impacted via ingestion of contaminated groundwater and/or via the inhalation of air emissions.
7. Contribution of metals (in household batteries) to MSW in case region is proportional to the population.

Battery Constituent	Total Annual Amount from Batteries into MSW Streams (National — million g)	Total Annual Amount from Batteries into MSW Streams (Ontario — g)	Total Annual Amount from Batteries into MSW Streams (Case Region 3 — g)	Total Annual Amount from Batteries into Landfill (Case Region 3 — mg)	Leachate	Potential Ingestion Intake (mg/person/year)	Potential Receptor Exposure Via Ingestion of Leachate Released into Ground Water (mg/kg/day)	
							Child	Adult
Cadmium (Cd)	4.73E + 01	1.70E + 07	1.70E + 05	8.52E + 07	4.28E + 04	4.28E − 01	7.29E − 05	1.67E − 05
Manganese (Mn)	1.61E + 03	5.60E + 08	5.80E + 06	2.90E + 09	1.45E + 06	1.45E + 01	2.48E − 03	5.67E − 04
Mercury (Hg)	1.04E + 00	3.80E + 05	3.80E + 03	1.90E + 06	9.50E + 02	9.50E − 03	1.63E − 06	3.72E − 07
Nickel (Ni)	5.07E + 01	1.83E + 07	1.83E + 05	9.13E + 07	4.54E + 04	4.56E − 01	7.81E − 05	1.79E − 05
Zinc (Zn)	8.09E + 02	2.91E + 08	2.91E + 06	1.46E + 09	7.28E + 05	7.28E + 00	1.25E − 03	2.85E − 04

Table 7.36 Exposure Modeling and Evaluation Results Case Model III — Combined Landfilling and Incineration Disposal for MSW (Continued)

Total Annual Amount from Batteries Into Incinerator (Case Region 3 — mg)	Bottom Ash (mg)	Incinerator Fly Ash Component (mg)	Total Fly Ash Captured by Pollution Control Devices (mg/yr)	Total Fly Ash Released Into Air (mg/yr)	Potential Inhalation Intake (mg/person/yr)	Potential Receptor Exposure Via Inhalation of Incinerator Releases Into Air (mg/kg/day)	
						Child	Adult
8.52E + 07	5.96E + 07	2.56E + 07	2.53E + 07	2.56E + 05	2.56E + 00	4.38E – 04	1.00E – 04
2.90E + 09	2.03E + 09	8.70E + 08	8.61E + 08	8.70E + 06	8.70E + 01	1.49E – 02	3.40E – 03
1.90E + 06	1.33E + 06	5.70E + 05	5.64E + 05	5.70E + 03	5.70E – 02	9.74E – 06	2.23E – 06
9.13E + 07	6.39E + 07	2.74E + 07	2.71E + 07	2.74E + 05	2.74E + 00	4.69E – 04	1.07E – 04
1.46E + 09	1.02E + 09	4.37E + 08	4.32E + 04	4.37E + 06	4.37E + 01	7.48E – 03	1.71E – 03

Table 7.37 Risk Characterization: Case Model III — Combined Landfilling and Incineration Disposal for MSW

Battery Constituent	Potential Ingestion Intake (mg/person/year)	Potential Receptor Exposure Via Ingestion of Leachate Releases into Ground-water (mg/kg/day)		Potential Inhalation Intake (mg/person/year)	Potential Receptor Exposure Via Inhalation of Incinerator Releases into Air (mg/kg/day)	
		Child	Adult		Child	Adult
Cadmium (Cd)[a]	4.26E – 01	7.29E – 05	1.67E – 05	2.56E + 00	4.38E – 04	1.00E – 04
Manganese (Mn)	1.45E + 01	2.48E – 03	5.67E – 04	8.70E + 01	1.49E – 02	3.40E – 03
Mercury (Hg)	9.50E – 03	1.63E – 06	3.72E – 07	5.70E – 02	9.76E – 06	2.23E – 06
Nickel (Ni)	4.56E – 01	7.81E – 05	1.79E – 05	2.74E + 00	4.69E – 04	1.07E – 04
Zinc (Zn)	7.28E + 00	1.25E – 03	2.85E – 04	4.37E + 01	7.48E – 03	1.71E – 03

[a] Cadmium is a B1 carcinogen (i.e., a probable human carcinogen) by the inhalation pathway only.

Table 7.37 Risk Characterization: Case Model III — Combined Landfilling and Incineration Disposal for MSW (Continued)

Oral RfD (mg/kg/day)	Oral SF (1/[mg/kg/day])	Inhalation RfD (mg/kg/day)	Inhalation SF (1/[mg/kg/day])	HI for Child Exposure	HI for Adult Exposure	Carcinogenic Risks for Child Exposure	Carcinogenic Risks for Adult Exposure
5.00E − 04	—	5.00E − 04	6.10E + 00	1.02	0.23	2.67E − 03	6.10E − 04
1.00E − 01	—	2.00E − 01	—	0.10	0.02	0.00E + 00	0.00E + 00
3.00E − 04	—	3.00E − 04	—	0.04	0.01	0.00E + 00	0.00E + 00
2.00E − 02	—	2.00E − 02	—	0.03	0.01	0.00E + 00	0.00E + 00
2.00E − 01	—	1.00E − 02	—	0.75	0.17	0.00E + 00	0.00E + 00
Total				1.9	0.4	2.67E − 03	6.10E − 04

[a] Cadium is a B1 carcinogen (i.e., a probable human carcinogen) by the inhalation pathway only.

Table 7.38 Risk Comparison for Disposal Alternatives

Disposal Option	Quantitative Risk Measure[a] HI	Carcinogenic Risk
Landfilling	0.4 (1.6)[b]	0.0 (0.0)
Incineration	3.5 (0.0)	5.3×10^{-3} (1.4×10^{-6})
Combined landfilling and incineration	1.9	2.7×10^{-3}

[a] Shows value for the most sensitive potential receptor, i.e., population indicating highest risk measure. Acceptable HI ≤ 1. Acceptable carcinogenic risk range is 10^{-4} to 10^{-7}, with 10^{-6} used as point of departure.

[b] Numbers in parentheses show values for typical/actual case studies for selected disposal options; these are represented by Waterloo Landfill Site (Waterloo) and Tricil SWARU incinerator facility (Hamilton), both in Ontario.

Table 7.39 Recommended Management Methods for Used Dry Cell Batteries

Battery Type	Preferred Management Option	Alternative Management Option	Comments
Alkaline (manganese)	Landfilling	Incineration	Neither landfilling nor incineration of even concentrated forms appear to present any significant risks
Zinc-carbon/ zinc-chloride	Landfilling	Incineration	Neither landfilling nor incineration of even concentrated forms appear to present any significant risks
Ni-Cads	Recycling	Landfilling	Separate collection and recycling of Ni-Cads preferred due to potential risks from Cd

In fact, batteries themselves are not the largest contributor to HHW, yet they are a ubiquitous and diffuse source of heavy metal waste. It is possible that managing their disposal efficiently could reduce the overall environmental threat posed by such metals. Indiscriminate policy decision aimed at all batteries, on the other hand, could be detrimental and would only result in ineffective and uneconomical programs at best and be potentially hazardous and environmentally unsound at worst.

7.4 CASE 4: DECISION BETWEEN REMEDIAL ACTIONS INVOLVING EXCAVATION OF SOILS

Air dispersion modeling results are particularly useful and necessary for predicting impacts from proposed remedial actions, since it often is not pragmatic to place air monitoring stations at actual offsite receptor locations of interest. However, it is necessary to characterize concentrations at such locations to be able to conduct a health and environmental impact assessment. Under such circumstances, dispersion patterns based on modeling results can be used to extrapolate concentrations monitored at a potentially contaminated site to offsite receptor locations. A screening assessment based on emission/dispersion modeling is performed to characterize hazardous air contaminants released from a facility.

7.4.1 Background Information and Objectives

Environmental site investigations have been conducted for a hypothetical property, ZZZ, located in southern U.S., which consists of several inactive wastewater evaporation ponds. This property is located within a residential and light industrial setting. The subject ponds have thick clay linings; it is therefore not expected that the underlying unconfined water supply aquifer has been impacted; this is confirmed by results of a field investigation. The waste ponds have periodically been dredged and sludge stockpiled on surrounding openland. The site assessment indicates the presence of several chemicals at the facility (Table 7.40).

To abate potential health risks from the ZZZ property, a number of remedial action alternatives to deal with the stockpiled waste have been proposed, including

- Transport for offsite disposal
- Landfilling onsite
- Stabilization by onsite treatment, and subsequent transport for offsite disposal

All these involve different degrees of excavation, loading, transporting, and unloading of wastes. A risk assessment is carried out to determine potential risks associated with each alternative proposed for the remediation of the stockpiled wastes at the site. Both short- (subchronic) and long-term (chronic) effects of the chemicals of concern are to be evaluated. The overall objective of the health risk assessment is to determine the magnitude and probability of actual or potential harm that the ZZZ site may pose to human health.

The critical and significant exposure pathway is determined to be due to wind erosion and mechanical resuspension of fugitive dust from contaminated soils in the evaporation ponds and waste stockpiles. The modeling procedure for fugitive dust emission that serves as input to the risk assessment process consists of a simplified air modeling methodology (U.S. EPA, 1985). This is a Gaussian dispersion model and comprises an analysis of aerodynamic fugitive dust particles 10 μm in size and less. The result of the fugitive dust modeling is the respirable concentration of each chemical that might produce exposures of potential receptors via inhalation. The quantification of exposure to wind-generated dust requires estimation of dust emissions and estimation of concentrations of particles downwind from the site of generation. The Gaussian model helps determine emission factors from the surface, based on inhalable particles <10 μm. These concentrations are further used in the risk characterization. The sequential steps for modeling of the fugitive dust dispersion are described further below.

Table 7.40 Data Evaluation for Selected Chemicals of Concern

Chemical of Concern	Soil Concentration Range (mg/kg)	Average Soil Concentration (mg/kg)
Antimony (Sb)	3–10	3.9
Arsenic (As)	1–7	1.6
Barium (Ba)	1.3–3600	523.0
Beryllium (Be)	0.3–0.6	0.3
Cadmium (Cd)	0.009–4.9	0.7
Chromium (Cr)	0.06–19	6.4
Cobalt (Co)	3–100	24.9
Copper (Cu)	0.06–18	8.4
Lead (Pb)	0.7–50	19.4
Mercury (Hg)	0.0006–0.4	0.2
Nickel (Ni)	3–110	31.8
Selenium (Se)	0.5–0.8	0.5
Silver (Ag)	0.21–4.9	1.5
Vanadium (V)	0.05–33	16.1
Zinc (Zn)	13–4680	740.0

7.4.2 Estimating/Modeling Particulate Emissions

The steps involved in estimating particulate emissions are (CDHS, 1986; U.S. EPA, 1985)

1. Determine soil particle size distribution. The particle size distribution mode is estimated at 0.01 mm for this site.
2. Determine the roughness height, Z_0, of the site terrain. This is the effective height associated with the surroundings which serve to dissipate the eroding winds. Based on the assumption that the critical receptor surroundings are suburban and residential, a value of $Z_0 = 5$ cm is assigned, using nomographs in the literature (e.g., CDHS, 1986; U.S. EPA, 1985)
3. Estimate the threshold friction velocity, U_f, and the threshold wind velocity, U_t. Using nomographs (CDHS, 1986; U.S. EPA, 1985), and based on particle size of 0.01 mm, U_f is determined to be 2.45 cm/sec. A further correcton factor is applied to the U_f value to account for nonerodible particles at the site (represented by soil particles of >1 cm). For each square meter of the site, about 10% of the area is estimated to be covered by nonerodible particles; using applicable nomographs (e.g., CDHS, 1986; U.S. EPA, 1985) based on the fraction of nonerodible particles, the ratio of corrected threshold friction velocity (U_{fc}) to the original uncorrected value (U_f) is

$$\frac{U_{fc}}{U_f} = 7.0 \qquad and \qquad U_{fc} = 17.15\, cm/s$$

4. Estimate the respirable particulate emission rates, E_{10}.
5. Estimate emission rate for chemicals of interest, R_{10}.
6. Determine the ambient air concentrations from the emission rates, using the Gaussian dispersion model; this gives an estimate of the ground-level ambient concentration.

It is conservatively assumed that no vegetation is present on site to limit erosion. Thus, the emission factor for 10-μm particles is estimated using the unlimited erosion potential model according to the following relationship (CDHS, 1986; U.S. EPA, 1985):

$$E_{10} = 0.036(1 - f_v)\left[\frac{U}{U_t}\right]^3 F(X)$$

where

- E_{10} = PM_{10} (i.e., particulate matter of 10 μm and less) emission factor, given as the annual average PM_{10} emission rate per unit area of contaminated surface (g/m^2-h)
- f_v = fraction of contaminated surface vegetative cover ($0 < f_v < 1$)
- U = mean annual wind speed (m/sec) (obtainable from literature, e.g., U.S. EPA, 1985, or from such sources as the U.S. National Oceanic and Atmospheric Administration, NOAA)
- U_t = usually U_7, which is value of wind speed at standardized elevation of 7 m, obtained from use of graphical relationships in the literature (e.g., U.S. EPA, 1985) in association with the corrected threshold friction velocity; also estimated from $U(z) / U_f = (1/0.4) \ln(Z/Z_0)$, where U(z) is the wind speed (m/sec) at height Z and Z is the height (cm) above surface (m/sec)
- F(X) = function of the ratio of threshold wind speed to mean annual wind speed, where $X = \{0.886 \times U_7/U\}$; value obtainable from standard nomographs in the literature (e.g., U.S. EPA, 1985) (dimensionless).

To evaluate short-term (subchronic) exposures, the worst-case erosion model is used, the emission factor of which is estimated by

$$E_{10wc} = 0.036(1 - f_v)(U_{6h})^3$$

where U_{6hr} is the expected maximum 6-h mean speed during the year (e.g., U.S. EPA, 1985) (m/sec).

The emission factor for mechanical agitation and resuspension of fugitive dust from vehicular traffic is calculated from the following equation (CDHS, 1986; U.S. EPA, 1985):

$$E_{10m} = 0.85\left(\frac{s}{10}\right)\left(\frac{V}{24}\right)0.8\left(\frac{W}{7}\right)0.3\left(\frac{N_w}{6}\right)1.2\left(\frac{[365\text{-}p]}{365}\right)$$

where

s = silt content of road surface material (%) — estimated at 50%
V = mean vehicle speed (km/h) — estimated at 10 km/h
W = mean vehicle weight (Mg) — estimated at 3, 15, and 26 Mg for 4-, 6-, and 10-wheeled vehicles, respectively
N_w = mean number of wheels for vehicles
p = number of days with at least 0.25 mm of precipitation per year (days) — conservatively estimated as 5 days

Contaminant Emission Rates

The contaminant emission rate is subsequently calculated for each of the emission factors. The following equation is used for the unlimited and worst-case models and modified for mechanical agitation and resuspension from vehicular traffic same:

$$R_{10} = (\alpha)(E_{10})(A)$$

where

R_{10} = the emission rate of contaminants as particulate matter of 10 μ and less, PM_{10}
α = mass fraction of contaminant/chemical in PM_{10} emissions, given by the ppm concentration of the chemical of concern multiplied by 10^{-6} g/gram of soil
A = source extent, represented by the contaminated area (m^2); for mechanical resuspension, this is a specified averaging time, given by the vehicle kilometers traveled per year, A_m, estimated by

$$A_m = (trips / month)(vehicles / trip)(meters / vehicle)(km / 1000\ m)$$
$$(12\ month / 1\ year)$$

The emission rates are converted into scaling factors according to:

$$Q_I = \frac{(R_{10})}{(P_R)} \quad or \quad Q_{II} = \frac{(R_{10m})}{(P_R)}$$

where

Q_I = the annual wind erosion scaling factor (for erosion values)
Q_{II} = the annual mechanical resuspension scaling factor

P_R = a parameter reflecting the percent of time that a defined climatic region has conditions promoting wind erosion — estimated at 0.152 for case region (*Source*: U.S. National Oceanic and Atmospheric Administration).

Determination of Respirable Concentrations

Finally, the annual scaling factor (in grams per second) can be converted into respirable concentrations (in micrograms per cubic meter) at various selected receptor locations from the source, using a matrix of unscaled concentrations available from the literature (e.g., U.S. EPA, 1985). The respective annual scaling factor for each contaminant is multiplied by the unscaled concentration to yield the modeled ambient contaminant concentration at the critical receptor locations. The respirable concentrations calculated for wind erosion and for mechanical resuspension are added to obtain the total respirable concentration of each chemical at selected receptor location impacted. A wind rose (i.e., a diagram or pictorial representation that summarizes pertinent statistical information about wind speed and direction at a specified location) for a nearby weather monitoring station is used to determine the predominant wind direction, which helps to identify impacted regions or sections of study area. A spreadsheet was set up for this computation and summary results are shown in Table 7.41.

7.4.3 Risk Characterization

Risks are calculated and summed for all of the chemicals of concern based on the ground-level concentration (GLC, in micrograms per cubic meter). Four different receptor groups, consisting of adult workers, adult residents, resident children between 6 and 12 years, and resident children under 6 years, are evaluated. Three exposure pathways are analyzed for each of the receptor groups:

- Particulate inhalation from fugitive dust
- Incidental and pica ingestion of soils
- Skin absorption from dermal contacts with soil

Risk parameters estimated for all categories of receptor groups and exposure parameters are

- Carcinogenic risks for carcinogens
- Chronic and subchronic HIs for noncarcinogenic effects of all chemicals

Table 7.42 summarizes results for the total carcinogenc risks and the chronic and subchronic HIs associated with the various receptor groups; this is obtained by applying the relevant equations presented in Appendix B. The subchronic HIs are based on exposures to the maximum concentrations of the chemicals present at the ZZZ site.

Offsite disposal scenario — For all receptor groups, risks resulting from exposure to known or suspected carcinogens present at the ZZZ site are within the acceptable range of 10^{-4} to 10^{-7}. Thus, no unacceptable carcinogenic risks are

Table 7.41 Summary of Soil Respirable Chemical Exposure Concentrations from the Air Dispersion Modeling for the ZZZ Site

Chemical of Concern	Maximum Soil Concentration (ppm)	Average Soil Concentration (ppm)	Offsite Disposal Option (mg/m^3)		OnsiteLandfilling Option (mg/m^3)		Onsite Treatment/ Offsite Disposal (mg/m^3)	
			Chronic	Subchronic	Chronic	Subchronic	Chronic	Subchronic
Antimony (Sb)	1.00E + 01	3.90E + 00	1.95E – 07	3.20E – 05	1.39E – 07	3.20E – 05	1.39E – 07	3.20E – 05
Arsenic (As)	7.00E + 00	1.64E + 00	8.13E – 08	1.35E – 05	5.86E – 08	1.34E – 05	5.86E – 08	1.34E – 05
Barium (Ba)	3.60E + 03	5.23E + 02	2.59E – 05	4.30E – 03	1.87E – 05	4.29E – 03	1.87E – 05	4.29E – 03
Beryllium (Be)	6.00E – 01	3.40E – 01	1.68E – 08	2.79E – 06	1.21E – 08	2.79E – 06	1.21E – 08	2.79E – 06
Cadmium (Cd)	4.90E + 00	7.30E – 01	3.62E – 08	5.99E – 06	2.61E – 08	5.98E – 06	2.61E – 08	5.98E – 06
Chromium (Cr)	1.90E + 01	6.44E + 00	3.19E – 07	5.29E – 05	2.30E – 07	5.28E – 05	2.30E – 07	5.28E – 05
Cobalt (Co)	1.00E + 02	2.49E + 01	1.24E – 06	2.05E – 04	8.91E – 07	2.04E – 04	8.91E – 07	2.04E – 04
Copper (Cu)	1.80E + 01	8.44E + 00	4.18E – 07	6.93E – 05	3.02E – 07	6.92E – 05	3.02E – 07	6.92E – 05
Lead (Pb)	5.00E + 01	1.94E + 01	9.64E – 07	1.60E – 04	6.95E – 07	1.59E – 04	6.95E – 07	1.69E – 04
Mercury (Hg)	4.00E – 01	2.00E – 01	9.91E – 09	1.64E – 06	7.15E – 09	1.64E – 06	7.15E – 09	1.64E – 06
Nickel (Ni)	1.10E + 02	3.18E + 01	1.57E – 06	2.61E – 04	1.14E – 06	2.60E – 04	1.14E – 06	2.60E – 04
Selenium (Se)	8.00E – 01	5.00E – 01	2.48E – 08	4.10E – 06	1.79E – 08	4.10E – 06	1.79E – 08	4.10E – 06
Silver(Ag)	4.90E + 00	1.47E + 00	7.28E – 08	1.21E – 05	5.25E – 08	1.20E – 05	5.25E – 08	1.20E – 05
Vanadium (V)	3.30E + 01	1.61E + 01	7.99E – 07	1.32E – 04	5.76E – 07	1.32E – 04	5.76E – 07	1.32E – 04
Zinc (Zn)	4.68E + 03	7.40E + 02	3.67E – 05	6.08E – 03	2.64E – 05	6.06E – 03	2.64E – 05	6.06E – 03

Table 7.42 Risk Characterization Summary

Population Group	Risk Class	Scenario/Action		
		OffsiteDisposal	Onsite Landfilling	Onsite Treatment/ Offsite Disposal
Child <6 years	Carcinogenic Risk (CR)	3.30E – 05	3.30E – 05	3.30E – 05
	Chronic Hazard Index (CHI)	2	2.02	2.02
	Subchronic Hazard Index(SHI)	0.8	0.8	0.8
Child 6–12 years	Carcinogenic Risk (CR)	1.20E – 05	1.20E – 05	2.00E – 05
	Chronic Hazard Index (CHI)	1.1	1.1	1.1
	Subchronic Hazard Index(SHI)	0.6	0.6	0.6
Adult resident	Carcinogenic Risk (CR)	5.00E – 05	6.00E – 05	6.00E – 05
	Chronic Hazard Index (CHI)	0.38	0.36	0.4
	Subchronic Hazard Index(SHI)	0.33	0.33	0.3
Adult worker	Carcinogenic Risk (CR)	4.00E – 05	4.00E – 05	4.00E – 05
	Chronic Hazard Index (CHI)	0.3	0.2	0.2
	Subchronic Hazard Index(SHI)	0.1	0.1	0.1

indicated due to choice of this remedial alternative. However, noncarcinogenic risks are indicated for some receptor groups with respect to chronic exposure, but not for subchronic exposures. Since the remedial process will be conducted over a short time period, the subchronic HI is the preferred indicator to use in making decisions here.

Onsite landfilling — For all receptor groups, risks resulting from exposure to known or suspected carcinogens present at the ZZZ site are within the acceptable range of 10^{-4} to 10^{-7}. Thus, no unacceptable carcinogenic risks are indicated due to choice of this remedial alternative. However, noncarcinogenic risks are indicated for some receptor groups with respect to chronic exposure, but not for subchronic exposures. Since the remedial process will be conducted over a short time period, the subchronic HI is the preferred indicator to use in making decisions here.

Onsite treatment/offsite disposal scenario — For all receptor groups, risks resulting from exposure to known or suspected carcinogens present at the ZZZ site are within the acceptable range of 10^{-4} to 10^{-7}. Thus, no unacceptable carcinogenic risks are indicated due to choice of this remedial alternative. However, noncarcinogenic risks are indicated for some receptor groups with respect to chronic exposure, but not for subchronic exposures. Since the remedial process will be conducted over a short time period, the subchronic HI is the preferred indicator to use in making decisions here.

It may be concluded that since chronic hazards are indicated for some receptor groups, the site need to be remediated and dust control measures should be used during remediation. Although no clear-cut, risk-based distinction can be made from this risk assessment, a more refined evaluation may indicate distinct variations in the levels of risk; else non-risk-based factors such as costs, etc. will form the basis for selecting between these alternatives.

7.5 CASE 5: EVALUATION OF TRANSPORTATION RISKS IN AN ACCIDENT CORRIDOR

To illustrate the potential application of PRA methodologies in hazardous waste management, a very simplified evaluation scheme is presented concerning transportation risks in an accident corridor. Consider, for instance, the transport of hazardous wastes 500 km in a specially equipped tanker/truck. About 200 shipments of such wastes are undertaken in a year. Accident frequencies have been evaluated; in the event of an accident along the transportation corridor, the container may be damaged. In such circumstances, the severity of public exposure to chemicals depends on several factors such as meteorological conditions, the likelihood of containment breach, extent of release due to spillage, and the dispersion pattern of released chemicals. The transportation risk data for the hypothetical case are as shown in Table 7.43 Based on these information, the following calculations are made:

$$\textit{Average annual number of trucks damaged}$$
$$= \left(2.0 \times 10^{-7}\right) \times (200) \times (500) = 2.0 \times 10^{-2}$$

Table 7.43 Hazardous Waste Transportation Risk and Health Impacts Data

Parameter	Risk Value	Population Impacted	Exposure Probability (EP)[a]
Average number of trucks damaged	2.0×10^{-7}/truck km	—	—
Release likelihood from accident	0.4	—	—
Release severity index:			
(1) Small (limited)	0.5	0	0.50
(2) Medium (average)	0.3	500	0.07
(3) Large (extensive)	0.1	1000	0.02

[a] EP refers to the likelihood of exposures to lethal concentrations of chemicals released into environment.

Table 7.44 Health Impact Data for F-N Relationship

Population Size	Average Annual Exposure Frequency
0	2.00×10^{-3}
500	1.68×10^{-4}
1000	1.60×10^{-5}

and,

(1) *Average annual frequency of release of limited severity from accident*

$$= \left(2.0 \times 10^{-2}\right) \times (0.4) \times (0.5) = 4 \times 10^{-3}$$

Average annual frequency of exposure

$$= \left(4 \times 10^{-3}\right) \times (0.5) = 2 \times 10^{-3}$$

(2) *Average annual frequency of release of medium severity from accident*

$$= \left(2.0 \times 10^{-2}\right) \times (0.4) \times (0.3) = 2.4 \times 10^{-3}$$

Average annual frequency of exposure

$$= \left(2.4 \times 10^{-3}\right) \times (0.07) = 1.68 \times 10^{-4}$$

(3) *Average annual frequency of release of extensive severity from accident*

$$= \left(2.0 \times 10^{-2}\right) \times (0.4) \times (0.1) = 8 \times 10^{-4}$$

Average annual frequency of exposure

$$= \left(8 \times 10^{-4}\right) \times (0.02) = 1.60 \times 10^{-5}$$

Potential health impacts of any toxic releases depends on the severity of releases, meteorological conditions, as well as the population density along the transportation corridor. Table 7.44 summarizes these results, and Figure 7.3 is a

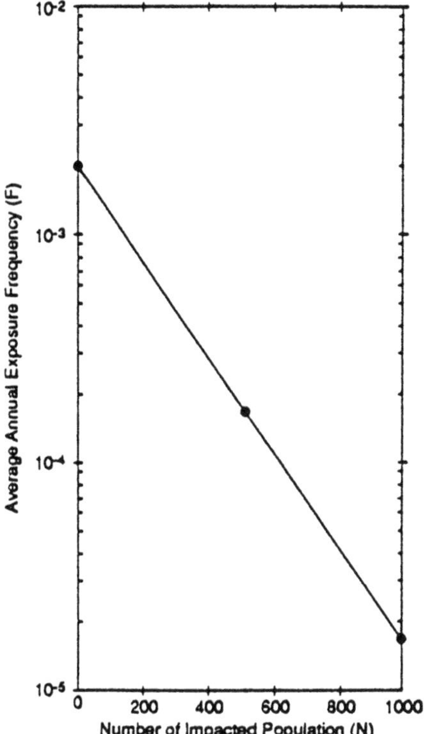

Figure 7.3 F-N/risk curve for hazardous waste transportation for a hypothetical example.

risk curve showing a plot of the annual frequency of exposure vs. the number of people affected; these results can be used in management decisions with respect to mode of transportation and traffic network design for the transportation of hazardous materials. Additionally, the average individual lifetime exposure probability can be estimated based on the average human life expectancy (say, 70 years), and these results compared to other activities that populations in the community are exposed to; subsequently, appropriate risk management and risk prevention programs can be developed to mitigate potential problems.

7.6 REFERENCES

Asante-Duah, D. K., S. P. Sayko, and R. E. Lees. "A Case Model for Assessment of Contaminant Concentration Leaching into an Aquifer," in Proc. of Hydrology Days Conf., Fort Collins, CO (April 1989).

Balfour, R. L. "Household and Other Batteries: Source Reduction and Recycling," presented at the 9th Annu. Resource Recovery Conf. of the U. S. Conference of Mayors, Washington, DC (1990).

Calabrese, E. J., et al. "How Much Soil Do Young Children Ingest: An Epidemiologic Study," *Regul. Toxicol. Pharmacol.* 10:123–137 (1989).

CAPCOA (California Air Pollution Control Officers Association). "Air Toxics Assessment Manual," California Air Pollution Control Officers Association, Draft Manual, August 1987 (amended, 1989), California (1989).

CDHS (California Department of Health Services). "The California Site Mitigation Decision Tree Manual," California Department of Health Services, Toxic Substances Control Division, Sacramento, (1986).

Christensen, T. H., R. Cossu, and R. Stegmann, Eds. "Sanitary Landfilling: Process, Technology and Environmental Impact," (London: Academic Press, 1989).

Clancy, K. M. and A. A. Jennings. "Environmental Verification of Multitransport Groundwater Contamination Predictions," *Water Resour. Bull.* 24(2):307–316 (1988).

CMU (Carnegie Mellon University). *Household Batteries: Is There a Need for Change in Regulation and Disposal Procedure?*, (Pittsburgh, PA: Carnegie Mellon University, 1989).

CRA. "Waterloo Landfill Site Hydrogeologic Investigation (Final Report, June 1990)," Prepared for the Regional Municipality of Waterloo (1990a).

CRA. "Waterloo Landfill Site Progress Report," Report to the MOE, Hamilton (1990b).

Denison, R. A. *The Hazards of Municipal Incinerator Ash and Fundamental Objectives of Ash Management*, (Washington, DC: Environmental Defense Fund, 1987).

DOE (U.S. Department of Energy). "The Remedial Action Priority System (RAPS): Mathematical Formulations," U.S. Dept. of Energy, Office of Environment, Safety, and Health, Washington, DC (1987).

Domenico, P. A. and V. V. Palciauskas. "Alternative Boundaries in Solid Waste Management," *Groundwater* 20(3):303–311 (1982).

Environment Canada. "The National Incinerator Testing and Evaluation Program: Air Pollution Control Technology," Report EPS 3/UP/2, Environment Canada, Ottawa (1986).

Environment Canada. "Used Batteries and the Environment: A Study on the Feasibility of Their Recovery," Report: EPS 4/CE/1 (May), Prepared by Eutrotech Inc. for Technology Development Branch, Environment Canada, Ottawa (1991).

Erickson, A. J. "Aids for Estimating Soil Erodibility — 'K' Value Class and Tolerance," U.S. Dept. of Agriculture, Soil Conservation Service, Salt Lake City, UT (1977).

Evans, L. J. "Chemistry of Metal Retention by Soils," *Environ. Sci. Technol.* 23(9):1047–1056 (1989).

Fochtman, E. G. and W. R. Haas. "Relationship of Spent Dry Batteries to the Heavy Metal Content of Solid Wastes," 2nd Natl. Conf. on Water Use, Chicago, May 4 to 8 (1975).

Forker, T. *Strategic Approaches to the Used Household Battery Problem. A Report on European Experiences and Their Implication for Action in the United States* (1989).

Goldman, S. J., K. Jackson, and T. A. Bursztynsky. *Erosion and Sediment Control Handbook* (New York: McGraw-Hill, 1986).

Goldstein, G. "Rechanneling the Waste Stream," *Mech. Eng. Mag.* August (1989).

Greghian, A. B., D. S. Ward, and R. W. Cleary. "A Finite Element Model for the Migration of Leachate from a Sanitary Landfill in Long Island, New York. Part 1. Application," *Water Resour. Bull.* 17(1):62–65 (1981).

Griffin, R. A. and N. F. Shimp. "Attenuation of Pollutants in Municipal Landfill Leachate by Clay Minerals," PB-287 140/8ST, August 1978, National Technical Information Service, U.S. Dept. of Commerce, Springfield, VA (1978).

Hawley, J. K. "Assessment of Health Risk from Exposure to Contaminated Soil," *Risk Anal.* 5(4):289–302 (1985).

Haith, D. A. "A Mathematical Model for Estimating Pesticide Losses in Runoff," *J. Environ. Qual.* 9(3):428–433 (1980).

Hughes, M. K. "Effect of Heavy Metal Pollution on Plants," *Cycling Trace Metals Ecosyst.* 2 (1981).

IRR (Institute for Risk Research). "Assessing the Environmental Effects of Disposal Alternatives for Household Batteries," Research Report, IRR, (Ontario: University of Waterloo, 1992).

Jones, C. J., P. J. McGugan and P. F. Lawrence. "An Investigation of the Degradation of Some Dry Cell Batteries Under Domestic Waste Landfill Conditions," *J. Hazard. Mater.* 2:259–289. (1977–1978).

Kellermeyer, D. A. "Quantitative Health Risk Assessment for Incinerator Ash Disposal," Paper presented at MSW Incineration Ash Conf., Orlando, FL, December (1989).

Kelly, H. G. "Pilot Testing for Combined Treatment of Leachate from a Domestic Waste Landfill Site," *Water Poll. Cont. Fed. J.* 59(5):254–261 (1987).

Kemper, J. M. and R. B. Smith. "Leachate Production by Landfill Processed Municipal Wastes," in Land Disposal Municipal Solid Waste. Proc. of the 7th Annu. Research Symp., EPA-600/9-81-0029, U.S. EPA, Cincinnati, OH (1981), pp. 18–36.

Kineman, R. and D. Natini. "Hazardous Household Waste in Sanitary Landfills," Fourth Annu. Conf. on Solid Waste Management and Materials Policy, January 30, (1988).

Klaassen, C. D., M. O. Amdur and J. Doull. *Casarett and Doull's Toxicology: The Basic Science of Poisons* (New York: Macmillan Publishing, 1986).

Lepow, M. L., et al. "Role of Airborne Lead in Increased Body Burden of Lead in Hartford Children," *Environ. Health Perspect.* 6:99–100 (1974).

Lepow, M. L., et al. "Investigation into Sources of Lead in the Environment of Urban Children," *Environ. Res.* 10:414–426 (1975).

Lindsay, W. L. *Chemical Equilibria in Soils* (New York: John Wiley & Sons, 1979).

Marcus, W. L. "Lead Health Effects in Drinking Water," *Toxicol. Ind. Health* 2(4):363–407 (1986).

Martin, W. H. and P. J. Coughtrey. Impact of Metals on Ecosystem Function and Productivity, vol. 2, Effect of Heavy Metal Pollution on Plants.

Miller, C. J. and M. Mishra, "Modeling of Leachate Through Cracked Clay Liners. 1. State of the Art," *Water Resour. Bull.* 25(3):551–555 (1989).

Mills, W. B., Dean, J. D., Porcella, D. B., et al. Water Quality Assessment: A Screening Procedure for Toxic and Conventional Pollutants: Parts 1, 2, and 3, Environmental Research Laboratory, Office of Research and Development, EPA-600/6-82/004 a. b. c., U.S. Environmental Protection Agency, Athens, GA (1982).

Mitchell, J. K. and G. D. Bubenzer. "Soil Loss Estimation," in *Soil Erosion*, Kirby, M. J. and R. P. C. Morgan, Eds. (New York: John Wiley & Sons, 1980).

Mockus, J. "Estimation of Direct Runoff from Storm Rainfall," in National Engineering Handbook, Section 4: Hydrology, U.S. Department of Agriculture Soil Conservation Service, Washington, DC (1972).

MPCA (Minnesota Pollution Control Agency). "Household Battery Recycling and Disposal Study," Report prepared by K. Arnold (St. Paul, MN: MPCA, 1991).

Oda, S. "The Disposal of Ni-Cd Batteries in Landfills and the Effect of Cadmium on the Human System," in Proc. of the 1st Int. Seminar on Battery Waste Management, Deerfield Beach, FL (1989).

Page, G. W. and M. Greenberg. "Maximum Contaminant Levels for Toxic Substances in Water: A Statistical Approach," *Water Resour. Bull.* 18(6):955–962, (1982).

Porter, J. W. "The Solid Waste Dilemma: An Agenda for Action," Final Report of the Municipal Solid Waste Task Force, February, (1989).

Rader, W. A. and J. E. Spaulding. Regulatory Elements of Trace Elements in the Environment, vol. 2 of Toxicity of Heavy Metals in the Environment, (New York).

Russell, G. M., M. Stewart, and A. L. Higer. "Examples of Landfill-Generated Plumes in Low Relief Areas, Southeastern Florida," *Water Resour. Bull.* 23(5):863–866 (1987).

Sax, N. I. and R. J. Lewis, Sr. *Hawley's Condensed Chemical Dictionary* (New York: Van Nostrand Reinhold, 1987).

Sedman, R. "The Development of Applied Action Levels for Soil Contact: A Scenario for the Exposure of Humans to Soil in a Residential Setting," *Environ. Health Perspect.* 79:291–313 (1989).

Shields, E. J. *Pollution Control Engineer's Handbook* (Des Plaines, IL: Cahners, 1985).

Tchobanoglous, G., H. Theisen, and R. Eliassen. *Solid Wastes: Engineering Principles and Management Issues* (New York: McGraw-Hill Inc., 1977).

Ujihara, A. M. and M. Gough. "Managing Ash from Municipal Waste Incinerators," A report, (Washington, DC: Center for Risk Management, Resources for the Future, 1989).

USBR (U.S. Bureau of Reclamation). "Design of Small Dams," U.S. Dept. of the Interior, Bureau of Reclamation, U.S. Govt. Printing Office, Washington, DC (1977).

U.S. EPA. "Hazardous Waste Land Treatment," Office of Solid Waste and Emergency Response, SW-874, Washington, DC (1983).

U.S. EPA. "Rapid Assessment of Exposure to Particulate Emissions From Surface Contamination Sites," EPA/600/8-85/002, NTIS PB85-192219, Washington, DC (1985).

U.S. EPA. "Superfund Exposure Assessment Manual," Report No. EPA/540/1-88/001, OSWER Directive 9285. 5-1, U. S. EPA, Office of Remedial Response, Washington, DC (1988a).

U.S. EPA. "The Solid Waste Dilemma: An Agenda for Action," Report of the Municipal Solid Waste Task Force, Office of Solid Waste,PB88-251145, Washington DC (1988b).

U.S. EPA. "Guidance on Metals and Hydrogen Chloride Controls for Hazardous Waste Incinerators," Office of Solid Waste, Washington, DC (1989a).

U.S. EPA. "Risk Assessment Guidance for Superfund, Vol. I: Human Health Evaluation Manual (Part A)," EPA/540/1-89/002, Office of Emergency and Remedial Response, Washington, DC (1989b).

U.S. EPA. "*Exposure Factors Handbook,*" EPA/600/8-89/043 (1989c).

U.S. EPA. "Characterization of Municipal Waste Combustion Ash, Ash Extracts, and Leachates," EPA 530-SW-90-029A. Office of Solid Waste and Emergency Response (0S-305), Washington, DC (1990).

Vogg, H., H. Braun, M. Metzger and J. Schneider. "The Specific Role of Cadmium and Mercury in Municipal Solid Waste Incineration," *Waste Manage. Res.* 4:65–74 (1986).

Walton, W. C. 1984. Practical Aspects of Groundwater Modeling. National Water Well Association.

Williams, J. R. "Sediment-Yield Prediction with the Universal Equation Using Runoff Energy Factor," in Present and Prospective Technology for Predicting Sediment Yields and Sources, U.S. Department of Agriculture, ARS-S-40, Washington, DC (1975).

Yakowitz, H. (1986). Fate of Small Quantities of Hazardous Wastes. OECD Environment Monographs.

Yakowitz, H. (1987). Waste Management Activities in Selected Industrialized Countries. Preliminary Report for the OECD, Paris.

CHAPTER 8

Epilogue

Some element of risk exists in all technological developments and activities, and these risks must be assessed and courses of action decided so as to minimize any consequences that are attributed to such risks. The bewildering array of risks caused by hazardous wastes has motivated society to develop systematic tools that will bring the situation under control less expensively. In recent years, focus seems to be on the use of risk assessment techniques to provide a structured, systematic framework for the evaluation of hazardous waste management programs. The development of a structured risk assessment framework will facilitate systematic decision making for the protection of public health and the environment from hazardous waste problems. Such a structure will provide an effective way to build a comprehensive and technically defensible information base for tackling potential health and environmental hazards.

The risk assessment process is intended to give the risk management team an effective tool for using scientific data to reach justifiable and defensible decisions on a wide range of issues. Risk assessment has become an important decision-making tool because it is a systematic, generally reproducible method of digesting and analyzing large bodies of complex information using a uniform set of rules. The concepts and techniques presented in this text should help managers of hazardous waste sites and/or facilities carry out hazardous waste management tasks effectively.

8.1 A RECAPITULATION

The primary objective of risk appraisal is the assessment of whether existing or future receptors are, or will be at risk of being harmed by exposure to potentially hazardous situations. This evaluation then serves as a basis for developing mitigation measures in a risk management and risk prevention program. Risk assessment will help define the level of risk as well as set performance goals for various response alternatives. The application of risk assessment can provide for prudent and technically feasible and scientifically justifiable decisions about corrective actions that will help protect public health and the environment in a most cost-effective manner.

The assessment of health and environmental risks play an important role in the RI/FS, the RAP development, and also the risk mitigation and risk management strategies in hazardous waste management. In particular, to reduce costs in planning for environmental cleanup of contaminated sites, it is important that the decision-making process involved be well defined. A major consideration in developing a RAP for a contaminated site is the level of cleanup to be achieved, i.e., "how clean is clean?" This could be the driving force behind remediation costs. It will therefore be prudent to allocate adequate resources for developing the appropriate cleanup criteria. The recommended site cleanup limit (RSCL) concept presented in this text will facilitate decisions as to the effective use of limited funds to clean a site to a level appropriate/safe for its intended use. In principle, the cleanup criteria selected for a potentially contaminated site may vary significantly from site to site due to the site-specific parameters. Similarly, mitigation measures may be case specific for various hazardous facilities and problems.

8.2 CONCLUDING STATEMENTS AND RECOMMENDATIONS

The following observations and recommendations are made in regards to the application of risk assessment to hazardous waste site management:

- It is important to adequately characterize the exposure and physical settings of a problem situation to allow for an effective application of appropriate risk assessment methods of approach.
- There are several complexities involved in real-life scenarios that are unique in characterizing hazardous waste sites and facilities. Careful development of exposure scenarios, guided by the use of the event tree structure is therefore encouraged.
- The populations potentially at risk from hazardous waste problems are usually heterogeneous; this can greatly influence the anticipated impacts/consequences. Critical receptors should therefore be carefully identified with respect to numbers, location (areal and temporal), sensitivities, etc., so that risks are neither underestimated or conservatively overestimated.
- Performing risk assessment to incorporate all scenarios envisaged rather than for the "worst-case" alone allows better comparison to be made between risk assessments performed by different scientists and analysts whose views on what represents a "worst case" may be very subjective and therefore may vary significantly.
- Exposure scenarios and chemical transport models may contribute significant uncertainty in the risk assessment. Uncertainties, heterogeneities, and similarities should be identified and well documented throughout the risk assessment.
- Whenever possible, the synergistic, antagonistic, and potentiation (i.e., the case of a nonhazardous situation becoming hazardous due to combination with others) effects

of chemicals and other hazardous situations should be carefully evaluated for inclusion in the risk assessment decisions.

- It is prudent to assess what the "baseline" (no action) risks are for a potentially hazardous waste site and/or facility. This will give a reflection of what the existing situation is, which can then be compared against future improved situations.
- It is recommended here that an evaluation of the "postremediation" risks (i.e., residual risks remaining after the implementation of corrective actions) for a potentially hazardous waste site and/or facility be carried out for alternative mitigation measures. This will give a reflection of what the anticipated improved situation is, compared with the prior conditions at the site and/or facility.
- In view of the nature of tasks involved in a risk assessment, multidisciplinary teams will generally be needed to effectively carry out a comprehensive risk assessment, and this will be case specific. Professionals normally involved may range from individuals with background in air modeling, ecology, economics, engineering (chemical, civil, environmental, geotechnical, etc.), environmental sciences, epidemiology, geology, health, hydrology, meteorology, systems analysis, toxicology, etc.

To further improve the potential applications of risk assessment methodologies to hazardous waste site management programs, several multidisciplinary research initiatives may be required, including those relating to the following:

- Rather limited information exists on exposure patterns and toxicological effects for mixtures and/or combination of chemicals. It is important to direct more future research in the development of appropriate analytical models for these purposes. This should include improvements in the toxicological evaluations for chemicals commonly encountered at potentially contaminated sites in particular and that found in the environment in general; this will help minimize uncertainties associated with the toxicity assessment during the dose-response modeling in a chemical risk assessment.
- In view of the several uncertainties that appear to surround risk assessment, more future research should consider investigations on the use of Monte Carlo simulation techniques for performing sensitivity and uncertainty analyses in the various stages, areas, and levels of risk assessment. This will ultimately help decide what the most sensitive parameters are and their relative effects on the results of a given risk assessment.
- Application of stochastic methods to the exposure assessment, including the modeling of contaminant migration, exposure concentration levels, and intake/dose estimations will help improve the risk characterization outputs.
- Investigation of the synergistic, antagonistic, and potentiation effects of chemical mixtures commonly encountered at potentially contaminated sites will help arrive at more reliable risk estimates.
- It is important to advance research for use in the practical consideration of human and ecological receptor exposures, especially in regards to receptor absorption fractions of chemicals contacted. This should include the effects of weather conditions and related factors on the absorption and bioaccumulation of chemicals contacted by potential receptors.

In general, the benefits of risk assessment outweigh the disadvantages, but it must be recognized that this process will not be without tribulations. Indeed, risk assessment is by no means a panacea. Its use, however, does widen the decision maker's knowledge base and thus improve the decision-making capability. Risk assessment can produce more efficient and consistent risk reduction policies. It can also be used as a screening device for setting priorities. It is apparent that, some form of risk assessment is inevitable if corrective actions are to be conducted in a sensible and deliberate manner. Indeed, the U.S. EPA and several federal, state, and local agencies require some form of risk assessment in their risk management and risk prevention programs.

Risk assessment seems to be gaining greater grounds in making public policy decisions in the control of risks associated with hazardous materials and installations. This is because the very process of performing a risk assessment does lead to a better understanding and appreciation of the nature of the risks inherent in a study and helps evolve steps that can be taken to reduce these risks. Overall, the method deserves the effort required for its continual refinement as a management tool.

Additional Bibliography

Ahmad, Y. J., S. E. El Serafy, and E. Lutz. *Environmental Accounting for Sustainable Development* (Washington, DC: World Bank, 1989).

Andelman, J. B. and D. W. Underhill. *Health Effects From Hazardous Waste Sites*, 2nd Printing (Chelsea, MI: Lewis Publishers, 1988).

Baasel, W. D. *Economic Methods for Multipollutant Analysis and Evaluation* (New York: Marcel Dekker, 1985).

Barnard, R. and G. Olivetti. "Rapid Assessment of Industrial Waste Production Based on Available Employment Statistics," *Waste Manage. Res.* 8(2):139–144 (1990).

Barnthouse, L. W., Suter, G. W., Bartell, S. M. Beauchamp, J. J., Gardener, R. H. Linder, E., O'Neill, R. V., and Rosen, A. E. "User's Manual for Ecological Risk Assessment," Environmental Sciences Division, Publication No. 2679, ORNL-6251 (1986).

Beck, L. W., A. W. Maki, N. R. Artman, and E. R. Wilson. "Outline and Criteria for Evaluating the Safety of New Chemicals," *Regul. Toxicol. Pharmacol.* 1:19–58 (1981).

Bentkover, J. D., V. T. Covello and J. Mumpower. *Benefits Assessment: The State of the Art* (Boston, MA: Reidel, 1986).

Binder, S., D. Sokal, and D. Maughan. "Estimating the Amount of Soil Ingested by Young Children through Tracer Elements," *Arch. Environ. Health* 41:341–345 (1986).

Bohnenblust, H. and S. Pretre. "Appraisal of Individual Radiation Risk in the Context of Probabilistic Exposures," *Risk Anal.* 10(2):247–253 (1990).

Boice, J. D. and J. F. Fraumeni, Eds. *Radiation Carcinogenesis: Epidemiology and Biological Significance* (New York: Raven Press, 1984).

Bretherick, L. *Handbook of Reactive Chemical Hazards*, 2nd ed. (Wolburn, MA: Butterworth Publishers, 1979).

Brown, C. "Statistical Aspects of Extrapolation of Dichotomous Dose Response Data," *J. Natl. Cancer Inst.* 60:101–108 (1978).

Brown, H. S., R. Guble and S. Tatelbaum. "Methodology for Assessing Hazards of Contaminants to Seafood," *Regul. Toxicol. Pharmacol.* 8:76–100 (1988).

Brown, K. W., Evans, G. B., Jr., and Frentrup, B. D., Eds. *Hazardous Waste and Treatment* (Boston, MA: Butterworth Publishers, 1983).

Buchel, K. H. *Chemistry of Pesticides* (New York: John Wiley & Sons, 1983).

Bullard, R. D. *Dumping in Dixie — Race, Class and Environmental Quality* (Boulder, Co: Westview Press, 1990).

Cairns, J., Jr., and T. V. Crawford, Eds. *Integrated Environmental Management* (Chelsea, MI: Lewis Publishers, 1991).

Calabrese, E. J. and P. T. Kostecki. 1991. Hydrocarbon Contaminated Soils, Vol. 1. (Chelsea, MI: Lewis Publishers, 1991).

Calabrese, E. J. and P. T. Kostecki. Petroleum Contaminated Soils, Vol. 2 (Chelsea, MI: Lewis Publishers, 1989).

Calabrese, E. J. and P. T. Kostecki. *Soils Contaminated by Petroleum: Environment and Public Health Effects* (New York: John Wiley & Sons, 1988).

Calabrese, E. J. *Principles of Animal Extrapolation* (New York: John Wiley & Sons, 1983).

Calabrese, E. J., R. Barnes, E. J. Stanek, III, H. Pastides, C. E. Gilbert, P. Veneman, X. Wang, A. Lasztity, and P. T. Kostecki. "How Much Soil Do Young Children Ingest: An Epidemiologic Study," *Regul. Toxicol. Pharmacol.* 10:123–137 (1989).

Carson, W. H., Ed. *The Global Ecology Handbook — What You Can Do About the Environmental Crisis,* The Global Tommorrow Coalition (Boston, MA: Beacon Press, 1990).

CCME (Canadian Council of Ministers of the Environment). Interim Canadian Environmental Quality Criteria for Contaminated Sites, Report CCME EPC–CS34, Winnipeg, Manitoba (1991).

Chatterji, M., Ed. *Hazardous Materials Disposal: Siting and Management* (England: Avebury, Gower Publishing, 1987).

Clark, T., K. Clark, S. Paterson, and R. D. Mackay. "Wildlife Monitoring, Modeling and Fugacity," *Environ. Sci. Technol.* 22:120–127 (1988).

Clausing, O., A. B. Brunekreef, and J. H. van Wijnen. "A Method for Estimating Soil Ingestion by Children," *Int. Arch. Occup. Environ. Health* 59:73–82 (1987).

Clayson, D. B., Krewski, D., and Munro, I., Eds. *Toxicological Risk Assessment,* Vols. 1 and 2 (Boca Raton, FL: CRC Press, Inc., 1985).

Clement Associates, Inc. "Multi–Pathway Health Risk Assessment Impact Guidance Document," South Coast Air Quality Management District, California (1988).

Clewell, H. J. and M. E. Andersen. "Risk Assessment Extrapolations and Physiological Modeling," *Toxicol. Ind. Health* 1:111–132 (1985).

CMA (Chemical Manufacturers Association). *Risk Management of Existing Chemicals* (Washington, DC: Chemical Manufacturers Association, 1984).

Cohen, Y. "Organic Pollutant Transport," *Environ. Sci. Technol.* 20(6):538–545 (1986).

Cohrssen, J. J. and V. T. Covello. "Risk Analysis: A Guide to Principles and Methods for Analyzing Health and Environmental Risks," National Technical Information Service, U.S. Dept. of Commerce, Springfield, VA (1989).

Conway, R. A., Ed. *Environmental Risk Analysis of Chemicals* (New York: Van Nostrand Reinhold, 1982).

Cooperrider, A. Y., R. J. Boyd, and H. R. Stuart. *Inventory and Monitoring of Wildlife Habitat* (Washington, DC: U. S. Government Printing Office, 1986).

Covello, V. T. and J. Mumpower. "Risk Analysis and Risk Management: An Historical Perspective," *Risk Anal.* 5:103–120 (1985).

Covello, V. T. et al. *Uncertainty in Risk Assessment, Risk Management, and Decision Making. Advances in Risk Analysis*, Vol. 4 (NewYork: Plenum Press, 1987).

Covello, V. T., J. Menkes, and J. Mumpower, Eds. *Risk Evaluation and Management. Contemporary Issues in Risk Analysis*, Vol. 1 (New York: Plenum Press, 1986).

Cowherd, C. M., G. E. Muleski, P. J. Englehart, and D. A. Gillette. *Rapid Assessment of Exposure to Particulate Emissions From Surface Contamination Sites* (Kansas City, MO: Midwest Research Institute, 1984).

Crandall, R. W. and B. L. Lave, Eds. *The Scientific Basis of Risk Assessment* (Washington, DC: Brookings Institution, 1981).

Crouch, E. A. C., R. Wilson, and L. Zeise. "The Risks of Drinking Water," *Water Resour. Res.* 19(6):1359–1375 (1983).

Crump, K. S. "An Improved Procedure for Low-Dose Carcinogenic Risk Assessment from Animal Data," *J. Environ. Toxicol.* 5:339–346 (1981).

Crump, K. S. and R. B. Howe. "The Multistage Model with Time-Dependent Dose Pattern: Applications of Carcinogenic Risk Assessment," *Risk Anal.* 4:163–176 (1984).

Daniels, S. L. "Environmental Evaluation and Regulatory Assessment of Industrial Chemicals," in 51st Annu. Conf. Water Pollution Control Federation, Anaheim, CA (1978).

DeCoufle, P., Thomas, T. L., and Pickle, L. W. Comparison of the Proportionate Mortality Ratio and Standardized Mortality Ratio Risk Measures," *Am. J. Epidemiol.* 111:2630–269.

Diesler, P. F., Ed. *Reducing the Carcinogenic Risks in Industry* (New York: Marcel Dekker, 1984).

Dunster, H. J. and W. Vinck. "The Assessment of Risk, its Value and Limitations," *European Nuclear Conf., Hamburg, FRG* (1979).

Eschenroeder, A., R. J. Jaeger, J. J. Ospital, and C. Doyle. "Health Risk Assessment of Human Exposure to Soil Amended with Sewage Sludge Contaminated with Polychlorinated Dibenzodioxins and Dibenzofurans," *Vet. Hum. Toxicol.* 28:356–442 (1986).

Ess, T. H. "Risk Acceptability," in Proc. Natl. Conf. on Risk and Decision Analysis for Hazardous Waste Disposal, August 24 to 27, Baltimore, MD, (1981a), pp.164–174.

Ess, T. H. "Risk Estimation," in Proc. Natl. Conf. on Risk and Decision Analysis for Hazardous Waste Disposal, August 24 to 27, Baltimore, MD,(1981b), pp.155–163.

Farrand, J. and J. Bull. *The Audubon Society Field Guide to North American Birds: Eastern Region*, 16th ed. (New York: Alfred A. Knopf, 1988).

Finkel, A. M. and J. S. Evans. "Evaluating the Benefits of Uncertainty Reduction in Environmental Health Risk Management," *J. Air Pollut. Control Assoc.* 37:1164–1171 (1987).

Fitchko, J. *Criteria for Contaminated Soil/Sediment Cleanup* (Northbrook, IL: Pudvan Publishing Co., 1989).

Forester, W. S. and J. H. Skinner Ed. *International Perspectives on Hazardous Waste Management — A Report from the International Solid Wastes and Public Cleansing Association (ISWA) Working Group on Hazardous Wastes* (London: Academic Press, 1987).

Frankiel, E. G. Systems Reliability and Risk Analysis (Boston, MA: Martinus Nijhoff Publishers, 1984).

Freese, E. (1973). Thresholds in Toxic, Teratogenic, Mutagenic, and Carcinogenic Effects," *Environ. Health Pers.* 6:171–178 (1973).

Frey A. E. "International Transport of Hazardous Wastes," *Environ. Sci.Technol.* 33(5):509 (1989).

GAO (U.S. General Accounting Office). *Probabilistic Risk Assessment: An Emerging Aid to Nuclear Power Plant Safety Regulation*, GAO/RCED-85-11 (June 19, 1985).

Gaylor, D. W. and Kodell, R. L. "Linear Interpolation Algorithm for Low Dose Risk Assessment of Toxic Substances," *J. Environ. Pathol. Toxicol.* 4:305–312 (1980).

Gaylor, D. W. and Shapiro, R. E. "Extrapolation and Risk Estimation for Carcinogenesis," in *Advances in Modern Toxicology, Vol. 1*, Mehlman, M. A., R. E. Shapiro, and H. Blumenthal, Eds. (New York: Hemisphere, 1979), pp. 65–85.

Gordon, S. I. *Computer Models in Environmental Planning* (New York: Van Nostrand Reinhold, 1985).

Haimes, Y. Y., Ed. *Risk/Benefit Analysis in Water Resources Planning and Management* (New York: Plenum Press, 1981).

Haimes, Y. Y., L. Duan, and V. Tulsiani. "Multiobjective Decision-Tree Analysis," *Risk Anal.* 10(1):111–129 (1990).

Haimes, Y. Y., V. Chankong, and C. Du. "Risk Assessment for Groundwater Contamination. I," in *Computer Applications in Water Resources.*, Torno, H. C., Ed. Proc. of the Specialty Conf., Buffalo, NY, June 10 to 12 (1985).

Hallenbeck, W. H. and K. M. Cunningham-Burns. *Pesticides and Human Health* (New York: Springer-Verlag, 1985).

Harr, M. E. *Reliability-Based Design in Civil Engineering* (New York: McGraw-Hill 1987).

Hawley, G. G. *The Condensed Chemical Dictionary*, 10th ed. (New York: Van Nostrand Reinhold, 1981).

Hawley, J. K. "Assessment of Health Risks from Exposure to Contaminated Soil," *Risk Anal.* 5(4):289–302 (1985).

Hayes, A. W., Ed. *Principles and Methods of Toxicology* (New York: Raven Press, 1982).

Henderson, M. *Living with Risk: The Choices, The Decisions.* British Medical Association, (Somerset, NJ: John Wiley & Sons, 1987).

Hertz, D. B. and H. Thomas. *Risk Analysis and Its Applications* (New York: John Wiley & Sons, 1983).

Hoel, D. G., D. W. Gaylor, R. L. Kirschstein, U. Saffiotti, and M. A. Schneiderman. "Estimation of Risks of Irreversible, Delayed Toxicity," *J. Toxicol. Environ. Health* 1:133–151 (1975).

Hoffman, D. J., B. A. Rattner, and R. J. Hall. "Wildlife Toxicology," *Environ. Sci. Technol.* 24:276 (1990).

Hogan, M. D. Extrapolation of Animal Carcinogenicity Data: Limitations and Pitfalls," *Environ. Health Pers.* 47:333–337 (1983).

Hohenemser, C. and J. X. Kasperson, Eds. *Risk in the Technological Society* (Denver, CO: Westview Press, 1982).

HSE (Health and Safety Executive). *Quantified Risk Assessment—Its Input to Decision Making* (London: HMSO, 1989b).

HSE (Health and Safety Executive). *Risk Criteria for Land-Use Planning in the Vicinity of Major Industrial Hazards* (London: HMSO, 1989b).

Hudson, R., R. Tucker, and M. Haegeli. Handbook of Toxicity of Pesticides to Wildlife, 2nd ed., USFWS Resources Publication No. 153 (1984).

IARC (International Agency for Research on Cancer). IARC Monographs on the Evaluation of Carcinogenic Risks to Humans, Overall Evaluations of Carcinogenicity, Suppl. 7 (Lyon, France: IARC, 1987).

IARC (International Agency for Research on Cancer). IARC Monographs on the evaluation of the carcinogenic risk of chemicals to man (multivolume work). (Geneva: IARC, World Health Organization, 1972–1985).

IARC. IARC Monographs on Evaluation of Carcinogenic Risks to Humans, Vol. 43 (Lyon, France: World Health Organization, 1988), pp. 15–32.

IJC (International Joint Commission). "Literature Review of the Effects of Persistent Toxic Substances on Great Lakes Biota," Report of the Health of Aquatic Communities Task Force, International Joint Commission (1986).

IRPTC. "Attributes for the Chemical Data Register of the International Register of Potentially Toxic Chemicals," Register Attribute Series, No. 1, International Register of Potentially Toxic Chemicals, UNEP, Geneva, Switzerland (1978).

IRPTC. Industrial Hazardous Waste Management, Industry and Environment Office and the International Register of Potentially Toxic Chemicals. United Nations Environment Programme, Geneva, Switzerland (1985).

IRPTC. "Treatment and Disposal Methods for Waste Chemicals: IRPTC File," Data Profile Series No. 5. International Register of Potentially Toxic Chemicals, United Nations Environment Programme, UNEP, Geneva, Switzerland (1985).

IRPTC. International Register of Potentially Toxic Chemicals, Part A, International Register of Potentially Toxic Chemicals, UNEP, Geneva, Switzerland (1985).

J. C. Consltancy Ltd., London. Risk Assessment for Hazardous Installations (Oxford: Pergamon Press, 1986).

Jennings, A. A. and P. Suresh. "Risk Penalty Functions for Hazardous Waste Management," *ASCE J. Environ. Eng.* 112(1):105–122 (1986).

Johnson, B. B. and Covello, V. T. *Social and Cultural Construction of Risk: Essays on Risk Selection and Perception* (Norwell, MA: Kluwer Academic, 1987).

Jorgensen, E. P., Ed. *The Poisoned Well—New Strategies for Groundwater Protection,* Sierra Club Legal Defense Fund (Washington, DC: Island Press, 1989).

Kamrin, M. A. *Toxicology: A Primer on Toxicology Principles and Applications* (Chelsea, MI: Lewis Publishers, 1988).

Kastenberg, W. E., T. E. McKone, and D. Okrent. "On Risk Assessment in the Absence of Complete Data," UCLA Rep. No. UCLA–ENG–677, School of Engineering and Applied Science, Los Angeles, CA (1976).

Kates, R. W. *Risk Assessment of Environmental Hazard,* SCOPE Report 8 (Chichester, NY: John Wiley & Sons, 1978).

Keeney, R. L. "Mortality Risks Induced by Economic Expenditures," *Risk Anal.* 10(1):147–159 (1990).

Kimmel, C. A. and D. W. Gaylor. "Issues in Qualitative and Quantitative Risk Analysis for Developmental Toxicology," *Risk Anal.* 8:15–20 (1988).

Kleindorfer, P. R. and H. C. Kunreuther, Eds. 1987. *Insuring and Managing Hazardous Risks: From Seveso to Bhopal and Beyond* (Berlin: Springer–Verlag, 1987).

Kostecki, P. T. and E. J. Calabrese, *Ed. Hydrocarbon Contaminated Soils and Groundwater,* Vol. 1. (Chelsea, MI: Lewis Publishers, 1991).

Kostecki, P. T. and E. J. Calabrese, Ed. *Petroleum Contaminated Soils,* Vols. 1, 2, and 3 (Chelsea, MI: Lewis Publishers, 1989).

Krewski, D. and C. Brown. "Carcinogenic Risk Assessment: A Guide to the Literature," *Biometrics* 37:353–366 (1981).

Krewski, D. and J. Van Ryzin. "Dose Response Models for Quantal Response Toxicity Data," in *Statistics and Related Topics,* Csorgo, M., D. A. Dawson, J. N. K. Rao, and A. K. E. Saleh, Eds. (New York: North-Holland, 1981), pp. 201–231.

Krewski, D., C. Brown, and D. Murdoch. "Determining Safe Levels of Exposure: Safety Factors or Mathematical Models," *Fundam. Appl. Toxicol.* 4:S383–S394 (1984).

Kunreuther, H. and M. V. Rajeev Gowda, Eds. *Integrating Insurance and Risk Management for Hazardous Wastes* (Boston, MA: Kluwer Academic, 1990).

LaGoy, P. K. "Estimated Soil Ingestion Rates for Use in Risk Assessment," *Risk Anal.* 7(3):355–359 (1987).

Lee, J. A. *The Environment, Public Health, and Human Ecology — Considerations for Economic Development,* A World Bank Publication (Baltimore, MD: Johns Hopkins University Press, 1985).

Leiss, W. *Prospects and Problems in Risk Communication,* Institute for Risk Research (Waterloo, Ontario, Canada: University of Waterloo Press, 1989).

Lind, N. C., J. Nathwani, and E. Siddall. "Measurement of Safety in Relation to Social Well-Being," Discussion Paper, Institute for Risk Research Paper No. 23, Waterloo, Ontario, Canada (1990).

Lindsay, W. L. *Chemical Equilibria in Soils* (New York: Wiley-Interscience, 1979).

Little, E. L. *The Audubon Society Field Guide to North American Trees: Eastern Region*, 2nd ed. (New York: Alfred A. Knopf, 1983).
Loehr, R. C., Ed. *Land as a Waste Management Alternative* (Ann Arbor, MI: Ann Arbor Science, 1976).
Long, F. A. and G. E. Schweitzer, Eds. *Risk Assessment at Hazardous Waste Sites*, (Washington, DC: American Chemical Society, 1982).
Long, W. L. "Economic Aspects of Transport and Disposal of Hazardous Wastes," *Marine Policy Int. J.* 14(3):198–204 (1990).
Lowrance, W. W. *Of Acceptable Risk: Science and the Determination of Safety* (Los Altos, CA: William Kaufman, 1976).
Lu, F. C. "Acceptable Daily Intake: Inception, Evolution, and Application," *Regul. Toxicol. Pharmacol.* 8:45–60 (1988).
Lu, F. C. "Safety Assessments of Chemicals with Threshold Effects," *Regul. Toxicol. Pharmacol.* 5:121–132 (1985).
Lu, F. C. *Basic Toxicology* (Washington, DC: Hemisphere, 1985).
MacDonald, D. W., Ed. *The Encyclopedia of Mammals* (Oxford: Equinox, 1984).
Madsen, H. L., S. Krenk, and N. C. Lind. *Methods of Structural Safety* (Englewood Cliffs, NJ: Prentice-Hall, 1987).
Mahmood, R. J. and R. C. Sims. Mobility of Organics in Land Treatment Systems, *J. Environ. Eng.* 112(2):236–245 (1986).
Mark, R. K. and D. E. Stuart-Alexander. "Disasters as a Necessary Part of Benefit-Cost Analyses," *Science* 197:1160 (1977).
Martin, E. J. and J. H. Johnson, Jr., Ed. *Hazardous Waste Management Engineering* (New York: Van Nostrand Reinhold, 1987).
Martin, L. R. G. and G. Lafond, Eds. *Risk Assessment and Management: Emergency Planning Perspectives,* Institute for Risk Research (Waterloo, Ontario, Canada: University of Waterloo Press, 1988).
Maurits la Riviere, J. W. "Threats to the World's Water," *Sci. Am.*, Managing Planet Earth, Special Issue, September (1989).
Mausner, J. S. and A. K. Bahn. *Epidemiology, An Introductory Text* (Philadelphia: W. B. Saunders, 1974).
Mayer, F. L. and M. R. Ellersick. "Manual of Acute Toxicity: Interpretation and Database for 410 Chemicals and 66 Species of Freshwater Animals," U.S. Dept. of Interior, Fish and Wildlife Service, Resource Publ. 160, Washington, DC (1986).
McColl, R. S. *Environmental Health Risks: Assessment and Management*, Institute for Risk Research (Waterloo, Ontario, Canada: University of Waterloo Press, 1987).
McCormick, N. J. *Reliability and Risk Analysis: Methods and Nuclear Power Applications* (New York: Academic Press, 1981).
McKone, T. E., W. E. Kastenberg, and D. Okrent. "The Use of Landscape Chemical Cycles for Indexing the Health Risks of Toxic Elements and Radionuclides," *Risk Anal.* 3(3):189–205 (1983).
McTernan, W. and E. Kaplan, Eds. Risk Assesssment for Groundwater Pollution Control, ASCE Monograph, New York (1990).
Merck. *The Merck Index: An Encyclopedia of Chemicals, Drugs and Biologicals,* 11th (Centennial) ed. (Rockway, NJ: Merck & Co., 1989).

Miller, D. W., Ed. *Waste Disposal Effects on Ground Water* (Berkeley, CA: Premier Press, 1980).

Mitruka, B. M., H. M. Rawnsley, and D. V. Vadehra. *Animals for Medical Research: Models for the Study of Human Disease* (New York: Wiley, 1976).

Monahan, D. J., "Estimation of Hazardous Wastes from Employment Statistics — Victoria, Australia." *Waste Manage. Res.* 8(2):145–149 (1990).

Moore, A. O. *Making Polluters Pay — A Citizen's Guide to Legal Action and Organizing* (Washington, DC: Environmental Action Foundation, 1987).

Mulkey, L. A. "Multimedia Fate and Transport Models: An Overview," *J. Toxicol. Clin. Toxicol.* 21(1–2):65–95 (1984).

Munro, I. C. and D. R. Krewski. "Risk Assessment and Regulatory Decision Making," *Food Cosmet. Toxicol.* 19:549–560, (1981).

NAS (National Academy of Sciences). "Drinking Water and Health," Safe Drinking Water Committee, Advisory Center on Toxicology, National Research Council, Washington, DC (1977).

Nathwani, J., N. C. Lind, and E. Siddall. "Risk-Benefit Balancing in Risk Management: Measures of Benefits and Detriments," Institute for Risk Research Paper No. 18, Waterloo, Ontario, Canada. Presented at the Annu. Meet. of the Soc. for Risk Anal., October 29 to November 6, 1989, San Francisco, CA (1990).

Nathwani, J., N. C. Lind, and E. Siddall. "Safety, Social Well-Being and Its Measurement," Institute for Risk Research Paper No. 21, Waterloo, Ontario, Canada. Presented at the 1st World Congr. on Safety Science, Cologne, Germany, September 24 to 26, 1990 (1990b).

Neely, W. B. *Chemicals in the Environment (Distribution, Transport, Fate, Analysis* (NewYork: Marcel Dekker, 1980).

Newell, A. J., S. W. Johnson, and L. K. Allen. "Niagara River Biota Contamination Project: Fish Flesh Criteria for Piscivorous Wildlife," New York State Dept. of Environmental Conservation, Div. of Fish and Wildlife, Technical Report 87-3, Albany, NY (1987).

NIOSH (National Institute for Occupational Safety and Health). "1979 Registry of Toxic Effects of Chemical Substances," Vol. 1" (NIOSH Publication No. 80-111 (1980).

NIOSH (National Institute of Occupational Safety and Health). "Registry of Toxic Effects of Chemical Substances," Tatken, R. L. and R. J. Lewis, Eds. U.S. Dept. Health and Human Services (DDSH), NIOSH, Cincinnati, OH, DHHS (NIOSH) Publ. No. 83-107 (1982).

NRC (National Research Council). *Risk and Decision-Making: Perspective and Research,* Committee on Risk and Decision-Making (Washington, DC: National Academy Press, 1982).

NRC (National Research Council). *Drinking Water and Health, Vol. 1* (Washington, DC: National Academy Press, 1977).

NRC (National Research Council). *Drinking Water and Health, Vol. 2* (Washington, DC: National Academy Press, 1980).

NRC (National Research Council). *Drinking Water and Health, Vol. 3* (Washington, DC: National Academy Press, 1980).

NRC (National Research Council). *Drinking Water and Health, Vol. 4* (Washington, DC: National Academy Press, 1982).

NRC (National Research Council). *Drinking Water and Health, Vol. 5* (Washington, DC: National Academy Press, 1983).

NRC (National Research Council). *Ground Water Models: Scientific and Regulatory Applications* (Washington, DC: National Academy Press, 1989).

NRC (National Research Council). *Hazardous Waste Site Management: Water Quality Issues* (Washington, DC: National Academy Press, 1988).

NRC (National Research Council). *Prudent Practices for Handling Hazardous Chemicals in Laboratories* (Washington, DC: National Academy Press, 1981).

NRC (National Research Council). *Specifications and Criteria for Biochemical Compounds,* 3rd ed. (Washington, DC: National Academy of Sciences, 1972).

NUREG (Nuclear Regulatory Commission). "PRA Procedures Guide — A Guide to the Performance of Probabilistic Risk Assessment for Nuclear Power Plants," Report NUREG/CR-2300, Office of Nuclear Regulatory Research, Washington, DC (1983).

OECD. "Existing Chemicals - Systematic Investigation, Priority Setting and Chemical Review," Organization for Economic Cooperation and Development, Paris, France (1986).

OECD. "Report of the OECD Workshop on Practical Approaches to the Assessment of Environmental Exposure," April 14 to 18, 1986, Vienna, (1986).

O'Hare, M., L. Bacow, and D. Sanderson. *Facility Siting and Public Opposition* (New York: Van Nostrand Reinhold, 1983).

Okrent, D. "A General Evaluation Approach to Risk-Benefit for Large Technological Systems and its Application to Nuclear Power," Final Report No. UCLA ENG-7777, Los Angeles, CA (1977).

Okrent, D. "Risk-Benefit Evaluation for Large Technological Systems," *Nucl. Safety* 20:148 (1979).

Onishi, Y., A. R. Olsen, M. A. Parkhurst, and G. Whelan. "Computer-Based Environmental Exposure and Risk Assessment Methodology for Hazardous Materials," *J. Hazard. Mater.* 10:389–417 (1985).

ORNL (Oak Ridge National Laboratory). "User's Manual for Ecological Risk Assessment," Environmental Sciences Division, Publication No. 2679, Oak Ridge, TN (1986).

OSHA (Occupational Safety and Health Administration). "Identification, Classification, and Regulation of Potential Occupational Carcinogens," *Fed. Regist.* 45:5002–5296 (1980).

OTA (Office of Technology Assessment). "Assessment of Technologies for Determining Cancer Risks from the Environment," Office of Technology Assessment, Washington, DC (1981).

OTA (Office of Technology Assessment). "Technologies and Management Strategies for Hazardous Waste Control," Congress of the U. S., Office of Technology Assessment, Washington, DC (1983).

Overcash, M. R. and D. Pal. *Design of Land Treatment Systems for Industrial Wastes — Theory and Practice* (Ann Arbor, MI: Ann Arbor Science, 1979).

Park, C. N. and R. D. Snee. "Quantitative Risk Assessment: State of the Art for

Carcinogenesis," *Am. Stat.* 37(4):427–441 (1983).

Peck, D. L., Ed. *Psychosocial Effects of Hazardous Toxic Waste Disposal on Communities* (Springfield, IL: Charles C Thomas Publishers, 1989).

Pedersen, J. "*Public Perception of Risk Associated with the Siting of Hazardous Waste Treatment Facilities,*" European Foundation for the Improvement of Living and Working Conditions, Dublin, Eire (1989).

Peterson, R. T. *A Field Guide to Birds: A Completely New Guide to All the Birds of Eastern and Central North America*, 4th ed. sponsored by the National Audubon Society and the National Wildlife Federation (Boston, MA: Houghton Mifflin, 1980).

Peterson, R. T. and G. A. Petrides. *A Field Guide to Trees and Shrubs: Northeastern and North Central United States and Southeastern and South Central Canada*, 2nd ed., sponsored by the National Audubon Society and the National Wildlife Federation (Boston, MA: Houghton Mifflin, 1986).

Pickering, Q. H. and C. Henderson. "The Acute Toxicity of Some Heavy Metals to Different Species of Warm Water Fish," *Air Water Pollut. Int. J.* 19–453 (1966).

Piddington K. W. "Sovereignty and the Environment," *Environment* 31(7):18–20, 35–39 (1989).

Postel, S. *Controlling Toxic Chemicals.* State of the World 1988. (New York: Worldwatch Institute, 1988).

Rail, C. C. *Groundwater Contamination: Sources, Control and Preventive Measures* (Lancaster, PA: Technomic Publishing Co., Inc., 1989).

Ricci, P. F., Ed. *Principles of Health Risk Assessment* (Englewood Cliffs, NJ: Prentice-Hall, 1985).

Rodricks, J. and M. R. Taylor. "Application of Risk Assessment to Food Safety Decision Making," *Regul. Toxicol. Pharmacol.* 3:275–307 (1983).

Rodricks, J. V. "Risk Assessment at Hazardous Waste Disposal Sites," *Hazard. Waste Hazard. Mater.* 1(3)333–362, (1984).

Rodricks, J. V. and R. G. Tardiff, Eds. *Assessment and Management of Chemical Risks*, ACS Symp. Ser. 239 (Washington, DC: American Chemical Society, 1984).

Royal Society of London. *Risk Assessment: A Study Group Report* (London: Royal Society, 1983).

Ruckelshaus, W. D. "Risk, Science, and Democracy," *Issues Sci. Technol.* Spring:19–38 (1985).

Russell, M. and M. Gruber. "Risk Assessment in Environmental Policy-Making," *Science* 236:286–290 (1987).

Sax, N. I. *Dangerous Properties of Industrial Materials*, 5th ed. (New York: Van Nostrand Reinhold, 1979).

Saxena, J. and F. Fisher, Eds. *Hazard Assessment of Chemicals* (New York: Academic Press, 1981).

Schramm, G. and J. J. Warford, Ed. *Environmental Management and Economic Development,* A World Bank Publication (Baltimore, MD: Johns Hopkins University Press, 1989).

Schroeder, R. L. "Habitat Suitability Index Models: Northern Bobwhite," U.S. Department of Interior, Fish and Wildlife Service, Washington, DC (1985).

Schwartz, S. I. and W. B. Pratt. *Hazardous Waste from Small Quantity Generators — Strategies and Solutions for Business and Government* (Washington, DC: Island Press, 1990).

Schwing, R. C. and W. A. Albers, Jr., Eds. *Societal Risk Assessment: How Safe is Safe Enough* (New York: Plenum Press, 1980).

Searle, C. E., Ed. *Chemical Carcinogens*, ACS Monograph 173 (Washington, DC: American Chemical Society, 1976).

Sebek, V., Ed. "Maritime Transport, Control and Disposal of Hazardous Waste." *Marine Policy Int. J.* Special Issue, 14(3), May (1990).

Shepard, T. H. *Catalog of Teratogenic Agents*, 3rd ed. (Baltimore, MD: Johns Hopkins University Press, 1980).

Shrader-Frechette, K. S. *Risk Analysis and Scientific Method* (Boston, MA: D. Reidel Publishing Co., 1985).

Sielken, R. L. "Some Issues in the Quantitative Modeling Portion of Cancer Risk Assessment," *Regul. Toxicol. Pharmacol.* 5:175–181 (1985).

Sims, R. C. and J. L. Sims. "Cleanup of contaminated soils," in *Utilization, Treatment, and Disposal of Waste on Land* (Madison, WI: Soil Science Society of America, 1986), pp. 257–278.

Sitnig, M. *Handbook of Toxic and Hazardous Chemicals and Carcinogens* (Park Ridge, NJ: Noyes Data Corp., 1985).

Smith, A. H. "Infant Exposure Assessment for Breast Milk Dioxins and Furans Derived from Waste Incineration Emissions," *Risk Anal.* 7:347–353 (1987).

Spencer, E. Y. "Guide to the Chemicals Used in Crop Protection," 7th ed., Research Institute, Agriculture Canada, Information Canada, Ottawa, Publ. No. 1093 (1982).

Sposito, G., C. S. LeVesque, J. P. LeClaire, and N. Sensi. "Methodologies to Predict the Mobility and Availability of Hazardous Metals in Sludge-Amended Soils. California Water Resource Center, Contribution No. 189. University of California (1984).

Starr, C., R. Rudman, and C. Whipple. "Philosophical Basis for Risk Analysis," *Annu. Rev. Energy* 1:629–662 (1976).

States, J. B., P. T. Hang, T. B. Shoemaker, L. W. Reed, and E. B. Reed. "A System Approach to Ecological Baseline Studies," FQS/DBS–78/21, USFWS, Washington, DC (1978).

Suess, M. J. and J. W. Huismans, Ed. "Management of Hazardous Wastes: Policy Guidelines and Code of Practice," WHO Regional Publication, European Series No. 14, Copenhagen, World Health Organization, Regional Office for Europe, (1983).

Tardiff, R. G. and J. V. Rodricks, Eds. *Toxic Substances and Human Risk* (New York: Plenum Press, 1987).

Tasca, J. J., M. F. Saunders, and R. S. Prann. Terrestrial Food-Chain Model for Risk Assessment. In Superfund '89: Proceedings 10th National Conference.

The Conservation Foundation. *Risk Assessment and Risk Control* (Washington, DC: The Conservation Foundation, 1985).

The World Bank. "Manual of Industrial Hazard Assessment Techniques," Office of Environment and Scientific Affairs, Washington, DC, October (1985).

The World Bank. *Striking a Balance — The Environmental Challenge of Develop-*

ment (Washington, DC: IBRD/The World Bank, 1989).

Theiss, J. C. "The Ranking of Chemicals for Carcinogenic Potency," *Regul. Toxicol. Pharmacol.* 3:320–328 (1983).

Thibodeaux, L. J. *Chemodynamics: Environmental Movement of Chemicals in Air, Water and Soil* (New York: John Wiley & Sons, 1979).

Tolba, M. K. "The Global Agenda and the Hazardous Waste Challenge," *Marine Policy Int. J.* 14(3):205–209(1990).

Tolba, M. K., Ed. *Evolving Environmental Perceptions: From Stockholm to Nairobi,* United Nations Environment Programme, UNEP, Nairobi (1988).

Travis, C. C. and H. A. Hattemer-Frey. "Determining an Acceptable Level of Risk," *Environ. Sci. Technol.* 22(8) (1988).

Travis, C. T. and A. D. Arms. "Bioconcentration of Organics in Beef, Milk, and Vegetation," *Environ. Sci. Technol.* 22:271 (1988).

U.S. EPA "User's Guide to the Contract Laboratory Program," Office of Emergency and Remedial Response, OSWER Dir. 9240. 0–1 (1989).

U.S. EPA (U.S. Environmental Protection Agency). "Appendix A: Uncontrolled Hazardous Waste Site Ranking System: A User's Manual," U. S. Environmental Protection Agency, *Fed. Regist.* 37(137):31219–31243 (1982).

U.S. EPA (U.S. Environmental Protection Agency). "Chemical, Physical, and Biological Properties of Compounds Present at Hazardous Waste Sites," Report Prepared by Clement Associates for the U.S. EPA, September (1985).

U.S. EPA (U.S. Environmental Protection Agency). "Ecological Information Resources Directory," Office of Information Resource Management (1989).

U.S. EPA (U.S. Environmental Protection Agency). "Exposure Assessment Methods Handbook," Office of Health and Environmental Assessment, U.S. EPA, Cincinnati, OH (1989).

U.S. EPA (U.S. Environmental Protection Agency). "Guidance on Remedial Actions for Contaminated Ground Water at Superfund Sites," Office of Emergency and Remedial Response, EPA/540/G–88/003 (1988).

U.S. EPA (U.S. Environmental Protection Agency). "Guidelines and Methodology Used in the Preparation of Health Effect Assessment Chapters of the Consent Decree Weater Criteria Documents," *Fed. Regist.* 45:79347–79357 (1980).

U.S. EPA (U.S. Environmental Protection Agency). "Handbook for Conducting Endangerment Assessments," U.S. EPA, Research Triangle Park, NC (1987).

U.S. EPA (U.S. Environmental Protection Agency). "Hazardous Waste Land Treatment," Revised ed., SW–8974. U.S. EPA, Cincinnati, OH (1983).

U.S. EPA (U.S. Environmental Protection Agency). "Registry of Toxic Effects of Chemical Substances," U.S. EPA, Research Triangle Park, NC (1986).

U.S. EPA (U.S. Environmental Protection Agency). "Superfund Public Health Evaluation Manual," EPA/540/1–86/060, Office of Emergency and Remedial Response, Washington, DC (1986).

U.S. EPA (U.S. Environmental Protection Agency). "Test Methods for Evaluating Solid Waste: Physical/Chemical Methods, 1st ed., SW–846, U.S. EPA (1982).

U.S. EPA (U.S. Environmental Protection Agency). Review of In-Place Treatment Techniques for Contaminated Surface Soils. Vols. 1 and 2, U. S. Environmental

Protection Agency, Hazardous Waste Engineering Research Laboratory, Cincinnati, OH, EPA-540/2–84–003a and b (1984).

UNCHE. The United Nations Conference on the Human Environment, Stockholm, June 5–16, 1972, Recommendation 74(e) (1972).

UNEP (United Nations Environment Programme). *The State of the Environment* (Nairobi: UNEP, 1988).

UNEP. "Report of the UNEP Expert Workshop on an International Register of Potentially Toxic Chemicals," Bilthoven, Netherlands, January 6–11, Document UNEP/WG. 1/4 (1975).

Van Ryzin, J. "Quantitative Risk Assessment," *J. Occup. Med.* 22:321–326 (1980).

Verschueren, K. *Handbook of Environmental Data on Organic Chemicals*, 2nd ed. (New York: Van Nostrand Reinhold, 1983).

Waller, R. A., and V. T. Covello, Eds. "Low-Probability/High-Consequence Risk Analysis: Issues, Methods, and Case Studies," in *Advances in Risk Analysis* (New York: Plenum Press, 1984).

WCED (The World Commission on Environment and Development). *Our Common Future* (Oxford, U.K.: WCED, 1987).

Weast, R. C., Ed. *Handbook of Chemistry and Physics*, 65th ed. (Boca Raton, FL: CRC Press, Inc., 1984).

Whitaker, J. O. *The Audubon Society Field Guide to North American Mammals*, 4th ed. (New York: Alfred A. Knopf, 1988).

WHO (World Health Organization). Management of Hazardous Waste, WHO Regional European Series No. 14 (1983).

Wilson, R. and E. A. C. Crouch. "Risk Assessment and Comparisons: An Introduction," *Science* 236–267(1987).

Zirm, K. L. and J. Mayer, Eds. *The Management of Hazardous Substances in the Environment* (London, England: Elsevier, 1990).

APPENDIX A

Abbreviations, Acronyms, and Glossary of Terms and Definitions

A.1 LIST OF SELECTED ABBREVIATIONS AND ACRONYMS

ACL alternate concentration limit
ADD average daily dose
ADI acceptable daily intake (or allowable daily intake)
AIC acceptable chronic intake
AL action level
ARAR applicable or relevant and appropriate requirement
ASC allowable soil concentartion
ATSDR Agency for Toxic Substances and Disease Registry
AWQC ambient water quality criteria

BAT best available technology
BCF bioconcentration (bioaccumulation) factor

CAA Clean Air Act
CAG Carcinogen Assessment Group
CAS Chemical Abstracts Service
CDI chronic daily intake
CERCLA Comprehensive Environmental Response, Compensation, and Liability Act (also Superfund)
CFR Code of Federal Regulations
CPF carcinogen potency factor
CSIN Chemical Substances Information Network

CWA	Clean Water Act
EA	endangerment assessment
EED	estimated exposure dose
EPA	Environmental Protection Agency
ERA	environmental (or ecological) risk assessment
ESA	Endangered Species Act
FDA	Food and Drug Administration
FIFRA	Federal Insecticide, Fungicide, and Rodenticide Act
FS	feasibility study
FWPCA	Federal Water Pollution Control Act
HEA(ST)	Health Effects Assessment (Summary Tables) (U.S. EPA)
HI	hazard index
HMTA	Hazardous Materials Transport Act
HRS	hazard ranking system
HSWA	Hazardous and Solid Waste Ammendments (of 1984, to RCRA)
IARC	International Agency for Research on Cancer
IRIS	Integrated Risk Information System
IRPTC	International Register of Potentially Toxic Chemicals
LADD	lifetime average daily dose
LC_{50}	mean lethal concentration
LD_{50}	mean lethal dose
LOAEL	lowest observed adverse effect level
LOEL	lowest observed effect level
MAC	maximum acceptable concentration
MCL	maximum contaminant level
MCLG	maximum contaminant level goal
MDD	maximum daily dose
MEL	maximum exposure level
MF	modifying factor
MOE	margin of exposure
MSW	municipal solid waste
(M)USLE	(modified) universal soil loss equation
NAAQS	National Ambient Air Quality Standards
NCP	National Contingency Plan
NESHAP	National Emission Standards for Hazardous Air Pollutants
NOAEL	no observable adverse effect level
NOEL	no observable effect level

NPL	National Priorities List
NTIS	National Technical Information Service
OSHA	Occupational Safety and Health Administration
PAR	population at risk
PEL	permissible exposure limit
ppb	parts per billion
ppm	parts per million
PRA	probabilistic risk assessment
PRP	potentially responsible party
PWP	pathway probability
RAP	remedial action plan
RCRA	Resource Conservation and Recovery Act
RFA	RCRA facility assessment
RfD	reference dose
RgD	regulatory dose
RFI	RCRA facility investigation
RI	remedial investigation
RI/FS	remedial investigation/feasibility study
RMCL	recommended maximum contaminant level (renamed MCLG)
RME	reasonable maximum exposure
RMPP	risk management and prevention program
RSD	risk-specific dose
RSCL	recommended soil cleanup limit
SARA	Superfund Amendments and Reauthorization Act
SDI	subchronic daily intake
SDWA	Safe Drinking Water Act
SF	slope factor
TBC	to-be-considered
TLV	threshold limit value
TSCA	Toxic Substances Control Act
TSDF	treatment, storage, and disposal facility
UCR	unit cancer risk
UF	uncertainty factor (also safety factor)
UR (F)	unit risk (factor)
VSD	virtually safe dose
WQA	Water Quality Act

A.2 GLOSSARY OF TERMS AND DEFINITIONS

Absorption — The uptake of a chemical by a cell or an organism, including the flow into the bloodstream following exposure through the skin, lungs, and/or gastrointestinal tract.

Absorbed dose — The amount of a chemical substance actually entering an exposed organism (i.e., the amount penetrating the exchange boundaries of an organism after contact). It is calculated from the intake and the absorption efficiency, expressed in mg/kg-day.

Absorption factor — The percent or fraction of a chemical in contact with an organism that becomes absorbed into the receptor.

Acceptable daily intake (ADI) — The amount of a chemical (in mg/kg body weight/day) that will not cause adverse effects following chronic exposure (i.e., lifetime daily exposure) to potential receptors (or population at risk); used interchangeably with RfD.

Action level (AL) — Level of a chemical concentration in selected media of concern above which there are potential adverse health and/or environmental effects, and above which corrective action needs to be taken.

Acute exposure — The single large exposure or dose to a chemical, generally occuring over a short period.

Acute toxicity — The development of symptoms of poisoning or the occurence of adverse health effects after exposure to a single dose or multiple doses of a chemical within a short period of time.

Adsorption — The physical process of attracting and holding molecules of other substances or particles to the surfaces of solid bodies with which the former are in contact.

Acceptable risk — A risk level which is considered by society as tolerable.

Antagonism — The interference or inhibition of the effects of one chemical substance by the action of other chemicals.

Arithmetic mean (also Average) — A measure of central tendency for data from a normal distribution, given for a set of n values, by the sum of values divided by n:

$$X_m = \frac{\sum_{i=1}^{n} X_i}{n}$$

Background level — The normal ambient environmental concentration levels of a chemical contaminant.

Bioaccumulation — The retention and concentration of a chemical in the tissues of an organism or biota.

Bioconcentration — The accumulation of a chemical in tissues of organisms to levels greater than levels in the surrounding media for the organism's habitat; often used synonymously with bioaccumulation.

Bioconcentration factor (BCF) — A measure of the amount of selected chemical substances that accumulates in humans or in biota. It is the ratio of the concentration of substances in an organism to the concentration of the substance in surrounding environmental media.

Cancer — A disease characterized by malignant, uncontrolled invasive growth of body tissue cells.
Carcinogen — A chemical or substance capable of producing cancer in living organisms.
Carcinogen Assessment Group (CAG) — The group within the U.S. EPA responsible for the evaluation of carcinogen bioassay results and estimates of the carcinogenic potency of various chemicals.
Carcinogenic — Tending to produce or incite cancer in living organisms.
Carcinogenicity —The ability of a chemical to cause cancer in a living organism.
Chronic — Pertaining to the long term (i.e., of long duration).
Chronic daily inatake (CDI) — The exposure, expressed in mg/kg-day, averaged over a long period of time.
Chronic exposure — The long-term, low-level exposure to chemicals, i.e., the repeated exposure or doses to a chemical over a long period of time. It may cause latent damage that does not appear until a later period in time.
Chronic toxicity — The occurence of symptoms, diseases, or other adverse health effects that develop and persist over time, after exposure to a single dose or multiple doses of a chemical delivered over a relatively long period of time.
Confidence interval (C.I.) — Pertaining to a range and the probability that an uncertain quantity falls within this range.
Confidence limits — The upper and lower boundary values of a range of statistical probability numbers that define the confidence interval.
Confidence limits, 95 percent — The limits of the range of values within which a single evaluation/analysis will be included 95% of the time. For large samples (i.e., $n > 30$),

$$95\%CL = X_m \pm \frac{1.96\sigma}{n^{0.5}}$$

where CL is the confidence level, and σ is the estimate of the standard deviation of the mean (X_m). For a limited number of samples ($n \leq 30$), a confidence limit or confidence interval may be be estimated from

$$CL = X_m \pm \frac{ts}{n^{0.5}}$$

where t is the value of the student t distribution (refer, e.g., *Introductory Statistics*, 2nd ed., by Wonnacott, T. H. and R. J. Wonnacott, John Wiley & Sons, New York, 1972) for the desired confidence level and degrees of freedom, (n–1).
Consequence — The impacts resulting from the response due to specified exposures or loading/stress conditions.

Decision analysis — A process of systematic evaluation of alternative solutions to a problem where the decision is made under uncertainty.

Degradation — The physical, chemical, or biological breakdown of a complex compound into simpler compounds and byproducts.

de Minimus — Legal doctrine dealing with levels associated with insignificant vs significant issues relating to human exposures to chemicals that present very low risk.

Dermal exposure — Exposure of an organism or receptor through skin absorption.

Dose — That amount of a chemical taken in by potential receptors on exposure; it is a measure of the amount of the substance received by the receptor, as a result of exposure, expressed as an amount of exposure (in mg) per unit body weight of the receptor (in kg).

Dose-response — The quantitative relationship between the dose of a chemical and an effect caused by exposure to such substance.

Dose-response curve — A graphical representation of the relationship between the degree of exposure to a chemical substance and the observed or predicted biological effects or reponse.

Dose-response evaluation — The process of quantitatively evaluating toxicity information and characterizing the relationship between the dose of a chemical administered or received and the incidence of adverse health effects in the exposed population.

Ecosystem — The interacting system of a biological community and its abiotic (i.e., nonliving) environment.

Ecotoxicity assessment — The measurement of effects of environmental toxicants on indigenuos populations of organisms within an ecosytem.

Effect (local) — The response produced due to a chemical that occurs at the site of first contact.

Effect (systemic) — The response produced due to a chemical that requires absorption and distribution of the chemical and tends to affect the receptor at sites away from the entry point(s).

Endangerment assessment — A site-specific risk assessment of the actual or potential danger to human health and welfare and also the environment, from the release of hazardous chemicals into various environmental media.

Endpoint — A biological effect used as index of the impacts of a chemical on an organism.

Environmental fate — The ultimate and intermediary destinies of a chemical after release into the environment, and following transport through various environmental compartments.

Exposure — Receiving a dose of a chemical substance (or physical agent) or coming in contact with a hazard.

Exposure assessment — The qualitative or quantitative estimation, or the measurement, of the dose or amount of a chemical to which potential receptors have been exposed or could potentially be exposed to; it comprises of determining the magnitude, frequency, duration, route, and extent of exposure (to the chemicals or hazards of potential concern).

Exposure conditions — Factors (such as location, time, etc) that may have an impact on the system response or adverse consequences.

Exposure pathway — The course a chemical or physical agent takes from a source to an exposed population or organism; it describes a unique mechanism by which an individual or population is exposed to chemicals or physical agents at or originating from a site.

Exposure point — A location of potential contact between an organism and a chemical or physical agent.

Exposure route — The avenue by which an organism contacts a chemical, such as inhalation, ingestion, and dermal contact.

Exposure scenario — A set of conditions or assumptions about hazard sources, exposure pathways, levels of chemicals, and potential receptors that aids in the evaluation and quantification of exposure in a given situation.

Extrapolation — The estimation of an unknown value by projecting from known values.

Event tree analysis (ETA) — A procedure often used to evaluate series of events which lead to an upset or accident scenario, using deductive logic.

Fault tree analysis (FTA) — A procedure often used to evaluate series of events which lead to an upset or accident scenario, using inductive logic.

Feasibility Study (FS) — The analysis and selection of alternative remedial or corrective actions for hazardous waste or potentially contaminated sites.

Fugitive dust — Atmospheric dust arising from disturbances of granular matter exposed to the air.

Geometric mean — A measure of the central tendency for data from a positively skewed distribution (lognormal), given by:

$$\sqrt[n]{(X_1)(X_2)(X_3)\ldots(X_n)}$$

or,

$$X_{gm} = anti\log\left\{\frac{\sum_{i=1}^{n}\log X_i}{n}\right\}$$

Hazard — The inherent adverse effect that a chemical or other object poses. It is that which has the potential for creating adverse consequences.

Hazard assessment — The evaluation of system performance and associated consequences over a range of operating and/or failure conditions. It involves gathering and evaluating data on types of injury or consequences that may be produced by a hazardous situation or substance.

Hazard identification — The systematic identification of potential accidents, upset conditions, etc. For a health risk assessment, it is the process of determining whether exposure to an agent can cause an increase in the incidence of a particular adverse health effect in receptors of interest.

Hazardous waste — Wastes that are ignitable, corrosive, reactive, or toxic; it is that by-product which has the potential of causing detrimental effects on human health and/or the environment if not managed efficiently.

Human health risk — The likelihood (or probability) that a given exposure or series of exposures to a hazardous substance will cause adverse health impacts on individual receptors experiencing the exposures.

Individual excess lifetime cancer risk — An upper-bound estimate of the increased cancer risk, expressed as a probability, that an individual receptor could expect from exposure over a lifetime; it is a statistical concept and is not dependent on the average residency time in an area.

Ingestion — An exposure type whereby chemical substances enter the body through the mouth and into the gastrointestinal system.

Inhalation — The intake of a substance by receptors through the respiratory tract system.

Intake — The amount of material inhaled, ingested, or dermally absorbed during a specified time period. It is a measure of exposure, expressed in mg/kg-day.

Integrated Risk Information System (IRIS) — A U.S. EPA database containing verified reference doses (RfDs) and slope factors (SFs) and up-to-date health risk and EPA regulatory information for numerous chemicals. It serves as an important source of toxicity information for health and environmental risk assessment.

K_d — Soil/water partition coefficient, provides a soil- or sediment-specific measure of the extent of chemical partitioning between soil or sediment and water, unadjusted for the dependence on organic carbon.

K_{oc} — Organic carbon adsorption coefficient, provides a measure of the extent of chemical partitioning between organic carbon and water at equilibrium.

K_{ow} — Octanol/water partition coefficient, provides a measure of the extent of chemical partitioning between water and octanol at equilibrium.

K_w — Water/air partition coefficient, provides a measure of the distribution of a chemical between water and air at equilibrium.

LC_{50} (mean lethal concentration) — The lowest concentration of a chemical in air or water that will be fatal to 50% of test organisms living in that media.

LD_{50} (mean lethal dose) — The single dose of a chemical (ingested or dermally absorbed) required to kill 50% of a test animal group; expressed in mg/kg body weight.

Leachate — A contaminated liquid resulting when water percolates, or trickles, through waste materials and collects components of those wastes; leaching usually occurs at landfills and may result in hazardous chemicals entering soils, surface water, or groundwater.

Life expectancy — The number of years remaining of an average person's life, for a specified attained age.

Lifetime risk — Risk which results from lifetime exposure to a chemical substance.

Lifetime average daily dose (LADD) — The exposure, expressed as mass of a substance contacted and absorbed per unit body weight per unit time, averaged over a lifetime.

Lifetime exposure — The total amount of exposure to a substance that a human receptor would be subjected to in a lifetime.

LOAEL (lowest observed adverse effect level) — That chemical dose rate causing statistically or biologically significant increases in frequency or severity of adverse effects between the exposed and control groups. It is the lowest dose level, expressed in mg/kg body weight/day, at which adverse effects are noted in the exposed population.

LOEL (lowest observed effect level) — The lowest exposure or dose level to a substance at which effects are observed in the exposed population; the effects may or may not be serious.

MCL (maximum contaminant level) — A legally enforceable maximum chemical concentration standard that is allowable in drinking water, issued by the U.S. EPA under the SDWA authorities.

MCLG (maximum contaminant level goal) — Nonenforceable health goals for public drinking water systems issued by the U.S. EPA under the SDWA authorities; it is also referred to as the recommended maximum contaminant level (RMCL).

Modeling — Use of mathematical algorithms to simulate and predict real events and processes.

Monitoring — Measurement of concentrations of chemicals in environmental media or in tissues of humans and other biological receptors/organisms.

Monte Carlo simulation — A technique in which outcomes of events or variables are determined by selecting random numbers based on a defined probability distribution.

NOAEL (No observed adverse effect level) — The highest chemical intakes at which there are no statistically or biologically significant increases in frequency or severity of adverse effects between the exposed and control groups (meaning statistically significant effects are observed at this level, but they are not considered to be adverse). It is a dose level, expressed in mg/kg body weight/day, at which no adverse effects are noted in the exposed population.

NOEL (no-observed-effect level) — That dose rate of chemical at which there are no statistically or biologically significant increases in frequency or severity of any effects between the exposed and control groups, (i.e., the highest level at which a chemical causes no observable changes in the species under investigation). It is a dose level, expressed in mg/kg body weight/day, at which no effects are noted in exposed populations.

Nonthreshold chemical — Also called zero threshold chemical, refers to a substance which is known, suspected, or assumed to potentially cause some adverse response at any dose above zero.

NPL (National Priorities List) — A list of waste sites for which the U.S. EPA has assessed the relative threat of a site contamination on air, surface and groundwater, soil, and the population potentially at risk; the site listing is found under CERCLA (Section 105).

Pathway — Any specific route by which a potential receptor or individual may be exposed to an environmental hazard, such as the release of a chemical material.

PEL (permissible exposure limit) — A maximum (legally enforceable) allowable level for a chemical in workplace air, expressed as ppm or mg/m^3 of substance in air.

Pica — The behavior in children and toddlers (usually under age 6 years) involving the intentional eating/mouthing of large quantities of dirt and other objects.

PM-10, PM_{10} — Particulate matter with physical/aerodynamic diameter <10 μm; represents the respirable particulate emissions.

Population at risk (PAR) — A population subgroup that is more susceptible to hazard or chemical exposures; it is that group which is more sensitive to a hazard or chemical than is the general population.

Population excess cancer burden — An upper-bound estimate of the increase in cancer cases in a population as a result of exposure to a carcinogen.

Potency — A measure of the relative toxicity of a chemical.

ppb (parts per billion) — An amount of substance in a billion parts of another material; also expressed by μg/kg or μg/L.

ppm (parts per million) — An amount of substance in a million parts of another material; also expressed by mg/kg or mg/L.

ppt (parts per trillion) — An amount of substance in a trillion parts of another material; also expressed by ng/kg or ng/L.

Probability — The likelihood of an event occuring.

PRP (potentially responsible party) — Refers to those identified by the U.S. EPA as potentially liable under CERCLA for cleanup costs at specified waste sites.

Qualitative — Referring to the occurrence of a situation without numerical specifications.

Quantitative — Describing the amounts in exact numerical terms.

Receptor — Refers to members of a potentially exposed population, e.g., persons or organisms that are potentially exposed to concentrations of a particular chemical compound.

Reference dose (RfD) — The maximum amount of a chemical that the human body can absorb without experiencing chronic health effects; it is expressed in mg of chemical per kg body weight per day. It is the estimate of lifetime daily exposure of a noncarcinogenic substance for the general human population which appears to be without an appreciable risk of deleterious effects; used interchangeably with ADI.

Remedial Investigation (RI) — The field investigations of hazardous waste sites to determine pathways, nature, and extent of contamination, as well as identify preliminary alternative remedial actions.

Residual risk — The risk of adverse consequences that remains after corrective actions have been implemented.

Response — The reaction of the body to a chemical substance or other physical, chemical, or biological agent.

Risk — The probability or likelihood of an adverse consequence from a hazardous situation or hazard, or the potential for the realization of undesirable adverse consequences from impending events.

Risk acceptance — The willingness of an individual, group, or society to accept a specific level of risk to obtain some gain or benefit.

Risk assessment — The determination of the potential adverse effects due to hazardous exposure in a particular situation; it is the total process of qualifying or quantifying risks and finding acceptable levels of the risks for an individual, group, or society. It may involve the characterization of the types of health and environmental effects expected from exposure to a chemical substance, estimation of the probability (risk) of occurrence of adverse effects, estimation of the number of cases, and a recommendation for corrective actions.

Risk estimate — A description of the probability that a potential receptor exposed to a specified dose of a chemical will develop an adverse response.

Risk estimation — The process of quantifying of the probability and consequence values for an identified risk.

Risk evaluation — The complex process of developing acceptable levels of risk to individuals or society.

Risk group — Refers to a real or hypothetical exposure group composed of the general population and/or workers.

Risk management — The steps and processes taken to reduce, abate, or eliminate the risk that has been revealed by a risk assessment; it is an activity concerned with decisions about whether an assessed risk is sufficiently high to present a public health concern and about the appropriate means for controlling the risks judged to be significant.

Risk perception — The magnitude of the risk as it is perceived by an individual or society; it consists of the measured risk and the preconceptions of the observer.

Risk reduction — The action of lowering the probability of occurrence and/or the value of a risk consequence, thereby reducing the magnitude of the risk.

Risk-specific dose (RSD) — The dose of a chemical associated with a specified risk level.

Sensitive receptor — Individual in a population who is particularly susceptible to health impacts due to exposure to a chemical pollutant.

Sensitivity analysis — A method used to examine the operation of a system by measuring the deviation of its nominal behavior due to pertubations in the performance of its components from their nominal values.

Slope factor (SF) — A plausible upper-bound estimate of the probability of a response per unit intake of a chemical over a lifetime. It is used to estimate an upperbound probability of an individual developing cancer as a result of a lifetime of exposure to a particular level of a potential carcinogen.

Standard — A general term used to describe legally established values above which regulatory action will be required.

Standard deviation — The most widely used measure to describe the dispersion of a data set, defined for a set of n values as follows:

$$s = \left[\frac{\sum_{i=1}^{n} \left(X_i - X_m \right)^2}{(n-1)} \right]^{0.5}$$

where X_m is the arithmetic mean for the data set of n values.

Subchronic — Relates to intermediate duration, usually used to describe studies or exposure levels spanning 5 to 90 days duration.

Subchronic exposure — The short-term, high-level exposure to chemicals, i.e., the maximum exposure or doses to a chemical over a portion of a lifetime.

Subchronic daily intake (SDI) — The exposure, expressed in mg/kg-day, averaged over a portion of a lifetime.

Superfund — A common name for the Comprehensive Environmental Response, Compensation, and Liability Act (CERCLA), also referred to as the Trust Fund.

Synergism — An interaction of two or more chemicals that results in an effect that is greater than the sum of their effects taken independently. More generally stated, this represents the effects from a combination of two or more events, efforts, or substances that are greater than would be expected from simply adding the individual effects.

Systemic — Relates to whole body, rather than individual parts of exposed individual or receptor.

Threshold — The lowest dose or exposure of a chemical at which a specified measurable effect is observed and below which such effect is not observed.

Threshold chemical — Also nonzero threshold chemical, refers to a substance which is known or assumed to have no adverse effects below a certain dose.

TLV (threshold limit value) — The maximal allowable workplace air concentration level for a chemical.

Tolerance limit — Level or concentration of a chemical residue in media of concern above which adverse health effects are possible and above which corrective action should be taken.

Toxic — Harmful or deleterious with respect to the effects produced by exposure to a chemical substance.

Toxicant — Refers to any synthetic or natural chemical with an ability to produce adverse health effects; it is a poisonous contaminant that may injure an exposed organism.

Toxicity — The harmful effects produced by a chemical substance. It is the quality or degree of being poisonous or harmful to human or ecological receptors.

Toxicity assessment — Evaluation of the toxicity of a chemical based on available human and animal data. It is the characterization of the toxicological properties and effects of a chemical substance, with special emphasis on the establishment of dose-response characteristics.

Uncertainty — The lack of confidence in the estimate of a variable's magnitude or probability of occurrence.

Uncertainty factor (UF) — Also called safety factor, refers to a factor that is used to provide a margin of error when extrapolating from experimental animals to estimate human health risks.

Unit cancer risk (UCR) — The excess lifetime risk of cancer due to a continuous lifetime exposure/dose of one unit of carcinogenic chemical concentration (caused by one unit of exposure in the low exposure region).

Unit risk (UR) — Measure of the carcinogenic potential of a substance when a dose is received through the inhalation pathway that is based on several assumptions; it is an upper-bound estimate of the probability of contracting cancer as a result of constant exposure over the individual lifetime to an ambient concentration of 1 $\mu g/m^3$.

Upper-bound estimate — The estimate not likely to be lower than the true value.

APPENDIX B

Relevant Equations Commonly Utilized in Human Health Risk Assessments

B.1 ESTIMATION OF RECEPTOR EXPOSURES TO CHEMICALS: EQUATIONS FOR CALCULATING CHEMICAL INTAKES AND DOSES

Introduction

An analysis of the potential exposures associated with potentially contaminated site problems generally involve a complexity of integrated evaluations and issues to be addressed (Figure B.1). The primary pathways of general concern include inhalation exposures, dermal exposures, soil ingestion, water ingestion, and crops ingestion (for crops contaminated from direct deposition of contaminants); secondary pathways of interest will generally comprise of ingestion of mother's milk, fish ingestion, poultry and eggs ingestion, meat and dairy products ingestion, and crops ingestion (for crops contaminated from root uptake of chemicals). Consumption of locally produced and homegrown food sources (i.e., animals and crops) should be determined and fully incorporated in all multipathway risk assessment. The methods by which each type of exposure is estimated are well documented in materials prepared under the auspices of various regulatory agencies (e.g., CDHS, 1986; U.S. EPA, 1989a,b; CAPCOA, 1990). Receptor exposures for the different primary routes of contact are presented.

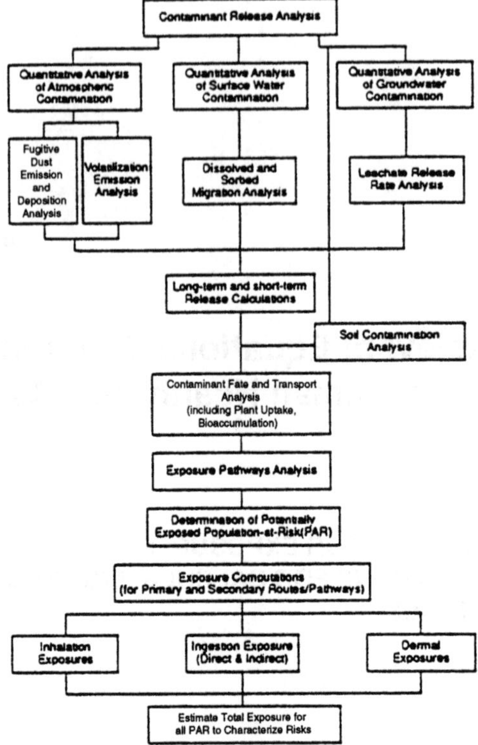

Figure B.1 Simplified schematic for the analysis of exposures associated with potentially contaminated site problems.

Inhalation

Chemical intake via the inhalation exposure pathway is conservatively estimated as follows:

$$Inhalation\ exposure\ (mg/kg\text{-}day) = \frac{\{GLC \times RR \times CF\}}{BW}$$

where

GLC = ground-level concentration (mg/m^3)
RR = respiration rate (m^3/day)
CF = conversion factor (= 1 mg/1000 μg = 1.0E–03 mg/μg)
BW = body weight (kg)

Ingestion

Chemical intake through water, soils, crops, and dairy/beef ingestion exposure pathways are conservatively estimated as follows:

$$\text{Water ingestion exposure } (mg/kg\text{-}day) = \frac{\{CW \times WIR \times GI\}}{BW}$$

$$\text{Soil ingestion exposure } (mg/kg\text{-}day) = \frac{\{CS \times SIR \times GI\}}{BW}$$

$$\text{Crop ingestion exposure } (mg/kg\text{-}day) = \frac{\{CS \times RUF \times CIR \times GI\}}{BW}$$

$$\text{Dairy and beef products ingestion exposure } (mg/kg\text{-}day) = \frac{\{CD \times FIR \times GI\}}{BW}$$

where

CW = chemical concentration in water (mg/L)
WIR = water consumption rate (L/day)
CS = chemical concentration in soil (mg/kg)
SIR = soil consumption rate (kg/day)
RUF = root uptake factor
CIR = crop consumption rate (kg/day)
CD = concentration of chemical in diet (mg/kg); for grazing animals, the concentration of chemicals in tissue, CT, is CT = BCF × F × CD, where BCF is the bioconcentration factor (fat basis) for the organism, expressed as {mg/kg fat}/{mg/kg of diet}, and F is the fat content of tissues (in kg fat/kg tissue)
FIR = food (meat and dairy) consumption (kg/day)
GI = gastrointestinal absorption factor
BW = body weight (kg)

The total dose received by the potential receptors from chemical ingestions will in general be dependent on the absorption of the chemical across the gastrointestinal (GI) lining. The scientific literature provides some estimates of such absorption factors for various chemical substances. It is worthwhile to note, though, that for some chemicals for which a carcinogenic SF is available, GI absorption are already implicitly accounted for in some cases. For chemicals without published absorption values and for which absorption factors are not implicitly accounted for in toxicological parameters, absorption may conservatively be assumed to be 100%.

Dermal

For dermal exposure, the calculation of chemical intakes are carried out as follows:

$$\text{Dermal exposure to soil } (mg/kg\text{-}day) = \frac{\{SS \times SA \times CS \times UF \times CF\}}{BW}$$

$$Dermal\ exposure\ to\ water\ (mg\,/\,kg\text{-}day) = \frac{\{WS \times SA \times CW \times UF\}}{BW}$$

where

SS = surface dust on skin ($mg/cm^2/day$)
CS = chemical concentration in soil (mg/kg)
CF = conversion factor (= 1.00E–06 kg/mg)
WS = water contacting skin ($L/cm^2/day$)
CW = chemical concentration in water (mg/L)
SA = exposed skin surface (cm^2)
UF = uptake factor
BW = body weight (kg)

Degradation Factor

Since exposure could be occurring over long time periods (up to an estimated human lifetime of 70 years or more), it is important in a detailed analysis to consider whether degradation or other transformation of the chemical at the source could occur. In such cases, the chemical and biological degradation properties of the contaminant should be reviewed. If significant degradation is likely to occur, exposure calculations become much more complicated. In that case, source contaminant levels must be calculated at frequent intervals and summed over the exposure period. For instance, assuming first-order kinetics, an approximation of the degradation effects can be obtained by multiplying the initial media concentration estimate by a degradation factor, DGF, defined by

$$DGF = \frac{\left(1 - e^{-kt}\right)}{kt}$$

where

k = chemical-specific degradation rate constant ($days^{-1}$)
t = time period over which exposure occurs (days)

For a first-order decaying substance, k is estimated from the following relationship:

$$t_{1/2}[days] = \frac{0.693}{k}$$

where $t_{1/2}$ is the half-life, which is the time after which the mass of a given substance will be one half its initial value. It should be recognized in carrying out all these manipulations, however, that in many cases when a substance is degraded, it produces an end product that could be of potentially equal or greater concern. Consequently, for simplicity, the decay factor will normally be ignored, except in situations where the end product is known to present no potential hazards to potential receptors.

Potentially Exposed Populations

An important step in the quantitative determination of the potential exposures involves the identification of the populations which may be potentially exposed to chemicals originating from a contaminated site. For instance, the difference in sensitivities between adults and children demands that they be treated separately in evaluating their exposure intakes and doses of chemicals/contaminants. Also, due to the variance in activity and behavior of children at different ages, child exposure of soils is usually broken down into two (or more) categories for such an evaluation, for example:

- Children aged up to 6 years (to include infants and preschool children)
- Children aged between 6 and 12 years (to include young children of school-going age)

For the purpose of a risk assessment and consistent with EPA guidance (U.S. EPA, 1989b), all population groups aged more than 12 years are normally included in the adult category.

Inhalation Exposures

Potential inhalation intakes are estimated based on the length of exposure, the inhalation rate of the exposed individual during the event, the concentration of contaminant in the air respired, and the amount retained in the lungs. Two major types of inhalation exposure pathways are generally considered (Figure B.2). The primary pathway is inhalation of airborne contaminants, in which all individuals within approximately 80 km (50 mi) radius of the site are potentially impacted. A secondary exposure pathway is inhalation of VOCs (i.e., airborne, vapor-phase chemicals) during domestic water use for showering. In fact, inhalation of VOCs may be considered for the groundwater sources only, since VOCs are not expected to remain in surface waters for the times required to reach service points of municipal/domestic water supply.

Inhalation of Volatile Compounds

Showering generally represents a system that promotes release of VOCs from water due to high turbulence, high surface area, and small droplets of water involved. Thus, the concentration of the contaminants in the shower air is assumed to be in equilibrium with the concentration in the water (DOE, 1987). In the case of volatile compounds released while bathing, the exposure relationship is defined by (U.S. EPA, 1988; 1989a,b)

$$INH = CW \times \left\{ \left[\frac{ET_1}{(VS \times 2)} \right] + \frac{ET_2}{VB} \right\} \times IR \times RR \times VW \times ABS_s \times EF$$

$$\times ED \times \frac{1}{BW} \times \frac{1}{AT}$$

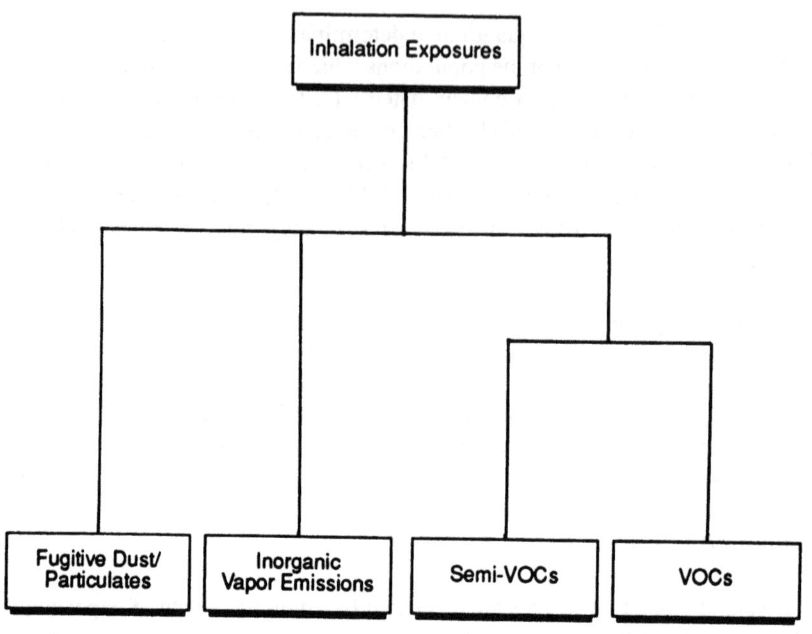

Figure B.2 Major inhalation exposure types.

where

- INH = inhalation intake whiles showering (mg/kg/day)
- CW = concentration of contaminant in water — adjusted for water treatment purification factor, T_f, which is the fraction remaining after treatment (i.e., $CW = CW_{source} \times T_f$) (mg/L)
- ET_1 = length of exposure in shower (h/day)
- ET_2 = length of additional exposure in enclosed bathroom (h/day)
- VS = volume of shower stall (m^3)
- VB = volume of bathroom (m^3)
- IR = breathing/inhalation rate (m^3/h)
- RR = retention rate of inhaled air (%)
- VW = volume of water used in shower (L)
- ABS_s = percent of chemical absorbed into the bloodstream (%)
- EF = exposure frequency (days/year)
- ED = exposure duration (years)
- BW = body weight (kg)
- AT = averaging time (period over which exposure is averaged, days)

The concentration of contaminants in water may further be adjusted for environmental degradation by multiplying by a factor of e^{-kt}, where k (in $days^{-1}$) is the environmental degradation constant of the chemical, and t (in days) is the average time of transit through the water distribution system; this yields a new CW value equal to $(CW)(e^{-kt})$ to be used for the intakes computation.

Particulate Inhalation Exposure — Fugitive Dust

The following relationship is used to calculate intakes as a result of the inhalation of wind-borne fugitive dust by potential receptors (CAPCOA, 1990; U.S. EPA, 1988, 1989a, 1989b):

$$INH = CA \times IR \times RR \times ABS_s \times ET \times EF \times ED \times 1 / BW \times 1 / AT$$

where

INH = inhalation intake (mg/kg/day)
CA = chemical concentration in air (mg/m^3)
IR = inhalation rate (m^3/h)
RR = retention rate of inhaled air (%)
ABS_s = percent of chemical absorbed into the bloodstream (%)
ET = exposure time (h/day)
EF = exposure frequency (days/year)
ED = exposure duration (years)
BW = body weight (kg)
AT = averaging time (period over which exposure is averaged, days)

The contaminant concentration in air, CA, is defined by the ground-level concentration (GLC), represented by the respirable (PM-10) particles, expressed in $\mu g/m^3$.

Ingestion Exposures

The major types of ingestion exposure pathways are shown in Figure B.3. Exposure through ingestion is a function of the concentration of the pollutant in the substance or material ingested (soil, water, or food), the GI absorption of the pollutant in solid or fluid matrix, and the amount ingested. The potential intake due to the ingestion of contaminants present in materials ingested (such as contaminated water or soils or sediments) is determined by multiplying the concentration of the chemical in the medium of concern by the amount of fluid or solids ingested per day and the degree of absorption. In general, exposure to contaminants via the ingestion of contaminated fluids or solids may be estimated according to the following relationship (U.S. EPA, 1988; 1989a,b):

$$ING = CONC \times IR \times CF \times FI \times ABS_s \times EF \times ED \times 1 / BW \times 1 / AT$$

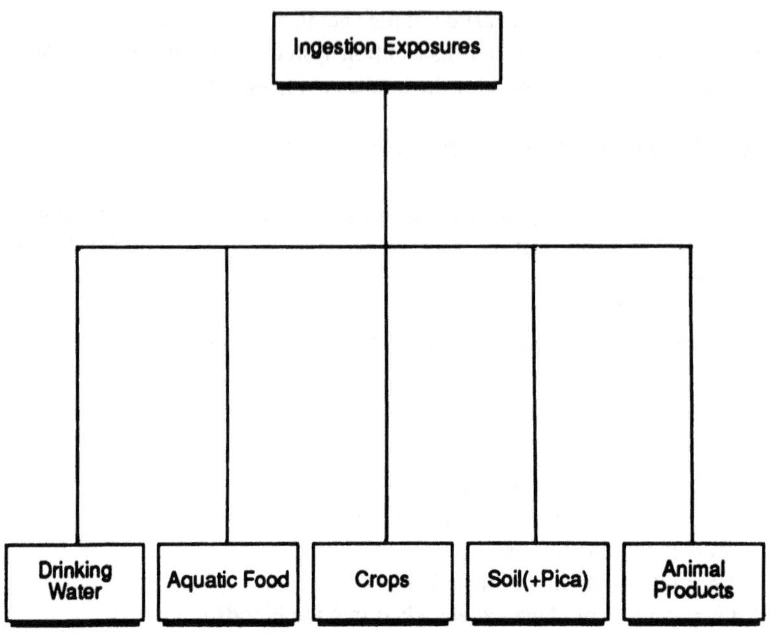

Figure B.3 Major ingestion exposure types.

where

- ING = ingestion intake, adjusted for absorption (mg/kg/day)
- CONC = chemical concentration in media of concern (mg/kg or mg/L)
- IR = ingestion rate (mg or L media material/day)
- CF = conversion factor (1.00E–06 kg/mg for solid media, or 1.00 for fluid media)
- FI = fraction ingested from contaminated source (unitless)
- ABS_s = bioavailability/gastrointestinal (GI) absorption factor (%)
- EF = exposure frequency (days/years)
- ED = exposure duration (years)
- BW = body weight (kg)
- AT = averaging time (period over which exposure is averaged, days)

Exposure Through Ingestion Of Drinking Water

The applicable relationship for the exposure intake through the ingestion of water is as follows:

$$ING_{dw} = CW \times IR \times FI \times ABS_s \times EF \times ED \times 1/BW \times 1/AT$$

where

ING_{dw} = ingestion intake, adjusted for absorption (mg/kg/day)
CW = chemical concentration in water (mg/L)
IR = average water ingestion rate (L/day)
FI = fraction ingested from contaminated source (unitless)
ABS_s = bioavailability/gastrointestinal (GI) absorption factor (%)
EF = exposure frequency (days/years)
ED = exposure duration (years)
BW = body weight (kg)
AT = averaging time (period over which exposure is averaged, days)

Exposure Through Ingestion of Chemicals During Swimming Activities

The applicable relationship for the exposure intake through the ingestion of chemicals in surface water during recreational activities is as follows:

$$ING_r = CW \times CR \times ABS_s \times ET \times EF \times ED \times 1 / BW \times 1 / AT$$

where

ING_r = ingestion intake, adjusted for absorption (mg/kg/day)
CW = chemical concentration in water (mg/L)
CR = contact rate (L/h)
ABS_s = bioavailability/gastrointestinal (GI) absorption factor (%)
ET = exposure time (h/event)
EF = exposure frequency (events/year)
ED = exposure duration (years)
BW = body weight (kg)
AT = averaging time (period over which exposure is averaged, days)

Exposure Through Ingestion Of Food

Exposure from the ingestion of food can be via the ingestion of plant products, fish, animal products, and mother's milk. The applicable relationship for the exposure intake through the ingestion of foods is as follows:

$$ING_f = CF \times IR \times CF \times FI \times ABS_s \times EF \times ED \times 1 / BW \times 1 / AT$$

where

ING_f = ingestion intake, adjusted for absorption (mg/kg/day)
CF = chemical concentration in food (mg/kg or mg/l)
IR = average food ingestion rate (mg or L/meal)
CF = conversion factor (1.00E–06 kg/mg for solids and 1.00 for fluids)

FI = fraction ingested from contaminated source (unitless)
ABS_s = bioavailability/gastrointestinal (GI) absorption factor (%)
EF = exposure frequency (meals/year)
ED = exposure duration (years)
BW = body weight (kg)
AT = averaging time (period over which exposure is averaged, days)

Ingestion of Plant Products. Exposure through ingesting plants, ING_p, is a function of the type of plant, GI absorption factor, and the fraction of plants ingested that are affected by pollutants. The calculation is done for each plant type according to the following relationship (CAPCOA, 1990):

$$ING_p = CP_z \times PIR_z \times FI_z \times ABS_s \times EF \times ED \times 1 / BW \times 1 / AT$$

where

ING_p = exposure intake from ingestion of plant products, adjusted for absorption (mg/kg/day)
CP_z = chemical concentration in plant type Z (mg/kg)
PIR_z = average consumption rate for plant type Z (kg/day)
FI_z = fraction of plant type Z ingested from contaminated source (unitless)
ABS_s = bioavailability/gastrointestinal (GI) absorption factor (%)
EF = exposure frequency (days/years)
ED = exposure duration (years)
BW = body weight (kg)
AT = averaging time (period over which exposure is averaged, days)

Bioaccumulation and Ingestion of Seafood. Exposure from the ingestion of fish from contaminated surface water bodies may be estimated by the following relation (U.S. EPA, 1987; 1988):

$$ING_{sf} = CW \times FIR \times CF \times BCF \times FI \times ABS_s \times EF \times ED \times 1 / BW \times 1 / AT$$

where

ING_{sf} = total exposure, adjusted for absorption (mg/kg/day)
CW = chemical concentration in surface water (mg/L)
FIR = average fish ingestion rate (g/day)
CF = conversion factor (= 1.00E–03 kg/g)
BCF = chemical-specific bioconcentration factor (L/kg)
FI = fraction ingested from contaminated source (unitless)
ABS_s = bioavailability/gastrointestinal (gi) absorption factor (%)
EF = exposure frequency (days/years)
ED = exposure duration (years)

BW = body weight (kg)
AT = averaging time (period over which exposure is averaged, days)

Ingestion of Animal Products. Exposure through ingestion of animal products, ING_a, is a function of the type of meat ingested (including animal milk products and eggs), GI absorption factor, and the fraction of animal products ingested that are affected by pollutants. The calculation is done for each animal product type according to the following relationship (CAPCOA, 1990):

$$ING_a = CAP_z \times APIR_z \times FI_z \times ABS_s \times EF \times ED \times 1 / BW \times 1 / AT$$

where

ING_a = exposure intake through ingestion of plant products, adjusted for absorption (mg/kg/day)
CAP_z = chemical concentration in food type Z (mg/kg)
$APIR_z$ = average consumption rate for food type Z (kg/day)
FI_z = fraction of product type Z ingested from contaminated source (unitless)
ABS_s = bioavailability/gastrointestinal (GI) absorption factor (%)
EF = exposure frequency (days/years)
ED = exposure duration (years)
BW = body weight (kg)
AT = averaging time (period over which exposure is averaged, days).

Ingestion of Mother's Milk. Exposure through ingestion of mother's milk, ING_m, is a function of the average chemical concentration in mother's milk, the amount of mother's milk ingested, and GI absorption factor. This is estimated according to the following relationship (CAPCOA, 1990):

$$ING_m = CMM \times IBM \times ABS_s \times EF \times ED \times 1 / BW \times 1 / AT$$

where

ING_m = exposure intake through ingestion of mother's milk, adjusted for absorption (mg/kg/day)
CMM = chemical concentration in mother's milk — which is a function of mother's exposure through all routes and the contaminant body half-life (mg/kg)
IBM = daily average ingestion rate for breast milk (kg/day)
ABS_s = bioavailability/gastrointestinal (GI) absorption factor (%)
EF = exposure frequency (days/years)
ED = exposure duration (years)
BW = body weight (kg)
AT = averaging time (period over which exposure is averaged, days)

Pica and Incidental Soil/Sediment Ingestion

Absorbed dose due to the incidental ingestion of contaminants sorbed on soils is determined by multiplying the concentration of the contaminant in the medium of concern by the amount of soil ingested per day and the degree of absorption, according to the following relationship (U.S. EPA 1988; 1989a,b; CAPCOA, 1990):

$$ING = CS \times IR \times CF \times FI \times ABS_s \times EF \times ED \times 1/BW \times 1/AT$$

where

ING = ingestion intake, adjusted for absorption (mg/kg/day)
CS = chemical concentration in soil (mg/kg)
IR = average ingestion rate (mg soil/day)
CF = conversion factor (1.00E–06 kg/mg)
FI = fraction ingested from contaminated source (unitless)
ABS_s = bioavailability/gastrointestinal (GI) absorption factor (%)
EF = exposure frequency (days/years)
ED = exposure duration (years)
BW = body weight (kg)
AT = averaging time (period over which exposure is averaged, days)

In general, it normally is assumed that all ingested soil during receptor exposures come from a contaminated source, so that FI becomes unity.

Dermal Exposures

The major types of dermal exposure pathways are shown in Figure B.4. Dermal intake is determined by the chemical concentration in the medium of concern, the body surface area in contact with the medium, the duration of the contact, the flux of the medium across the skin surface, and the absorbed fraction.

Dermal Exposure — Absorption/Soils Contact

The dermal exposures to chemicals in soils and sediments from a site may be estimated by the following relationship (U.S. EPA 1989a,b; 1988; CAPCOA, 1990):

$$DEX = CS \times CF \times SA \times AF \times ABS_s \times SM \times EF \times ED \times 1/BW \times 1/AT$$

where

DEX = absorbed dose (mg/kg/day)
CS = chemical concentration in soil (mg/kg)
CF = conversion factor (1.00E–06 kg/mg)
SA = skin surface area available for contact, i.e., surface area of exposed skin (cm^2/event)

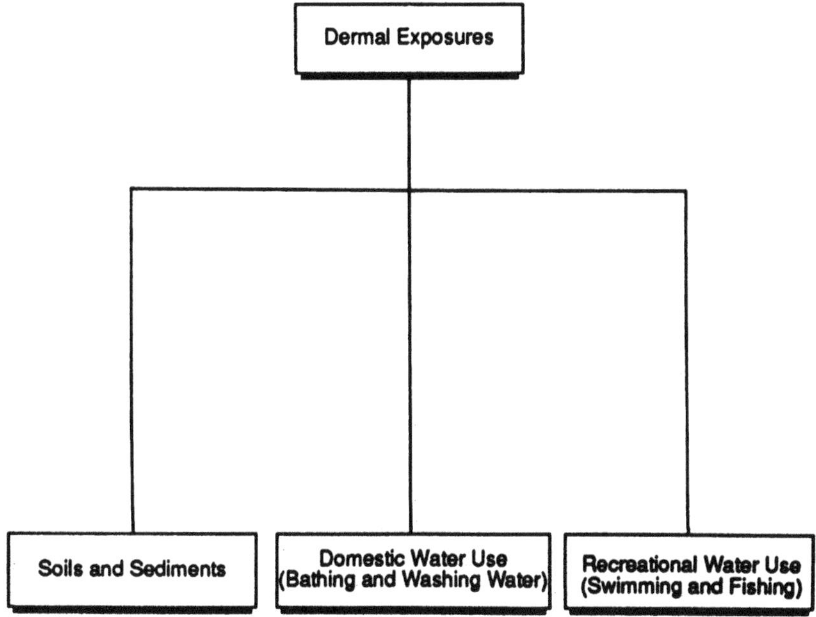

Figure B.4 Major dermal contact exposure types.

AF = soil to skin adherence factor, i.e., soil loading on skin (mg/cm^2)
ABS_s = skin absorption factor for chemicals in soil (%)
SM = factor for soil matrix effects (%)
EF = exposure frequency (events/year)
ED = exposure duration (years)
BW = body weight (kg)
AT = averaging time (period over which exposure is averaged, days)

Dermal Exposure to Waters and Seeps

Dermal exposures to chemicals may occur during domestic use (such as bathing and washing) or through recreational activities (such as swimming or fishing). The dermal intakes of chemicals in ground- or surface water and/or from seeps from a site may be estimated by the following relationship (U.S. EPA, 1988; 1989a,b):

$$DEX = CW \times SA \times PC \times ABS_s \times CF \times ET \times EF \times ED \times 1/BW \times 1/AT$$

where

DEX = total exposure (mg/kg/day)
CW = chemical concentration in water (mg/L)

SA = skin surface area available for contact, i.e., surface area of exposed skin (cm^2)
PC = chemical-specific dermal permeability constant (cm/h)
ABS_s = skin absorption factor for chemicals in water (%)
CF = volumetric conversion factor for water (1 L/1000 cm^3)
ET = exposure time (h/day)
EF = exposure frequency (days/year)
ED = exposure duration (years)
BW = body weight (kg)
AT = averaging time (period over which exposure is averaged, days)

Typical Computations

In general, default values may be obtained from literature for some of the parameters used in the estimation of intakes and doses. Table B.1 gives typical parameters commonly used; this is by no means complete. More detailed information can be obtained from several sources (e.g., U.S. EPA, 1987, 1988, 1989a; CAPCOA, 1990). A spreadsheet for automatically calculating intake factors for site-specific problems may be designed as shown in Table B.2. Numerical examples for potential receptor groups expected to be exposed through inhalation, soil ingestion (i.e., incidental or pica behavior), and dermal contact are discussed below. The same set of units are maintained throughout as given above.

Inhalation Exposures

The daily inhalation intake of fugitive dust for various population groups are calculated for both carcinogenic and noncarcinogenic effects. The assumed variables used in the numerical demonstration are given in Table B.1.

Carcinogenic Effects from a Contaminated Site — Estimation of LADD. For the fugitive dust inhalation pathway, the carcinogenic CDI (also the LADD) is estimated for the different population groups identified to represent the critical receptors.

The carcinogenic CDI for children aged up to 6 years is calculated to be

$$
\begin{aligned}
&CInh_{(1\text{-}6)} \\
&= CA \times IR \times RR \times ABS_s \times ET \times EF \times ED \times 1/BW \times 1/AT \\
&= [CA] \times 0.25 \times 1 \times ABS_s \times 12 \times 365 \times 5 \times 1/16 \times 1/(70 \times 365) \\
&= 1.34E\text{-}02 \times ABS_s \times [CA]
\end{aligned}
$$

The carcinogenic CDI for children aged 6 to 12 years is calculated to be

$$
\begin{aligned}
&CInh_{(6\text{-}12)} \\
&= CA \times IR \times RR \times ABS_s \times ET \times EF \times ED \times 1/BW \times 1/AT \\
&= [CA] \times 0.46 \times 1 \times ABS_s \times 12 \times 365 \times 6 \times 1/29 \times 1/(70 \times 365) \\
&= 1.63E\text{-}02 \times ABS_s \times [CA]
\end{aligned}
$$

The carcinogenic CDI for adult residents is calculated to be

$CInh_{(adultR)}$
$= CA \times IR \times RR \times ABS_s \times ET \times EF \times ED \times 1/BW \times 1/AT$
$= [CA] \times 0.83 \times 1 \times ABS_s \times 12 \times 365 \times 58 \times 1/70 \times 1/(70 \times 365)$
$= 1.18E\text{-}01 \times ABS_s \times [CA]$

The carcinogenic CDI for adult workers is calculated to be

$CInh_{(adultW)}$
$= CA \times IR \times RR \times ABS_s \times ET \times EF \times ED \times 1/BW \times 1/AT$
$= [CA] \times 0.83 \times 1 \times ABS_s \times 8 \times 260 \times 58 \times 1/70 \times 1/(70 \times 365)$
$= 5.60E\text{-}02 \times ABS_s \times [CA]$

Noncarcinogenic Effects from a Contaminated Site — Estimation of ADD. For the fugitive dust inhalation pathway, the noncarcinogenic CDI (also, the ADD) is estimated for the different population groups identified to represent the critical receptors.

The noncarcinogenic CDI for children aged up to 6 years is calculated to be

$NCInh_{(1\text{-}6)}$
$= CA \times IR \times RR \times ABS_s \times ET \times EF \times ED \times 1/BW \times 1/AT$
$= [CA] \times 0.25 \times 1 \times ABS_s \times 12 \times 365 \times 5 \times 1/16 \times 1/(5 \times 365)$
$= 1.88E\text{-}01 \times ABS_s \times [CA]$

The noncarcinogenic CDI for children aged 6 to 12 years is calculated to be

$NCInh_{(6\text{-}12)}$
$= CA \times IR \times RR \times ABS_s \times ET \times EF \times ED \times 1/BW \times 1/AT$
$= [CA] \times 0.46 \times 1 \times ABS_s \times 12 \times 365 \times 6 \times 1/29 \times 1/(6 \times 365)$
$= 1.90E\text{-}01 \times ABS_s \times [CA]$

The noncarcinogenic CDI for adult residents is calculated to be

$NCInh_{(adultR)}$
$= CA \times IR \times RR \times ABS_s \times ET \times EF \times ED \times 1/BW \times 1/AT$
$= [CA] \times 0.83 \times 1 \times ABS_s \times 12 \times 365 \times 58 \times 1/70 \times 1/(58 \times 365)$
$= 1.42E\text{-}01 \times ABS_s \times [CA]$

The noncarcinogenic CDI for adult workers is calculated to be

$NCInh_{(adultW)}$
$= CA \times IR \times RR \times ABS_s \times ET \times EF \times ED \times 1/BW \times 1/AT$
$= [CA] \times 0.83 \times 1 \times ABS_s \times 8 \times 260 \times 58 \times 1/70 \times 1/(58 \times 365)$
$= 6.76E\text{-}02 \times ABS_s \times [CA]$

Table B.1 Example Case-Specific Parameters for Exposure Assessment

Parameter	Children Aged Up to 6	Children Aged 6–12	Adult	Reference Sources
Physical characteristics				
Average body weight	16 kg	29 kg	70 kg	a,b,c
Average total skin surface area	6980 cm^2	10,470 cm^2	18,150 cm^2	a,b,e,h
Average lifetime			70 years	a,b,c,e
Average lifetime exposure period	5 years	6 years	58 years	b,e
Activity characteristics				
Inhalation rate	0.25 m^3/h	0.46 m^3/h	0.83 m^3/h	b,e
Retention rate of inhaled air	100%	100%	100%	e
Frequency of fugitive dust inhalation				
Offsite residents, schools, and passers-by	365 days/year	365 days/year	365 days/year	b,e
Offsite workers	—	—	260 days/year	b,e
Duration of fugitive dust inhalation (outside)				
Offsite residents, schools, and passers-by	12 h/day	12 h/day	12 h/day	b,e
Offsite workers	—	—	8 h/day	b,e
Amount of soil ingested incidentally	200 mg/day	100 mg/day	50 mg/day	a,b,c,e,h,i
Frequency of soil contact				
Offsite residents, schools, and passers-by	330 days/year	330 days/year	330 days/year	b,e
Offsite workers	—	—	260 days/year	b,e
Activity characteristics				
Duration of soil contact				
Offsite residents, schools, and passers-by	12 h/day	8 h/day	8 h/day	b,e
Offsite workers	—	—	8 h/day	b,e
Percentage of skin area contacted by soil	20%	20%	10%	b,e,h

Material characteristics				
Soil to skin adherence factor	0.75 mg/cm^2	0.75 mg/cm^2	0.75 mg/cm^2	a,b,e,f,g
Soil to matrix attenuation factor	15%	15%	15%	d

Note: The exposure factors represented here are for potential maximum exposures (for conservative estimates) and could be modified as appropriate to reflect the most reasonable exposure patterns anticipated. For instance, soil exposure will be reduced by snow cover and rainy days, thus reducing potential exposures for children playing in contaminated areas.

[a] U.S. EPA (1989b).
[b] U.S. EPA (1989a).
[c] U.S. EPA (1988).
[d] Hawley (1985).
[e] Estimate based on site-specific conditions.
[f] Lepow et al. (1975).
[g] Lepow et al. (1974).
[h] Sedman (1989).
[i] Calabrese et al. (1989).

Table B.2 Example Spreadsheet for Calculating Case-Specific Intake Factors in an Exposure Assessment[a]

	Fugitive Dust Inhalation Pathway								Soil Ingestion Pathway							
Group	IR	RR	ET	EF	ED	BW	AT	INH Factor	IR	CF	FI	EF	ED	BW	AT	ING Factor
C(1-6)@NCarc	0.25	1	12	365	5	16	1825	1.88E–01	200	1.00E–06	1	330	5	16	1825	1.13E–05
C(1-6)@Carc	0.25	1	12	365	5	16	25550	1.34E–02	200	1.00E–06	1	330	5	16	25550	8.07E–07
C(6-12)@NCarc	0.46	1	12	365	6	29	2190	1.90E–01	100	1.00E–06	1	330	6	29	2190	3.12E–06
C(6-12)@Carc	0.46	1	12	365	6	29	25550	1.63E–02	100	1.00E–06	1	330	6	29	25550	2.67E–07
ResAdult@NCarc	0.83	1	12	365	58	70	21170	1.42E–01	50	1.00E–06	1	330	58	70	21170	6.46E–07
ResAdult@Carc	0.83	1	12	365	58	70	25550	1.18E–01	50	1.00E–06	1	330	58	70	25550	5.35E–07
JobAdult@NCarc	0.83	1	8	260	58	70	21170	6.76E–02	50	1.00E–06	1	260	58	70	21170	5.09E–07
JobAdult@Carc	0.83	1	8	260	58	70	25550	5.60E–02	50	1.00E–06	1	260	58	70	25550	4.22E–07

Soil Dermal Contact Pathway								
SA	CF	AF	SM	EF	ED	BW	AT	DEX Factor
1396	1.00E–06	0.75	0.15	330	5	16	1825	8.87E–06
1396	1.00E–06	0.75	0.15	330	5	16	25550	6.34E–07
2094	1.00E–06	0.75	0.15	330	6	29	2190	7.34E–06
2094	1.00E–06	0.75	0.15	330	6	29	25550	6.30E–07
1815	1.00E–06	0.75	0.15	330	58	70	21170	2.64E–06
1815	1.00E–06	0.75	0.15	330	58	70	25550	2.19E–06
1815	1.00E–06	0.75	0.15	330	58	70	21170	2.08E–06
1815	1.00E–06	0.75	0.15	330	58	70	25550	1.72E–06

[a] Notations and units are as defined in the text.
INH Factor = inhalation factor for calculation of doses and intakes.
ING Factor = soil ingestion factor for calculation of doses and intakes.
DEX Factor = dermal exposure/skin adsorption factor for calculation of doses and intakes.
C(1-6)NCarc = noncarcinogenic effects for children aged 1 to 6 years.
C(1-6)@Carc = carcinogenic effects for children ages 1 to 6 years.
C(6-12)@NCarc = noncarcinogenic effects for children aged 6 to 12 years.
C(6-12)@Carc = carcinogenic effects for children aged 6 to 12 years.
ResAdult@NCarc = noncarcinogenic effects for resident adults.
ResAdult@Carc = carcinogenic effects for resident adults.
JobAdult@NCarc = noncarcinogenic effects for adult workers.
JobAdult@Carc = carcinogenic effects for adult workers.

Ingestion Exposures

The daily ingestion intake of soils for various population groups are calculated for both carcinogenic and noncarcinogenic effects. The assumed variables used in the numerical demonstration are given in Table B.1.

Carcinogenic Effects from a Contaminated Site — Estimation of LADD. For the soil ingestion pathway, the carcinogenic CDI (also the LADD) is estimated for the different population groups identified to represent the critical receptors.

The carcinogenic CDI for children aged up to 6 years is calculated to be

$$CIng_{(1\text{-}6)}$$
$$= CS \times IR \times CF \times FI \times ABS_s \times EF \times ED \times 1/BW \times 1/AT$$
$$= [CS] \times 200 \times 1.00E\text{-}06 \times 1 \times ABS_s \times 330 \times 5 \times 1/16 \times 1/(70 \times 365)$$
$$= 8.07E\text{-}07 \times ABS_s \times [CS]$$

The carcinogenic CDI for children aged 6 to 12 years is calculated to be

$$CIng_{(6\text{-}12)}$$
$$= CS \times IR \times CF \times FI \times ABS_s \times EF \times ED \times 1/BW \times 1/AT$$
$$= [CS] \times 100 \times 1.00E\text{-}06 \times 1 \times ABS_s \times 330 \times 6 \times 1/29 \times 1/(70 \times 365)$$
$$= 2.67E\text{-}07 \times ABS_s \times [CS]$$

The carcinogenic CDI for adult residents is calculated to be

$$CIng_{(adultR)}$$
$$= CS \times IR \times CF \times FI \times ABS_s \times EF \times ED \times 1/BW \times 1/AT$$
$$= [CS] \times 50 \times 1.00E\text{-}06 \times 1 \times ABS_s \times 330 \times 58 \times 1/70 \times 1/(70 \times 365)$$
$$= 5.35E\text{-}07 \times ABS_s \times [CS]$$

The carcinogenic CDI for adult workers is calculated to be

$$CIng_{(adultW)}$$
$$= CS \times IR \times CF \times FI \times ABS_s \times EF \times ED \times 1/BW \times 1/AT$$
$$= [CS] \times 50 \times 1.00E\text{-}06 \times 1 \times ABS_s \times 260 \times 58 \times 1/70 \times 1/(70 \times 365)$$
$$= 4.22E\text{-}07 \times ABS_s \times [CS]$$

Noncarcinogenic Effects from a Contaminated Site — Estimation of ADD. For the soil ingestion pathway, the noncarcinogenic CDI (also ADD) is estimated for the different population groups identified to represent the critical receptors.

The noncarcinogenic CDI for children aged up to 6 years is calculated to be

$$NCIng_{(1\text{-}6)}$$
$$= CS \times IR \times CF \times FI \times ABS_s \times EF \times ED \times 1/BW \times 1/AT$$
$$= [CS] \times 200 \times 1.00E\text{-}06 \times 1 \times ABS_s \times 330 \times 5 \times 1/16 \times 1/(5 \times 365)$$
$$= 1.13E\text{-}05 \times ABS_s \times [CS]$$

The noncarcinogenic CDI for children aged 6 to 12 years is calculated to be

$NCIng_{(6\text{-}12)}$
$= CS \times IR \times CF \times FI \times ABS_s \times EF \times ED \times 1/BW \times 1/AT$
$= [CS] \times 100 \times 1.00E\text{-}06 \times 1 \times ABS_s \times 330 \times 6 \times 1/29 \times 1/(6 \times 365)$
$= 3.12E\text{-}06 \times ABS_s \times [CS]$

The noncarcinogenic CDI for adult residents is calculated to be

$NCIng_{(adultR)}$
$= CS \times IR \times CF \times FI \times ABS_s \times EF \times ED \times 1/BW \times 1/AT$
$= [CS] \times 50 \times 1.00E\text{-}06 \times 1 \times ABS_s \times 330 \times 58 \times 1/70 \times 1/(58 \times 365)$
$= 6.46E\text{-}07 \times ABS_s \times [CS]$

The noncarcinogenic CDI for adult workers is calculated to be

$NCIng_{(adultW)}$
$= CS \times IR \times CF \times FI \times ABS_s \times EF \times ED \times 1/BW \times 1/AT$
$= [CS] \times 50 \times 1.00E\text{-}06 \times 1 \times ABS_s \times 260 \times 58 \times 1/70 \times 1/(58 \times 365)$
$= 5.09E\text{-}07 \times ABS_s \times [CS]$

Dermal Exposures

The daily dermal intake of soils for various population groups are calculated for both carcinogenic and noncarcinogenic effects. The assumed variables used in the numerical demonstration are given in Table B.1.

Carcinogenic Effects from a Contaminated Site — Estimation of LADD. For the soil dermal contact pathway, the carcinogenic CDI (also LADD) is estimated for the different population groups identified to represent the critical receptors.

The carcinogenic CDI for children aged up to 6 years is calculated to be

$CDEX_{(1\text{-}6)}$
$= CS \times CF \times SA \times AF \times ABS_s \times SM \times EF \times ED \times 1/BW \times 1/AT$
$= [CS] \times 1.00E\text{-}06 \times 1396 \times 0.75 \times ABS_s \times 0.15 \times 330 \times 5 \times 1/16 \times 1/(70 \times 365)$
$= 6.34E\text{-}07 \times ABS_s \times [CS]$

The carcinogenic CDI for children aged 6 to 12 years is calculated to be

$CDEX_{(6\text{-}12)}$
$= CS \times CF \times SA \times AF \times ABS_s \times SM \times EF \times ED \times 1/BW \times 1/AT$
$= [CS] \times 1.00E\text{-}06 \times 2094 \times 0.75 \times ABS_s \times 0.15 \times 330 \times 6 \times 1/29 \times 1/(70 \times 365)$
$= 6.30E\text{-}07 \times ABS_s \times [CS]$

The carcinogenic CDI for adult residents is calculated to be

$CDEX_{(adultR)}$
$= CS \times CF \times SA \times AF \times ABS_s \times SM \times EF \times ED \times 1/BW \times 1/AT$
$= [CS] \times 1.00E\text{-}06 \times 1815 \times 0.75 \times ABS_s \times 0.15 \times 330 \times 58 \times 1/70 \times 1/(70 \times 365)$
$= 2.19E\text{-}06 \times ABS_s \times [CS]$

The carcinogenic CDI for adult workers is calculated to be

$CDEX_{(adultW)}$
$= CS \times CF \times SA \times AF \times ABS_s \times SM \times EF \times ED \times 1/BW \times 1/AT$
$= [CS] \times 1.00E\text{-}06 \times 1815 \times 0.75 \times ABS_s \times 0.15 \times 260 \times 58 \times 1/70 \times 1/(70 \times 365)$
$= 1.72E\text{-}06 \times ABS_s \times [CS]$

Noncarcinogenic Effects from a Contaminated Site — Estimation of ADD. For the soil dermal contact pathway, the noncarcinogenic CDI (also ADD) is estimated for the different population groups identified to represent the critical receptors.

The noncarcinogenic CDI for children aged up to 6 years is calculated as follows

$NCDEX_{(1\text{-}6)}$
$= CS \times CF \times SA \times AF \times ABS_s \times SM \times EF \times ED \times 1/BW \times 1/AT$
$= [CS] \times 1.00E\text{-}06 \times 1396 \times 0.75 \times ABS_s \times 0.15 \times 330 \times 5 \times 1/16 \times 1/(5 \times 365)$
$= 8.87E\text{-}06 \times ABS_s \times [CS]$

The noncarcinogenic CDI for children aged 6 to 12 years is calculated to be

$NCDEX_{(6\text{-}12)}$
$= CS \times CF \times SA \times AF \times ABS_s \times SM \times EF \times ED \times 1/BW \times 1/AT$
$= [CS] \times 1.00E\text{-}06 \times 2094 \times 0.75 \times ABS_s \times 0.15 \times 330 \times 6 \times 1/29 \times 1/(6 \times 365)$
$= 7.34E\text{-}06 \times ABS_s \times [CS]$

The noncarcinogenic CDI for adult residents is calculated to be

$NCDEX_{(adultR)}$
$= CS \times CF \times SA \times AF \times ABS_s \times SM \times EF \times ED \times 1/BW \times 1/AT$
$= [CS] \times 1.00E\text{-}06 \times 1815 \times 0.75 \times ABS_s \times 0.15 \times 330 \times 58 \times 1/70 \times 1/(58 \times 365)$
$= 2.64E\text{-}06 \times ABS_s \times [CS]$

The noncarcinogenic CDI for adult workers is calculated to be

$NCDEX_{(adultW)}$
$= CS \times CF \times SA \times AF \times ABS_s \times SM \times EF \times ED \times 1/BW \times 1/AT$
$= [CS] \times 1.00E\text{-}06 \times 1815 \times 0.75 \times ABS_s \times 0.15 \times 260 \times 58 \times 1/70 \times 1/(58 \times 365)$
$= 2.08E\text{-}06 \times ABS_s \times [CS]$

B.2 RISK CHARACTERIZATION: EQUATIONS FOR CALCULATING CARCINOGENIC RISKS AND NONCARCINOGENIC HAZARD INDICES

Estimation of Carcinogenic Risks

The methodology for calculating carcinogenic risks of chemicals in the environment is well documented in materials prepared under the auspices of various regulatory agencies (e.g., CDHS, 1986; U.S. EPA, 1989b; CAPCOA, 1990). Cancer risk is a function of the LADD and the chemical-specific potency slope. For inhalation, cancer risk can be calculated using unit risk factors (URFs) or unit risk (UR) values and ground-level concentrations (GLCs). Thus,

$$\text{Risk (for noninhalation pathways)} = \text{dose} \times \text{potency slope}$$

and

$$\text{Risk (for inhalation pathway)} = \text{GLC} \times \text{UR}$$

where dose (or sum of doses from all routes of exposure) is expressed in mg/kg/day; the chemical-specific potency slope is given in units of $(\text{mg/kg/day})^{-1}$; GLC is expressed in $\mu g/m^3$; and the chemical-specific UR is in $(\mu g/m^3)^{-1}$.

More generally, the carcinogenic effects of the chemicals of concern are calculated according to the following relationship (CDHS, 1986; U.S. EPA, 1989b):

$$CR = CDI \times SF$$

where

CR = probability of an individual developing cancer (unitless)
CDI = chronic daily intake averaged over a lifetime, say 70 years (mg/kg/day)
SF = slope factor (1/[mg/kg/day])

This represents the linear low-dose cancer risk model and is valid only at low risk levels (i.e., below estimated risks of 0.01). For sites where chemical intakes are high (i.e., potential risks above 0.01), the one-hit model is used. The one-hit equation for high carcinogenic risk levels is given by the following relationship:

$$CR = \left[1 - \exp(-CDI \times SF)\right]$$

where the terms are the same as defined above for the low-dose model.

The aggregate cancer risk equation for multiple substances is subsequently obtained by summing the risks calculated for the individual chemicals using the above relationship(s). Thus, for multiple compounds,

$$Total\ risk = \sum_{i=1}^{n}\left(CDI_i \times SF_i\right)$$

for the linear low-dose model for low risk levels, or

$$Total\ risk = \sum_{i=1}^{n}\left(1 - \exp\left(-CDI_i \times SF_i\right)\right)$$

for the one-hit model used at high carcinogenic risk levels,

where
CDI_i = chronic daily intake for the ith contaminant
SF_i = slope factor for the ith contaminant
n = total number of carcinogens

For multiple compounds and multiple pathways, the overall total cancer risk for all exposure pathways and all contaminants considered in the risk evaluation will be

$$Overall\ total\ risk = \sum_{j=1}^{p}\sum_{i=1}^{n}\left(CDI_{ij} \times SF_{ij}\right)$$

for the linear low-dose model for low risk levels, or

$$Overall\ total\ risk = \sum_{j=1}^{p}\sum_{i=1}^{n}\left\{1 - \exp\left(-CDI_{ij} \times SF_{ij}\right)\right\}$$

for the one-hit model used at high carcinogenic risk levels,

where
CDI_{ij} = chronic daily intake for the ith contaminant and j^{th} pathway
SF_{ij} = slope factor for the ith contaminant and j^{th} pathway
n = total number of carcinogens
p = total number of pathways or exposure routes

The CDIs are estimated from the equations previously discussed in Appendix B.1 for calculating chemical intakes. The SF values are obtained from various sources or databases, including the IRIS and the HEAST, maintained by the U.S. EPA and other regulatory agencies. As a rule of thumb, incremental risks of between 10^{-7} and 10^{-4} are generally perceived as acceptable levels for the protection of human health.

Estimation of Noncarcinogenic Hazards

The methodology for calculating noncarcinogenic hazards of chemicals in the environment is well documented in materials prepared under the auspices of various regulatory agencies (e.g., CDHS, 1986; U.S. EPA, 1989b; CAPCOA, 1990). The noncarcinogenic effects of the chemicals of concern are calculated according to the following relationship (CDHS, 1986; U.S. EPA, 1989b):

$$\text{Hazard quotient, } HQ = \frac{E}{RfD}$$

where

E = chemical exposure level or intake (mg/kg/day)
RfD = reference dose (mg/kg/day)

The sum total of the hazard quotients for all chemicals of concern gives the HI for a given exposure pathway. The applicable relationship is

$$\text{Total hazard index, } HI = \sum_{i=1}^{n} \frac{E_i}{RfD_i}$$

where

E_i = exposure level (or intake) for the ith contaminant
RfD_i = acceptable intake level (or reference dose) for ith contaminant
n = total number of chemicals presenting noncarcinogenic effects

For multiple compounds and multiple pathways, the overall total noncancer risk for all exposure pathways and all contaminants considered in the risk evaluation will be

$$\text{Overall total hazard index} = \sum_{j=1}^{p} \sum_{i=1}^{n} \frac{E_{ij}}{RfD_{ij}}$$

where

E_{ij} = exposure level (or intake) for the ith contaminant and jth pathway
RfD_{ij} = acceptable intake level (or reference dose) for ith contaminant and jth pathway

The E values are estimated from the equations previously discussed in Appendix

B.1 for calculating chemical intakes. The RfD values are obtained from databases such as IRIS and HEAST maintained by the U.S. EPA and other regulatory agencies. In accordance with the U.S. EPA guidelines on the interpretation of HIs, for any given chemical, there may be potential for adverse health effects if the HI exceeds unity. For HI values greater than unity, the higher the value, the greater is the likelihood of adverse noncarcinogenic health impacts. In a comprehensive evaluation, it becomes necessary to introduce the idea of physiologic endpoints in the calculation process, in which case chemicals affecting the same target organs (i.e., chemicals determined to have the same physiologic endpoint) are grouped together in the calculation of total HI.

B.3 DEVELOPMENT OF HEALTH-BASED SITE CLEANUP CRITERIA: EQUATIONS FOR CALCULATING SOIL CLEANUP LEVELS FOR REMEDIAL ACTION PLANS

Introduction

The site cleanup level is a site-specific criterion that a remedial action would have to satisfy in order to keep exposures of potential receptors to levels at or below an AL. The ALs tend to drive the cleanup process for a contaminated site. The ALs are calculated for both the systemic toxicants and for the carcinogens; the more stringent of the two, where both exist, is selected as the site cleanup limit. This would represent the maximum acceptable contaminant level for site cleanup.

ALs for Systemic Toxicants

The governing equation for calculating action levels for noncarcinogens and noncarcinogenic effects of carcinogens present at a contaminated site is given by (U.S. EPA, 1987)

$$C_m = \frac{(RfD \times BW \times CF)}{(I \times A)}$$

where

C_m = AL in medium of concern (e.g., soil @ mg/kg)
RfD = reference dose (mg/kg/day)
BW = body weight (kg)
CF = conversion factor (e.g., 10^6 mg/kg for soil ingestion exposures)
I = intake assumption (e.g., soil ingestion rate @ mg/day)
A = absorption factor (dimensionless)

ALs for Carcinogenic Constituents

The governing equation for calculating ALs for carcinogenic constituents present at a contaminated site is given by (U.S. EPA, 1987):

$$C_m = \frac{(R \times BW \times LT \times CF)}{(SF \times I \times A \times ED)}$$

where

C_m = AL (equal to the RSD or VSD) in medium of concern (e.g., soil @ mg/kg)
R = specified (acceptable) risk level (dimensionless)
BW = body weight (kg)
LT = assumed lifetime (years)
CF = conversion factor (e.g., 10^6 mg/kg for soil ingestion exposures)
SF = cancer slope factor (1/[mg/kg/day])
I = intake assumption (e.g., soil ingestion rate @ mg/day)
A = absorption factor (dimensionless)
ED = exposure duration (years)

Allowable Soil Concentrations (ASCs)

To determine ASCs, the following relationships are used based on an algebraic manipulations of the HI or carcinogenic risk equations and the exposure estimation equations.

Noncarcinogenic Effects

The HI is given by

$$HI = \Sigma \left\{ \sum_{i=1}^{p} \frac{CDI}{RfD_p} \right\} = \frac{CDI_{inh}}{RfD_{inh}} + \frac{CDI_{ing}}{RfD_{ing}} + \frac{CDI_{der}}{RfD_{der}}$$

Assuming there is only one toxic constitiuent present in soils and that exposures via the dermal contact and ingestion routes only contribute to the total HI of 1 (a conservative assumption), then:

$$\Sigma CDI = RfD$$

or

$$\left(CDI_{ing} + CDI_{der}\right) = RfD_{oral}$$

$$\frac{\left(ASC \times IR \times CF \times FI \times ABS_{si} \times EF \times ED\right)}{(BW \times AT)} + \frac{\left(ASC \times CF \times SA \times AF \times ABS_{sd} \times SM \times EF \times ED\right)}{(BW \times AT)} = RfD_{oral}$$

or

$$ASC = \frac{(BW \times AT) \times \left(RfD_{oral}\right)}{(CF \times EF \times ED)\left\{\left(IR \times FI \times ABS_{si}\right) + \left(SA \times AF \times ABS_{sd} \times SM\right)\right\}}$$

where

CDI = chronic daily intake, adjusted for absorption (mg/kg/day)
ASC = allowable chemical concentration in soil (mg/kg)
RfD_{oral} = oral reference dose (mg/kg/day)
IR = ingestion rate (mg/day)
CF = conversion factor (1.00E–06 kg/mg)
FI = fraction ingested from contaminated source (unitless)
ABS_{si} = bioavailability absorption factor for ingestion exposures (%)
ABS_{sd} = bioavailability absorption factor for dermal exposures (%)
SA = skin surface area available for contact, i.e., surface area of exposed skin (cm^2/event)
AF = soil to skin adherence factor, i.e., soil loading on skin (mg/cm^2)
SM = factor for soil matrix effects (%)
EF = exposure frequency (days/years)
ED = exposure duration (years)
BW = body weight (kg)
AT = averaging time (period over which exposure is averaged, days)

Carcinogenic Effects

The cancer risk is given by

$$CR = \Sigma\left\{\sum_{i=1}^{p} CDI \times SF_p\right\} = \left(CDI_{inh} \times SF_{inh}\right) + \left(CDI_{ing} \times SF_{ing}\right) + \left(CDI_{der} \times SF_{der}\right)$$

Assuming there is only one toxic constitiuent present in soils and that exposures via the dermal and ingestion routes only contribute to the total CR of R ($=10^{-6}$, for example) (a conservative assumption), then

$$\sum CDI = \frac{R}{SF_{oral}} = RSD$$

or

$$\left(CDI_{ing} + CDI_{der}\right) = \frac{R}{SF_{oral}}$$

$$\frac{\left(ASC \times IR \times CF \times FI \times ABS_{si} \times EF \times ED\right)}{(BW \times AT)} + \frac{\left(ASC \times CF \times SA \times AF \times ABS_{sd} \times SM \times EF \times ED\right)}{(BW \times AT)} = \frac{R}{SF_{oral}}$$

or

$$ASC = \frac{(BW \times AT) \times (RSD)}{(CF \times EF \times ED)\left\{\left(IR \times FI \times ABS_{si}\right) + \left(SA \times AF \times ABS_{sd} \times SM\right)\right\}}$$

where

CDI = chronic daily intake, adjusted for absorption (mg/kg/day)
RSD = risk-specific dose (mg/kg/day)
ASC = allowable chemical concentration in soil (mg/kg)
SF_{oral} = oral slope factor (1/mg/kg/day)
IR = ingestion rate (mg/day)
CF = conversion factor (1.00E–06 kg/mg)
FI = fraction ingested from contaminated source (unitless)
ABS_{si} = bioavailability absorption factor for ingestion exposures (%)
ABS_{sd} = bioavailability absorption factor for dermal exposures (%)
SA = skin surface area available for contact, i.e., surface area of exposed skin (cm^2/event)
AF = soil to skin adherence factor, i.e., soil loading on skin (mg/cm^2)
SM = factor for soil matrix effects (%)
EF = exposure frequency (days/years)
ED = exposure duration (years)
BW = body weight (kg)
AT = averaging time (period over which exposure is averaged, days)

RSCLs

The RSCL is estimated in the same way as the soil AL or other equivalent methods, but with a consideration for the aggregation of the individual chemicals present at the case site. Then, assuming each compound contributes proportionately to the total carcinogenic risk and/or HI, the RSCL is estimated according to the following simplistic relationship:

$$RSCL = \frac{C_m}{N}$$

where

C_m	=	AL in medium of concern (e.g., soil @ mg/kg)
N	=	number of chemical contributors to overall HI or cancer risk, as appropriate

Also, the RSCL may alternatively be estimated by proportionately aggregating — or rather disaggregating — the target cancer risk (for carcinogens) or noncancer HI (for noncarcinogenic effects) between the chemicals of potential concern. This is carried out according to the following approximate relationships:

$$RSCL = \frac{(\% \times R \times BW \times LT \times CF)}{(SF \times I \times A \times ED)}$$

for carcinogens, and

$$RSCL = \frac{(\% \times RfD \times BW \times CF)}{(I \times A)}$$

for noncarcinogens. All the terms are the same as defined above, and % represents the proportionate contribution from a specific constituent to the target/acceptable risk level (for carcinogens) or HI (for systemic toxicants).

The assumption used here for allocating estimated excess carcinogenic risk is that all carcinogens have the same mode of biological actions and target organs; otherwise, excess carcinogenic risk is not allocated among carcinogens, but rather each assumes the same value in the computational efforts. Similarly, for the noncarcinogenic effects, the total HI is apportioned only between chemicals with the same toxicological endpoint.

B.4 REFERENCES

CAPCOA (California Air Pollution Control Officers Association). "Air Toxics 'Hot Spots' Program. Risk Assessment Guidelines," California Air Pollution Control Officers Association, California (1990).

CDHS (California Department of Health Services). "The California Site Mitigation Decision Tree Manual," California Department of Health Services, Toxic Substances Control Division, Sacremento, CA (1986).

CDHS (California Department of Health Services). "The California Site Mitigation Decision Tree Manual," California Department of Health Services, Toxic Substances Control Division, Sacremento, CA (1986).

Calabrese et al. "How Much Soil Do Young Children Ingest? An Epidemiological Study," *Regul. Toxicol. Pharmacol.* 10:123–137 (1989).

DOE (U.S. Department of Energy). "The Remedial Action Priority System (RAPS): Mathematical Formulations," U.S. Dept. of Energy, Office of Environment, Safety, and Health, Washington, DC (1987).

Hawley, J. K. "Assessment of Health Risk from Exposure to Contaminated Soil," *Risk Anal.* 5(4):289–302 (1985).

Lepow, M. L., et al. "Role of Airborne Lead in Increased Body Burden of Lead in Hartford Children," *Environ. Health Perspect.* 6:99–101 (1974).

Lepow, M. L., et al. "Investigation into Sources of Lead in the Environment of Urban Children," *Environ. Res.* 10:415–426 (1975).

Sedman, R. "The Development of Applied Action Levels for Soil Contact: A Scenario for the Exposure of Humans to Soil in a Residential Setting," *Environ. Health Perspect.* 79:291–313 (1989).

U.S. EPA (U.S. Environmental Protection Agency). "RCRA Facility Investigation (RFI) Guidance," EPA/530/SW-87/001, Washington, DC (1987).

U.S. EPA (U.S. Environmental Protection Agency). "Superfund Exposure Assessment Manual," Report No. EPA/540/1-88/001, OSWER Directive 9285.5-1, U.S. EPA, Office of Remedial Response, Washington, DC (1988).

U.S. EPA (U.S. Environmental Protection Agency). Exposure Factors Handbook, EPA/600/8-89/043 (1989a).

U.S. EPA (U.S. Environmental Protection Agency). "Risk Assessment Guidance for Superfund. Vol. I, Human Health Evaluation Manual (Part A)," EPA/540/1-89/002, Office of Emergency and Remedial Response, Washington, DC (1989b).

APPENDIX C

Carcinogen Classification and Identification Systems

C.1 CARCINOGENIC CLASSIFICATION SYSTEMS

Identifying Carcinogens

Carcinogens may be categorized into the following identifiable groupings:

- "Known human carcinogens," defined as those chemicals for which there is sufficient evidence of carcinogenicity from studies in humans to indicate a causal relationship between the agent and human cancer.
- "Reasonably anticipated to be carcinogens," those chemical substances for which there is limited evidence for carcinogenicity in human and/or sufficient evidence of carcinogenicity in experimental animals. Sufficient evidence in animals is demonstrated by positive carcinogenicity findings in multiple strains and species of animals, in multiple experiments, or to an unusual degree with regard to incidence, site or type of tumor, or age of onset.
- "Sufficient evidence" and "limited evidence" of carcinogenicity, used in the criteria (for judging the adequacy of available data for identifying carcinogens), refer only to the amount and adequacy of the available evidence and not to the potency of carcinogenic effect on the mechanisms involved (IARC, 1982; USDHS, 1989).

An important issue in chemical carcinogenesis concerns initiators and promoters. A *promoter* is defined as an agent which results in an increase in cancer induction when it is administered some time after the receptor has been exposed to an *initiator*. A *cocarcinogen* differs from a promoter only in that it is administered at the same time as the initiator. Initiators, cocarcinogens, and promoters do not usually induce tumors when administered separately. *Complete* carcinogens act as both initiator and promoter (OSTP, 1985). Federal regulatory agencies do not distinguish between initiators and promoters because it is very difficult to confirm that a given chemical acts by promotion alone (OSTP, 1985; OSHA, 1980; U.S. EPA, 1984).

Carcinogen Classification Systems/Schemes

A chemical's potential for human carcinogenicity is inferred from the available information relevant to the potential carcinogenicity of the chemical and from judgments as to the quality of the available studies. Two evaluation philosophies, one based on weight of evidence and the other on strength of evidence, have found common acceptance and usage. Systems that employ the weight-of-evidence evaluations consider and balance the negative indicators of carcinogenicity with those showing carcinogenic activity; schemes using the strength-of-evidence evaluations consider combined strengths of all positive animal tests (human epidemiology studies and genotoxicity) to rank a chemical without evaluating negative studies, nor considering potency or mechanism (Huckle, 1991).

A weight-of-evidence approach is used by the U.S. EPA to classify the likelihood that the agent in question is a human carcinogen. This is a classification system for characterizing the extent to which available data indicate that an agent is a human carcinogen (or for some other toxic effects such as developmental toxicity). A three-stage procedure is followed:

- Stage 1 — the evidence is characterized separately for human studies and for animal studies.
- Stage 2 — the human and animal evidence are integrated into a presumptive overall classification.
- Stage 3 — the provisional classification is modified (i.e., adjusted upwards or downwards), based on analysis of the supporting evidence.

The result is that each chemical is placed into one of five categories, according to the U.S. EPA Carcinogen Assessment Group (CAG) weight of evidence categories for potential carcinogens. Proposed guidelines for the classification of the weight-of-evidence for human carcinogenicity has been published by the U.S. EPA (U.S. EPA, 1984). These guidelines are adapted from those developed by the International Agency for Research on Cancer (IARC, 1984); they consist of the categorization of the weight of evidence into five groups (Groups A to E) as follows:

EPA Group/Category	Criterion/Basis for Category
A	Human carcinogen (i.e., known human carcinogen)
B	Probable human carcinogen: B1 indicates limited human evidence B2 indicates sufficient evidence in animals and inadequate or no evidence in humans
C	Possible human carcinogen
D	Not classifiable as to human carcinogenicity
E	No evidence of carcinogenicity in humans (or evidence of noncarcinogenicity for humans)

Group A — Human Carcinogen

This group is used to represent only those agents for which there is sufficient evidence from epidemiologic studies to support a causal association between exposure to the agent and human cancer. The following three criteria must be satisfied before a causal association can be inferred between exposure and cancer in humans (Hallenbeck and Cunningham, 1988):

- No identified bias which could explain the association
- Possibility of confounding factors (i.e., variables other than chemical exposure level which can affect the incidence or degree of the parameter being measured) has been considered and ruled out as explaining the association
- Association is unlikely to be due to chance

This group is used only when there is sufficient evidence from epidemiologic studies to support a causal association between exposure to the agent and cancer.

Group B — Probable Human Carcinogen

This group includes agents for which the weight-of-evidence of human carcinogenicity based on epidemiologic studies is "limited", i. e., ranges from almost sufficient to inadequate, and also includes agents for which the weight-of-evidence of carcinogenicity based on animal studies is "sufficient." The group is divided into two subgroups, reflecting higher (Group B1) and lower (Group B2) degrees of evidence. Usually, Group B1 is reserved for agents for which there is limited evidence of carcinogenicity to humans from epidemiologic studies; limited evidence of carcinogenicity indicates that a causal interpretation is credible, but that alternative explanations such as chance, bias, or confounding could not be excluded. Inadequate evidence indicates that one of two conditions prevailed (Hallenbeck and Cunningham, 1988):

- There were few pertinent data.
- The available studies, while showing evidence of association, did not exclude chance, bias, or confounding.

When there are inadequate data for humans, it is reasonable to regard agents for which there is sufficient evidence of carcinogenicity in animals as if they will present carcinogenic risk to humans. Therefore, agents for which there is "sufficient" evidence from animal studies and for which there is "inadequate" evidence from human (epidemiological) studies or "no data" from epidemiologic studies would usually result in a classification as B2 (CDHS, 1986; U.S. EPA, 1986; Hallenbeck and Cunningham, 1988).

Group C — Possible Human Carcinogen

This group is used for agents with limited evidence of carcinogenicity in animals in the absence of human data. Limited evidence means that the data suggest a carcinogenic effect, but are limited because of any of the following reasons (Hallenbeck and Cunningham, 1988):

- The studies involve a single species, strain, or experiment.
- The experiments are restricted by inadequate dosage levels, inadequate duration of exposure to the agent, inadequate period of follow-up, poor survival, too few animals, or inadequate reporting.
- An increase in the incidence of benign tumors only.

Group C includes a wide variety of evidence, including (U.S. EPA, 1986; Hallenbeck and Cunningham, 1988)

- Definitive malignant tumor response in a single, well-conducted experiment that does not meet conditions for "sufficient" evidence.
- Tumor response of marginal statistical significance in studies having inadequate design or reporting.
- Benign but not malignant tumors, with an agent showing no response in a variety of short-term tests for mutagenicity.
- Responses of marginal statistical significance in a tissue known to have a high and variable background rate.

Group D — Not Classifiable as to Human Carcinogenicity

This group is generally used for agents with inadequate human (epidemiological) and animal evidence of carcinogenicity or for which no data are available. Inadequate evidence indicates that because of major qualitative or quantitative limitations, the studies cannot be interpreted as showing either the presence or absence of a carcinogenic effect.

Group E — No Evidence of Carcinogenicity for Humans (or Evidence of Noncarcinogenicity for Humans)

This group is used for agents for which there is evidence of noncarcinogenicity for humans, with no evidence of carcinogenicity in at least two adequate animal tests in different species, or in both adequate animal and human (epidemiological) studies. The designation of an agent as being in this group is based on the available evidence and should not be interpreted as a definitive conclusion that the agent will not be a carcinogen under any circumstances.

Other varying schemes exist for other regulatory and legislative agencies in Europe and elsewhere. The IARC does its classification based on the strength-of-evidence philosophy. The corresponding IARC classification system, equivalent to the U.S. EPA description presented above is as follows:

IARC Group	Category
1	Human carcinogen (i.e., known human carcinogen)
2	Probable or Possible human carcinogen:
	2A indicates limited human evidence (i.e., probable)
	2B indicates sufficient evidence in animals and inadequate or no evidence in humans (i.e., possible)

3	Not classifiable as to human carcinogenicity
4	No evidence of carcinogenicity in humans

Group 1 — Known Human Carcinogen. This group is generally used for agents with sufficient evidence from human (epidemiological) studies as to human carcinogenicity.

Group 2A — Probable Human Carcinogen. This group is generally used for agents for which there is sufficient animal evidence, evidence of human carcinogenicity, or at least limited evidence from human (epidemiological) studies. These are probably carcinogenic to humans, with (usually) at least limited human evidence.

Group 2B — Possible Human Carcinogen. This group is generally used for agents for which there is sufficient animal evidence and inadequate evidence from human (epidemiological) studies, or there is limited evidence from human (epidemiological) studies in the absence of sufficient animal evidence. These are probably carcinogenic to humans, but (usually) have no human evidence.

Group 3 — Not Classifiable. This group is generally used for agents for which there is inadequate animal evidence and inadequate evidence from human (epidemiological) studies. There is sufficient evidence of carcinogenicity in experimental animals.

Group 4 — Noncarcinogenic to Humans. This group is generally used for agents for which there is evidence for lack of carcinogenicity. They are probably not carcinogens.

Typical Carcinogenic Chemicals Encountered at Hazardous Waste Sites

The evaluation of carcinogenicity in a risk assessment involves two basic steps:

- Identification of potential carcinogens from among the contaminants of potential concern present at the problem site
- Quantitative determination of the carcinogenic potency of the chemicals of potential concern; this is represented by the SF

Evidence of possible carcinogenicity in humans comes primarily from epidemiological studies and long-term animal exposure studies at high doses which have subsequently been extrapolated to humans. Results from these studies are supplemented with information from short-term tests, pharmacokinetic studies, comparative metabolism studies, molecular structure-activity relationships, and other relevant information sources. Table C.1 lists typical chemicals that may be encountered at hazardous waste sites that are known to be carcinogenic; Table C.2

Table C.1 Example Chemicals that Are Known to Be Carcinogenic[a]

4-Aminobiphenyl
Arsenic and certain arsenic compounds
Asbestos
Benzene
Benzidine
Bis(chloromethyl)ether (BCME)
1,4-Butanediol dimethylsulfonate (Myleran)
Chloromethyl methyl ether (CMME)
Chromium and certain chromium compounds
Diethylstilbestrol (DES)
2-Napthylamine
Vinyl chloride

[a] "Known carcinogens" are defined as those substances for which the evidence from human studies indicates that there is a causal relationship between exposure to the substance and human cancer (USDHS, 1989).

Table C.2 Example Chemicals which May Reasonably Be Anticipated to Be Carcinogens[a]

2-Acetylaminofluorene	4,4′-Methylene bis(2-chloroaniline)(MBOCA)
Benzotrichloride	4,4′-Methylene bis(*N,N*-dimethyl)benzenamine
Beryllium and certain beryllium compounds	Mirex
Bischloroethyl nitrosourea	Nickel and certain nickel com pounds
1,3-Butadiene	N-Nitrosodi-n-butylamine
Cadmium and certain cadmium compounds	N-Nitrosodiethanolamine
Carbon tetrachloride	N-Nitrosodiethylamine
Chlorinated paraffins	N-Nitrosodimethylamine (NDMA)
1-(2-Chloroethyl)-3-cyclohexyl-1-nitrosourea (CCNU)	p-Nitrosodiphenylamine
Chloroform	N-Nitrosodi-n-propylamine
3-Chloro-2-methylpropene	N-Nitroso-N-ethylurea
4-Chloro-o-phenylenediamine	N-Nitroso-N-methylurea
p-Cresidine	N-Nitrosomethylvinylamine
Dichlorodiphenyl trichloroethane (DDT)	N-Nitrosomorpholine
2,4-Diaminoanisole sulfate	N-Nitrosonornicotine
2,4-Diaminotoluene	N-Nitrosopiperidine
1,2-Dibromo-3-chloropropane (DBCP)	N-Nitrosopyrrolidine
1,2-Dibromoethane (EDB)	N-Nitrososarcosine
1,4-Dichlorobenzene	Polybrominated biphenyls (PBBs)
3,3′-Dichlorobenzidine	Polychlorinated biphenyls (PCBs)
1,2-Dichloroethane	Polycyclic aromatic hydrocarbons (PAHs)
Dichloromethane (Methylene Chloride)	Benz(a)anthracene
1,3-Dichloropropene	Benzo(b)fluoranthene
Diepoxybutane	Benzo(j)fluoranthene
Di(2-ethylhexyl)phthalate (DEHP)	Benzo(k)fluoranthene
Diethyl sulfate	Benzo(a)pyrene
3,3′-Dimethoxybenzidine	Dibenz(a,h)acridine
4-Dimethylaminoazobenzene	Dibenz(a,j)acridine
3,3′-Dimethylbenzidine	Dibenz(a,h)anthracene
1,1-Dimethylhydrazine (UDMH)	7H-Dibenzo(c,g)carbazole
Dimethyl sulfate	Dibenzo(a,e)pyrene
Dimethylvinyl chloride	Dibenzo(a,h)pyrene
1,4-Dioxane	Dibenzo(a,i)pyrene
Ethyl acrylate	Dibenzo(a,l)pyrene

Table C.2 (continued)

Ethylene oxide	Indeno(1,2,3-cd)pyrene
Ethylene thiourea	5-Methylchrysene
Hexachlorobenzene	Selenium sulfide
Hexamethylphosphoramide	Sulfallate
Hydrazine and hydrazine sulfate	2,3,7,8-Tetrachlorodibenzo-p-dioxin (TCDD)
Hydrazobenzene	Tetrachloroethylene (perchloroeth ylene)
Lead acetate and lead phosphate	Thioacetamide
Lindane and other hexachlorocyclohexane isomers	Toxaphene
	2,4,6-Trichlorophenol
	Tris(1-aziridinyl)phosphine sulfide
	Tris(2,3-dibromopropyl)phosphate

[a] Substances "which may reasonably be anticipated to be carcinogens" are defined as those for which there is a limited evidence of carcinogenicity in humans or sufficient evidence of carcinogenicity in experimental animals (USDHS, 1989).

contains a listing of those chemicals which may reasonably be anticipated to be carcinogenic.

C.2 REFERENCES

CDHS (California Department of Health Services). "The California Site Mitigation Decision Tree Manual. California Department of Health Services, Toxic Substances Control Division, Sacremento, CA (1986).

Hallenbeck, W. H. and K. M. Cunningham. *Quantitative Risk Assessment for Environmental and Occupational Health* (Chelsea, MI: Lewis Publishers, 1988).

Huckle, K. R. *Risk Assessment — Regulatory Need or Nightmare* (Shell Center, London: Shell Publications, 1991).

IARC (International Agency for Research on Cancer). IARC Monographs on the Evaluation of the Carcinogenic Risk of Chemicals to Humans. Chemicals, Industrial Processes and Industries Associated with Cancer in Humans, Suppl. 4 (Lyon, France: IARC, 1982).

IARC (International Agency for Research on Cancer). IARC Monographs on the Evaluation of the Carcinogenic Risk of Chemicals to Humans, Vol. 33 (Lyon, France: World Health Organization, 1984).

OSHA (Occupational Safety and Health Administration). "Identification, Classification, and Regulation of Potential Occupational Carcinogens," *Fed. Regist.* 45:5002–5296 (1980).

OSTP (Office of Science and Technology Policy). "Chemical Carcinogens: A Review of the Science and Its Associated Principles," *Fed. Regist.* 50:10372–442 (1985).

USDHS (U.S. Department of Health and Human Services). Public Health Service, Fifth Annual Report on Carcinogens, Summary (1989).

U.S. EPA (U.S. Environmental Protection Agency). "Proposed Guidelines for Carcinogen, Mutagenicity, and Developmental Toxicant Risk Assessment," *Fed. Regist.* 49:46294–46331 (1984).

U.S. EPA (U.S. Environmental Protection Agency). "Guidelines for Carcinogen Risk Assessment," *Fed. Regist.* 51(185):33992–34003 (1986).

APPENDIX D

Selected Databases and Information Library With Important Risk Information Used in Risk Assessment

D.1 THE INTEGRATED RISK INFORMATION SYSTEM (IRIS)

Overview of IRIS

IRIS, prepared and maintained by the Office of Health and Environmental Assessment of the U.S. EPA, is an electronic database containing health risk and regulatory information on several specific chemicals. It is an on-line database of chemical-specific risk information; it is also a primary source of EPA health hazard assessment and related information on chemicals of environmental concern. IRIS was developed for EPA staff in response to a growing demand for consistent risk information on chemical substances for use in decision making and regulatory activities. The information in IRIS is accessible to those without extensive training in toxicology, but with some rudimentary knowledge of health and related sciences.

To aid users in accessing and understanding the data in the IRIS chemical files, the following supportive documentation is provided:

- Alphabetical list of the chemical files in IRIS and list of chemicals by CAS (Chemical Abstracts Service) number
- Background documents describing the rationales and methods used in arriving at the results shown in the chemical files
- A user's guide that presents step-by-step procedures for using IRIS to retrieve chemical information
- An example exercise in which the use of IRIS is demonstrated

- Glossaries in which definitions are provided for the acronyms, abbreviations, and specialized risk assessment terms used in the chemical files and in the background documents

The information in IRIS is intended for use in protecting public health through risk assessment and risk management. More detailed information, and access to IRIS, may be obtained from the EPA (IRIS User Support, U.S. EPA, Environmental Criteria and Assessment Office, Cincinnati, OH 45268) or from other independent/private organizations (such as Chemical Information System, CIS, in Maryland).

Types of Information in IRIS

The IRIS system consists of a collection of computer files covering individual chemicals. These chemical files contain descriptive and numerical information on several subjects, including

- Oral and inhalation RfDs for chronic noncarcinogenic health effects
- Oral and inhalation SFs and unit risks for chronic exposures to carcinogens
- Summaries of drinking water health advisories from U.S. EPA Office of Drinking Water
- U.S. EPA regulatory action summaries
- Supplementary data on acute health hazards and physical/chemical properties of the chemicals

The IRIS is a computerized library of current information that is updated periodically. An alphabetical and CAS number listing of chemicals in IRIS is included in the system.

The Role of IRIS in Risk Assessment and Risk Management

IRIS is a tool which provides hazard identification and dose-response assessment information, but does not provide problem-specific information on individual instances of exposure. Combined with specific exposure information, the data in IRIS can be used to characterize the public health risks of an identified chemical under specific scenarios, which can then lead to a risk management decision designed to protect public health. The information contained in Section I (Chronic Health Hazard Assessment for Noncarcinogenic Effects) and Section II (Carcinogenicity Assessment for Lifetime Exposure) of the chemical files represents a consensus opinion of EPA's Reference Dose Work Group or Carcinogen Risk Assessment Verification Endeavor Work Group, respectively. These two work groups include high-level scientists from EPA's program offices and Office of Research and Development. Individual EPA offices have conducted comprehensive scientific reviews of the literature available on the particular chemical, and have performed the first two steps of risk assessment: hazard evaluation and dose-response assessment. These assessments have been summarized in the IRIS format and reviewed and revised by the appropriate work group. As new information becomes available, these work groups reevaluate their work and revise IRIS files accordingly.

D.2 THE INTERNATIONAL REGISTER OF POTENTIALLY TOXIC CHEMICALS (IRPTC)

Overview of the IRPTC

In 1972, the United Nations Conference on the Human Environment, held in Stockholm, recommended the setting up of an international registry of data on chemicals likely to enter and damage the environment. Subsequently, in 1974, the Governing Council of the United Nations Environment Programme (UNEP) decided to establish both a chemicals register and a global network for the exchange of information the register would contain. The definition of the register's objectives was subsequently elaborated as follows:

- Make data on chemicals readily available to those who need it.
- Identify and draw attention to the major gaps in the available information and encourage research to fill those gaps.
- Help identify the potential hazards of using chemicals and improve people awareness of such hazards.
- Assemble information on existing policies for control and regulation of hazardous chemicals at national, regional, and global levels.

In 1976, a central unit for the register, named the International Register of Potentially Toxic Chemicals, was created in Geneva, Switzerland, with the main function of collecting, storing, and disseminating data on chemicals, and also to operate a global network for information exchange. IRPTC network partners, the designation assigned to participants outside the central unit, consist of National Correspondents appointed by governments, national and international institutions, national academies of science, industrial research centers, and specialized research institutions.

Chemicals examined by the IRPTC have been chosen from national and international priority lists. The selection criteria used include the quantity of production and use, the toxicity to humans and ecosystems, persistence in the environment, and the rate of accumulation in living organisms. IRPTC stores information that would aid in the assessment of the risks and hazards posed by a chemical substance to human health and environment. The major types of information collected include that relating to the behavior of chemicals and information on chemical regulation. Information on the behavior of chemicals is obtained from various sources such as national and international institutions, industries, universities, private databanks, libraries, academic institutions, scientific journals, and United Nations bodies such as the International Programme on Chemical Safety (IPCS). Regulatory information on chemicals is largely contributed by IRPTC National Correspondents. Specific criteria are used in the selection of information for entry into the databases. Whenever possible, IRPTC uses data sources cited in the secondary literature produced by national and international panels of experts to maximize reliability and quality. The data are then extracted from the primary literature. Validation is performed prior to data entry and storage on a computer at the United Nations International Computing Centre (ICC).

Following the successful implementation of the IRPTC databases, a number of countries went into creating National Registers of Potentially Toxic Chemicals (NRPTCs) that would be completely compatible with the IRPTC system. The IPRTC, with its carefully designed database structure, provides a sound model for national and regional data systems. More importantly, it brings consistency to information exchange procedures within the international community. The IPRTC is serving as an essential international tool for chemical hazards assessment, as well as a mechanism for information exchange on several chemicals. More detailed information, and access to the IRPTC, may be obtained from the National Correspondent to the IRPTC, the U.S EPA, the National Technical Information Service (NTIS), the Agency for Toxic Substances and Disease Registry (ATSDR), or the National Academy of Sciences (NAS).

Types of Information in the IRPTC Databases

The complete IRPTC file structure consists of the following:

- Legal database
- Mammalian and special toxicity studies
- Chemobiokinetics and effects on organisms in the environment
- Environmental fate tests and environmental fate and pathways into the environment
- Identifiers, production, processes, and waste

The IRPTC Legal database contains national and international recommendations and legal mechanisms related to chemical substances control in media such as air, water, wastes, soils, sediments, biota, foods, drugs, consumer products, etc. This organization allows for rapid access to the regulatory mechanisms of several nations and to international recommendations for safe handling and use of chemicals. The Mammalian Toxicity database provides information on the toxic behavior of chemical substances in humans; toxicity studies on laboratory animals are included as a means of predicting potential human effects. The Special Toxicity databases contain information on particular effects of chemicals on mammals, such as mutagenicity and carcinogenicity, as well as data on nonmammalian species when relevant for the description of a particular effect. The Chemobiokinetics and Effects on Organisms in the Environment databases provide data that will permit the reliable assessment of the hazards posed by chemicals present in the environment to humans. The absorption, distribution, metabolism, and excretion of drugs, chemicals, and endogenous substances are described in the Chemobiokinetics databases. The Effects on Organisms in the Environment databases contain toxicological information regarding chemicals in relation to ecosystems and to aquatic and terrestrial organisms at various nutritional levels. The Environmental Fate Tests and Environment Fate and Pathways into the Environment databases assess the risk presented by chemicals to the environment. The Identifiers, Production, Processes, and Waste databases contain miscellaneous information about chemicals, including physical and chemical properties; hazard classification for chemical production and trade statistics of chemicals on worldwide or regional basis; information on production methods; information on uses and quantities of use for chemicals; data on persistence of chemicals in various environmental compartments or media; information on the intake of chemicals by humans in different geographical areas; sampling methods for various media and species, as well as analytical protocols for obtaining reliable data; recommendable methods for the treatment and disposal of chemicals; etc.

D.3 THE CHEMICAL SUBSTANCES INFORMATION NETWORK (CSIN)

CSIN is not a database, but rather an interactive network system that links together a number of databases relating to several chemical substances. CSIN accesses data on chemical nomenclature, composition, structure, properties, toxicological information, health and environmental effects, production and uses, regulations, etc. CSIN and the databases it accesses are in the public domain. However, users have to make independent arrangements with vendors of those databases in the network that needs to be used for specific assignments.

The Superfund Risk Assessment Information Directory (U.S. EPA, Office of Emergency and Remedial Response, Washington, DC, EPA/540/1-86/061; OSWER Directive 9285.6-1, November 1986) identifies and describes several sources of information that should prove useful in the conduct of risk assessments. It also identifies, references, and provides guidance and documentation on several EPA and non-EPA databases and information libraries that may be useful in performing risk assessments.

D.4 SELECTED ENVIRONMENTAL MODELS POTENTIALLY APPLICABLE TO RISK ASSESSMENT STUDIES

Table D.1 consists of selected models that may be applied to some aspect of hazardous waste risk assessment. Choice of model will be problem specific. This listing is by no means complete and exhaustive.

D.5 DESCRIPTION OF THE MATHEMATICAL STRUCTURE OF AIR PATHWAY MODELS

In a manner similar to most aspects of the environment, the ability exists to measure substances in the air and emission sources but such efforts are costly. In addition, it would be highly undesirable to first construct facilities (such as pollution control facilities) and then monitor the impact of the control of the emissions control, identify changes in air quality, and understand the fate and transport of substances emitted into the atmosphere. Instead, evaluation prior to construction of such facilities is essential. In response, we utilize mathematical models to predict how the emissions behave in the atmosphere.

Table D.1 Listing of Selected Environmental Models Potentially Applicable to Risk Assessment Studies

Model Name	Environmental Compartment	Model Description	Model Use	Sources/Refs.
AERAM (air emission risk assessment model)	Air	AERAM is a methodology for assessing the human health risks associated with hazardous air pollutants emitted by coal-fired (and gas-fired) power plants.	AERAM estimates the excess cancer risks and other health effects expected to affect human populations near a power plant as a result of air pollutants emitted by the plant.	EPRI (Electric Power Research Institute) (1988). Air Emission Risk Assessment Model (AERAM) Manager, User's Guide, Version 1.0. EPRI Report No. EA-5886-CCML.
AIRTOX (air toxics risk management framework)	Air	AIRTOX is a decision analysis model for air toxics risk management. The framework consists of a structural model that relates emissions of air toxics to potential health effects, and a decision tree model that organizes scenarios evaluated by the structural model.	AIRTOX evaluates the magnitude of health risks to a population, a specific source's contribution to the total health risk, and the cost effectiveness of current and future emission control measures.	EPRI (Electric Power Research Institute), Palo Alto, CA.
TOXIC	Air	TOXIC is a microcomputer program that calculates the incremental risk to the hypothetical maximum exposed individual (MEI) from hazardous waste incineration. It calculates exposure to each pollutant individually using a specified dispersion coefficient (which is the ratio of pollutant concentration in air in $\mu g/m^3$ to pollutant emission rate in gm/sec).	TOXIC is used in hazardous waste facility risk analysis. It is a flexible and convenient tool for performing inhalation risk assessments for hazardous waste incinerators. It gives point estimates of inhalation risks.	Rowe Research and Engineering Associates, Alexandria, VA.
BOXMOD (the Box model)	Air	BOXMOD is an interactive steady-state simple atmospheric area source box model for screening chemicals. It is applicable to regions containing many diffuse emission sources within	BOXMOD calculates a single annual average concentration applicable to the entire region based on a uniform area emission rate. It is used for detailed screening.	U.S. EPA, Office of Air Quality Planning and Standards, Research Triangle Park, NC.

		its boundaries, such as in an urban area. It is available through GEMS (see description below).		
GAMS (the GEMS atmospheric modeling subsystem)	Air	GAMS is an integrated atmospheric modeling system that can be used to estimate annual average concentrations, annual exposure, and lifetime and annual incidence of excess cancer cases.	GAMS is used for refined analysis. Exposure and risk calculations can also be carried out using this model.	Office of Air Quality Planning and Standards (OAQPS), 1988. Guidelines on Air Quality Models. NTIS No. PB88-150958.
SHORTZ and LONGZ (atmospheric dispersion models)	Air	SHORTZ is a computer algorithm designed to use sequential short-term meteorological inputs to calculate chemical concentrations for averaging times. It is applicable in areas of both flat and complex terrain and can accommodate receptors that are both above and below source elevations.	SHORTZ calculates the short-term ground-level pollutant concentrations produced at large number of receptors by emissions from various sources.	U.S. EPA (1982). User's Instructions for the SHORTZ and LONGZ Computer Programs, Vol. I & II, EPA 903/9-82-004a+b.
		LONGZ is the long-term average version of SHORTZ. It is an area-source computer program designed for application to square, ground-level area sources.	LONGZ, used in near-field concentration estimates, calculates vertical dispersion based on the integral of the vertical term in the Gaussian equation between the downwind edge and the upwind edge.	
ISCLT (industrial source complex long-term model)	Air	ISCLT is a long-term sector-averaged environmental model that uses statistical wind summaries to calculate annual ground-level concentrations or dispersion values.	ISCLT is an air model that calculates annual ground-level concentrations or deposition values and estimates risk and exposure level using these values. It is used for modeling long-term air exposures associated with point and areal sources of air emissions.	U.S. EPA, Office of Air Quality Planning and Standards, Research Triangle Park, NC U.S. EPA (1986). Industrial Source Comples (ISC) Dispersion Model User's Guide. NTIS, No. PB86-234259/LP.

Table D.1 (continued)

Model Name	Environmental Compartment	Model Description	Model Use	Sources/Refs.
ISCST (industrial source complex short-term model)	Air	ISCST is a short-term model that uses finite-line source approach to model area sources. Each square area source is modeled as a single line segment oriented normal to the wind direction. The model does not accurately account for source-receptor geometry.	ISCST algorithm is used to model short-term air exposures associated with air emissions.	U.S. EPA, Office of Air Quality Planning and Standards, Research Triangle Park, NC. U.S. EPA (1986). Industrial Source Complex (ISC) Dispersion Model User's Guide, NTIS, No. PB86-234259/LP.
PTPLU (the point plume model)	Air	PTPLU is a Gaussian plume dispersion model that estimates the location of the maximum short-term concentration in the atmosphere from a single point source (as a function of stability and wind speed).	PTPLU is a point-source Gaussian plume dispersion algorithm, used for detailed screening. It is used for estimating worst-case hourly concentrations for steady-state conditions.	Environmental Sciences Research Laboratory, Office of Research and Development, U.S. EPA, Research Triangle Park, NC. Pierce, T. E. and D. B. Turner (1982). PTPLU — A Single Source Gaussian Dispersion Algorithm, "EPA-600/8-82-014.
FDM (fugitive dust model)	Air	FDM is a computerized analytical air quality model specifically designed for computing concentration and deposition impacts from fugitive dust sources. The sources may be point, line, or areal; it contains no plume rise algorithm. The model is generally based on Gaussian Plume formulation for computing concentrations, with improved gradient-transfer deposition algorithm. FDM accounts for deposition losses as well as pollutant dispersion.	FDM models both short- and long-term average particulate emissions from surface mining and similar sources. A primary use is for computation of concentrations and depositions rates resulting from emission sources such as hazardous waste sites where fugitive dust is a concern. Concentration and deposition are computed at all user-selected receptor locations.	Support Center for Regulatory Air Models, Office of Air Quality Planning and Standards (OAQPS), U.S. EPA, Research Triangle Park, NC. U.S. EPA (1991). User's Guide for the Fugitive Dust Model (FDM) (Revised): User's Instructions, EPA-910/9-88-202R.

PAL (point, area, line-source model)	Air	PAL is a steady-state Gaussian plume model, recommended for source dimensions of tens to hundreds of meters. It uses Gaussian model equations for a finite line segment.	PAL calculates concentrations, including user-specified accuracy levels.	Office of Air Quality Planning and Standards OAQPS), U.S. EPA, Research Triangle Park, NC.
CREAMS (the chemical, run-off, and erosion from agricultural management systems model)	Surface water	CREAMS is a numerical, finite-difference method of solution, used to estimate sorption and degradation of chemicals and erosion of single land segments. Inputs required include watershed characteristics and chemical properties.	CREAMS is used to estimate sorption and degradation of chemicals and erosion of single land segments.	U.S. Department of Agriculture (USDA). U.S. Army Corps of Engineers.
QUAL2E (enhanced stream water quality model)	Surface water	QUAL2E is a steady-state numerical, finite-difference model for conventional pollutants in branching streams and well-mixed lakes. It includes conservative substances, temperature, coliform bacteria, BOD, DO, N, P, and algae.	QUAL2E is widely used for waste load allocation and permitting.	Center for Water Quality Modeling, Environmental Research Laboratory , Office of Research and Development, U.S. EPA, Athens, GA..
EXAMS-II (exposure analysis modeling system)	Surface water	EXAMS is a steady-state contaminant fate and transport numerical, finite-difference model, offering one-, two-, or three-dimensional compartmental solutions in surface water bodies. It is based on a series of equations which account for interactions between the canonical aquatic environment into which a chemical is released, the chemistry of a given chemical and the toxicant loading quantities. It includes process models of the physical, chemical, and biological phenomena governing the transport and fate of compounds.	EXAMS simulates the fate of organic chemicals in surface water bodies (i.e., rivers, lakes, reservoirs, estuaries). Model can estimate the time-varying and/or steady-state concentrations of the chemical in the water body in various phases (dissolved, sediment, sorbed, biosorbed). It has been designed to evaluate the consequences of longer term, primarily time-averaged chemical loadings that ultimately result in trace-level contamination of aquatic systems. Suitable for modeling synthetic organic chemicals for freshwater, nontidal aquatic systems.	Environmental Research Laboratory Office of Research and Development, U.S. EPA, Athens, GA. Burns, L. A., D. M. Cline, and R. R. Lassiter (1982). Exposure Analysis Modeling System (EXAMS): User Manual and System Documentation, NTIS, No. PB82-258096 ; EPA-600/3-82/023. Burns, L. A. and D. M. Cline (1985). Exposure Analysis Modeling System: Reference Manual for EXAMS II, NTIS, No. PB85-228138/AS; EPA/600/3-85 /038.

Table D.1 (continued)

Model Name	Environmental Compartment	Model Description	Model Use	Sources/Refs.
REACHSCA (reach scan model)	Surface water	REACHSCA is a simple dilution model used to estimate steady-state chemical concentration in surface water bodies due to continuous loading from a single discharging facility.	REACHSCA is used to estimate steady-state chemical concentration in surface water bodies (mainly river reaches).	Office of Toxic Substances, U.S. EPA, Washington, DC.
WTRISK (waterborne toxic risk assessment model)	Surface water	WTRISK provides a framework that employs a risk assessment methodology in which mathematical models simulating chemical fate and transport processes can be linked to determine pollutant concentrations in all appropriate environmental media. These predicted concentrations are then used as input values for modeling nearby population exposures and potential health risks.	WTRISK provides a flexible framework for the risk assessment of toxics in surface water. It aids the estimation of the source terms or quantities of toxics emitted into the environment.	EPRI (Electric Power Research Institute), Palo Alto, CA.
MINTEQA1 (metals exposure analysis modeling system)	Surface water	MINTEQA1 is an interactive complex computer program, consisting of a steady-state 3-dimensional numerical, finite-difference compartmental model.	MINTEQA1 is designed for modeling the equilibrium states for metal loadings in freshwater, nontidal aquatic systems.	Battelle Pacific Northwest Laboratories, Richland, WA.
SARAH (surface water back-calculation procedure	Surface water	SARAH is a semianalytical steady-state, 1-D surface water exposure assessment model. It models contaminated leachate plume feeding the downgradient surface waterbody (stream or river). It employs a Monte Carlo-simulated generic environment.	SARAH is used to model contaminated leachate plume feeding the downgradient stream or river.	Environmental Research Laboratory , U.S. EPA, College Station Rd., Athens, GA.

		Bioaccumulation in fish, degradation, sorption, dilution, and volatilization are included.		
MEXAMS (metals exposure analysis modeling system)	Surface water	MEXAMS is a steady-state 3-D compartmental model. It is a combination of two models (MINTEQ and EXAMS) designed for modeling metal loadings. The model links a complex speciation model with an aquatic transport/fate model and should help discriminate between the fraction of metal that is dissolved and in bioavailable form, and the fraction that is complexed and rendered relatively nontoxic.	MEXAMS is designed for modeling metals loadings and is suitable for freshwater, nontidal aquatic systems.	Battelle Pacific North west Laboratories, Richland, WA. Center for Water Quality Modeling, Environmental Research Laboratory, U.S. EPA, College Station Rd., Athens, GA.
PDM (probabilistic dilution model)	Surface water	PDM models exceedances of specified concentration levels in streams. The model estimates are based on statistical distribution of daily volume flow and on solution of mass balance dilution evaluation.	PDM estimates the percentage of time that a given concentration level may be exceeded in receiving surface water bodies.	Office of Toxic Substances, Exposure Evaluation Division, U.S. EPA, Washington, DC.
TOXIWASP	Surface water	TOXIWASP is a dynamic (time-varying) 3-D model for simulating transport and fate of toxic chemicals in water bodies. Both organic chemicals and sediments are simulated by the model. The model attempts to account for the full array of chemical transformations and sediment-chemical exchange processes through a set of transport, chemical transformation, and mass loading submodels. The model accounts for advection.	TOXIWASP calculates the concentrations for every segment of modeled waterbody, including surface water, subsurface water, surface bed, and subsurface bed. It is a chemical and stream quality model.	U.S. EPA (1983). User's Manual for the Chemical Transport and Fate Model (TOXIWASP), Version 1, EPA-600/3-83-005, Center for Water Quality Modeling, Environmental Research Laboratory, U.S. EPA, College Station Rd., Athens, GA. Center for Water Quality Modeling, Environmental Research Laboratory, U.S. EPA, College Station Rd., Athens, GA.

Table D.1 (continued)

Model Name	Environmental Compartment	Model Description	Model Use	Sources/Refs.
		dispersion, bed sedimentation and erosion, pore water diffusion, hydrolysis, photolysis, oxidation, biodegradation, and volatilization.		
SWIP (the survey waste injection program)	Ground-water	SWIP is a saturated zone 3-D numerical model which simulates contaminant (chemical or thermal) movement in an aquifer, including well-pumping effects. It incorporates finite difference approximation with options for several matrix solution techniques.	SWIP simulates chemical and/or thermal contaminant movement in an aquifer.	USGS (U.S. Geological Survey). U.S. EPA (1985). User's Guide to SWIP Model Execution Using Data Management Supporting System, Prepared by GSC for EPA's Office of Pesticides & Toxic Substances, Washington, DC.
SOLUTE (solute transport model)	Ground-water	SOLUTE is a PC basic program package of analytical models for solute transport in groundwater. The package includes several subprograms, including UNITS (for conversion of units); ERFC (to calculate error functions and complimentary error functions); ONED1 (for solute transport in 1-D); WMPLUME and SLUG (for solute transport in 2-D); RADIAL and LTIRD (for 2-D radial flow); and PLUME3D and SLUG3D (for 3-D transport).	SOLUTE is used for solute transport modeling, to estimate receptor exposure concentration distributions.	International Ground Water Modeling Center (IGWMC), Holcomb Research Institute, Butler University, Indianapolis, IN. Beljin, M. S. (1985). A Program Package of Analytical Models for Solute Transport in Groundwater; SOLUTE. IGWMC, Indianapolis, IN.
AT123D (analytical transient 1-2-3 dimensional simulation model)	Ground-water	AT123D is a saturated zone analytical transient 1-2-3-dimensional simulation model that is used to simulate chemical movement and	AT123D is an environmental model that predicts spread of a contaminant plume(chemical, thermal or radioactive) through groundwater	Oak Ridge National Laboratory (ORNL), Environmental Sciences Division.

		waste transport in the aquifer system. It predicts spread of contaminant plume through groundwater and can handle constant as well as time-varying chemical release to groundwater. Chemical processes include both adsorption and degradation. Output concentrations are time varying. It is available through GEMS.	(saturated zone) and estimates the chemical concentration within groundwater at positions on a user-specified three-dimensional grid.	Yeh, G. T. (1981). AT123D: Analytical Transient One-Two-and Waste Transport in the Aquifer System. ORNL-5602, Oak Ridge, TN.
MYGRT (migration of solutes in the subsurface environment)	Ground water	MYGRT is a 1-, 2-D fate model that provides a method for computing the fate of reacting or nonreacting inorganic chemicals released to the groundwater environment. The simulation is based on quasianalytical solutions to the conservation of mass equations, including advection, dispersion, and retardation. Simulation of both continuous and finite-duration solute releases are possible. It gives results that are ball-park approximations to the dynamic and complex problem of solute migration predictions.	MYGRT allows users to calculate concentration of solute at an elapsed time from the release time. Calculations can also be made to estimate time taken for a chemical to travel the distances of interest.	EPRI (Electric Power Research Institute), Palo Alto, CA.
SUTRA (saturated-unsaturated transport model)	Ground water	SUTRA is a 2-D numerical, finite-element and integrated finite-element solution technique. It is a solute transport simulation model that may be used to model natural or human-induced chemical species transport including processes of solute sorption, production, and decay.	SUTRA may be applied to analyze groundwater contaminant transport problems and aquifer restoration designs. It predicts fluid movement and the transport of either energy or dissolved substances in a subsurface environment.	USGS, Water Resources Department. Voss, C. I. (1984). A Finite-Element Simulation Model for Saturated-Unsaturated, Fluid-Density-Dependent Groundwater Flow with Energy Transport or Chemically-Reactive Single Species Solute Transport, USGS Water Resources Investigation Report No. 84-4369.

Table D.1 (continued)

Model Name	Environmental Compartment	Model Description	Model Use	Sources/Refs.
TRANS	Groundwater	TRANS is a 2-D numerical model that uses random walk solution technique. Concentration distribution in aquifer represents a vertically averaged value over the saturated thickness of the aquifer.	TRANS is used to predict groundwater pollution problems.	Illinois State Water Survey. Prickett, T. A., T. G. Namik, and C. S. Lonnquist (1981). A Random-Walk Solute Transport Model for Selected Groundwater Quality Evaluations, Illinois State Water Survey Bulletin No. 65.
SHALT (solute, heat, and liquid transport)	Groundwater	SHALT is a 2-D numerical, finite-element model for predicting liquid flow, heat transport, and solute transport in a regional groundwater flow system. Fractured media may be modeled by treating the fractured rock as a continuum. Model output consists of the pressure, concentration, and temperature distribution at each time step.	SHALT is used to predict liquid flow, heat transport, and solute transport in a regional groundwater flow system.	Inland Water Directorate, Environment Canada. Atomic Energy of Canada, Ltd., Report TR-81 (September, 1979).
FEMWASTE1 (finite-element model of waste transport)	Groundwater (unsaturated and saturated zones)	FEMWASTE1 is a 2-D transient model for predicting waste transport through saturated-unsaturated porous media under dynamic groundwater conditions. Solution is by finite-element weighted residual method. It models interzone transfer; incorporates convection, dispersion; simulates degradation of nonconservative substances; accounts for adsorption; and is capable of modeling layered,	FEMWASTE1 is used to simulate the fate/transport of contaminants in saturated and unsaturated porous media.	Oak Ridge National Laboratory (ORNL), Environmental Sciences Division, Oak Ridge, TN.

		heterogeneous soil zones. The model is applied to a single chemical species without considering the effects of other chemicals that may be present in the porous media.		
RITZ (regulatory and investigative treatment zone model)	Unsaturated zone	RITZ is a 1-D unsaturated zone analytical model, which models pollutant transport. It can consider soil solution, volatilization and atmospheric losses, and biological degradation of chemicals. It considers the effect of an oil phase on pollutant transport. The model was designed to predict fate of contaminants in a land treatment scenario and considers downward movement of chemicals. Output includes mass transport to groundwater that can become input to groundwater dilution model.	RITZ is useful in predicting contaminant transport of residual constituents in setups similar to a land treatment area with respect to contaminant transport.	Office of Environmental Processes and Effects Research, Oak Ridge, TN.
VHS (vertical and horizontal spread model)	Groundwater	VHS is a steady-state, analytical model used to simulate the dispersion of contaminants and calculate the contaminant concentrations at a receptor point or well directly downgradient of the waste disposal area.	VHS is a groundwater dilution model used for predicting steady-state contaminant concentrations at receptor locations.	U.S. EPA, Research Triangle Park, NC.
MOC (method-of-characteristics model for solute transport)	Groundwater	MOC is a finite-difference computer model that is applicable to 1- or 2-D problems involving steady-state or transient flow. Model computes changes in concentration over time caused by the processes of convective transport, hydrodynamic dispersion,	MOC is used for calculating changes in the concentration of dissolved (non-reactive) chemical species in flowing groundwater, by the model simulating solute transport in flowing groundwater. The purpose of the simulation is to compute the	USGS The Holcomb Research Institute, IN. Konikow, L. F. and J. D. Bredehoeft (1984). Computer Model of Two-Dimensional Solute Transport and

Table D.1 (continued)

Model Name	Environmental Compartment	Model Description	Model Use	Sources/Refs.
		and mixing (or dilution) from fluid sources. Model assumes that solute is nonreactive and that gradients of fluid density, viscosity, and temperature do not affect the velocity distribution; aquifer may be heterogeneous and/or anisotropic. Model couples the groundwater flow equation with the solute transport equation. T he model is based on a rectangular, block-centered, finite-difference grid. The method of characteristics is used in the model to solve the solute transport equation. By coupling the flow equation with the solute-transport equation, the model can be applied to both steady-state and transient flow problems.	concentration of a dissolved chemical species in an aquifer at any specified place and time.	Dispersion in Ground Water, Water Resources Investigations, Book 7, chap. C2, USGS, Reston, VA.
SESOIL (seasonal soil compartment model)	Soil (vadose zone)	SESOIL is a seasonal soil compartment model that estimates the rate of vertical chemical transport and transformation in the soil column in terms of mass and concentration distributions among the soil, water, and air phases in the unsaturated soil zone. It is designed for long-term environmental fate simulations of pollutants in the vadose zone. It is a 1-D unsaturated zone model for both organics and inorganics. It is integrated into GEMS.	Estimates the rate of vertical chemical transport and transformation in the soil column. Also has capability to simulate contaminant transport in washload at soil surface and volatilization rates to the atmosphere.	Office of Toxic Substances, Exposure Evaluation Division, U.S. EPA, Washington, DC. Bonazountas, M. and J. M. Wagner (1986). SESOIL: A Seasonal Soil Compartment Model, Report prepared by Arthur D. Little, Inc., for U.S. EPA, OTS, Washington, DC.

PRZM (pesticide root zone model)	Soil (unsaturated zone)	PRZM is an unsaturated zone dynamic numerical, finite-difference, 1-D transport model which simulates the vertical movement of pesticides in unsaturated soil within and below the plant root zone. Time-varying transport, including advection and dispersion, is represented by model. It has two major components — hydrology and chemical transport. It accommodates various release rates and schedules.	PRZM provides pollutant velocity distribution and concentration data for organic substances. Degradation is also simulated.	U.S. EPA, Environmental Research Laboratory, Athens, GA. U.S. EPA (1984). User's Manual for the Pesticide Root Zone Model (PRZM), EPA-600/3-84-109, ERL, Athens, GA.
SWAG (simulated waste access to groundwater)	Soil (to groundwater)	SWAG is a three-compartment analytical computer model for organic pollutant transport that considers transformations in the soil-geological matrix.	SWAG predicts organic pollutant transport to groundwater.	Office of Environmental Processes and Effects Research, Oak Ridge, TN.
HELP (hydrologic evaluation of landfill performance model)	Soil (to groundwater)	HELP is a quasi-two-dimensional, deterministic, numerical, finite-difference model that computes a daily water budget for a landfill represented as a series of horizontal layers. It models leaching from landfills to unsaturated soil beneath a landfill. It models both organics and inorganics, using rainfall and waste solubility to model leachate concentrations leaving the landfill.	HELP is used for water balance computation and for the estimation of chemical emissions, volatile emissions, and leachate quality.	U.S. EPA, National Computer Center, Research Triangle Park, NC. U.S. Army Corps of Engineers
SITES (the contaminated sites risk management system)	Soil (to multiple compartments)	SITES is a flexible interactive PC computerized decision-support tool for organizing relevant information and conducting risk management analyses for contaminated sites. It has the	SITES is a computer-based integrating framework used to help evaluate and compare site investigation and remedial action alternatives in terms of health and environmental effects	EPRI (Electric Power Research Institute), Palo Alto, CA. EPRI (1988). User Guide to the Contaminated Sites Risk Management

Table D.1 (continued)

Model Name	Environmental Compartment	Model Description	Model Use	Sources/Refs.
		dimensionality to model multiple chemicals, pathways, population groups, health effects, and remedial actions. The user completely defines the scope of the analyses. The model uses information from diverse sources, such as site investigations, transport and fate modeling, behavioral and exposure estimates, and toxicology. It explicitly addresses the many uncertainties and allows for quick sensitivity analyses. Both deterministic and probabilistic analysis method possible.	and total economic costs/impacts. The decision tree structure in SITES allows for explicit examination of key uncertainties and the efficient evaluation of numerous scenarios. The model's design and computer implementation facilitates extensive sensitivity analysis.	System. Prepared by Decision Focus, Inc. for EPRI, CA.
GEMS (graphical exposure modeling system)	Multimedia	GEMS is an operating environment that houses a variety of programs for performing exposure assessment studies. It is an interactive computer system for environmental models, physiochemical property estimation, and statistical analysis. GEMS has graphic display capabilities.	GEMS is an interactive management tool that allows quick and meaningful analysis of environmental problems. It allows user to estimate chemical properties; assess fate of chemicals in receiving environments; model resulting chemical concentrations; determine the number of people potentially exposed; and estimate the resultant human exposure and risk.	EPA, Research Triangle Park, NC. Office of Pesticide and Toxic Substances, U.S. EPA, Washington, DC. U.S. EPA (1989). GEMS User's Guide, Prepared by General Sciences Corp., MD for U.S. EPA, Office of Toxic Substances, Washington, DC.
PCGEMS (personal computer version of the graphical exposure modeling system)	Multimedia	PCGEMS is a complete information management tool designed to help in performing exposure assessment studies. It is able to work interactively with GEMS on the VAX Clusters.	PCGEMS can be used to estimate chemical properties and perform simulation studies of chemical release in air, soil, and groundwater systems. The environmental modeling programs allows for the simulation of	Office of Pesticides and Toxic Substances, Exposure Evaluation Division, U.S. EPA, Washington, DC. U.S. EPA (1989). PCGEMS User's Guide, Prepared by General

			the migration and transformation of chemicals through the air, water, soil, and groundwater subsystems.	Sciences Corp., MD for U.S. EPA, Office of Pesticide and Toxic Substances, Washington, DC.
ENPART (Environmental partitioning model)	Multimedia	ENPART uses simple physical chemical data to estimate equilibrium concentration ratios of a chemical between the environmental compartments of air, water, and soil. More of a screening tool. The model is available on GEMS.	ENPART is a multimedia model that estimates equilibrium concentration ratios of a chemical between the environmental compartments of air, water, and soil. It also provides a second level of concentration and mass partitioning called dynamic partitioning. Subsequently, potential exposure in each environmental compartment to the chemical are estimated.	Office of Toxic Substances, Exposure Evaluation Division, U.S. EPA, Washington, DC.
POSSM (PCB onsite spill model)	Multimedia	POSSM is an exposure assessment methodology. It consists of a chemical transport and fate model capable of considering all of the key processes controlling chemical losses from a spill site (including volatilization, leaching to groundwater, and chemical washoff from the land surface due to runoff/erosion). Contains several relevant subprograms, including PTDIS (air model), RIVLAK (surface water model), GROUND (ground water model), and EXPOSE (exposure intake model).	POSSM provides a quantitative framework for estimating general public exposure levels associated with spills from utility electrical equipment. Onsite environmental concentrations can be estimated with POSSM; offsite concentrations can be estimated with one of three relatively simple transport and fate models for air (PTDIS), surface water (RIVLAK), and groundwater (GROUND) that are also incorporated into the methodology. Given estimates of onsite and/or offsite concentrations and the characteristics and activity patterns of the receptors of concern, inhalation, ingestion, and dermal exposure levels can be calculated using the program, EXPOSE. Methodology was developed primarily for PCBs, but is applicable to a wide range of organic chemicals.	EPRI (Electric Power Research Institute), Palo Alto, CA.

Table D.1 (continued)

Model Name	Environmental Compartment	Model Description	Model Use	Sources/Refs.
INPOSSM (interactive PCB onsite spill model)	Multimedia	INPOSSM is an interactive PCB exposure assessment model. It includes a chemical transport and fate model capable of considering key processes controlling chemical losses from a PCB or organic chemical spill site, including volatilization, leaching to groundwater, and chemical washoff from the land surface due to runoff/erosion.	INPOSSM uses a Monte Carlo simulation model for evaluating the variability of predicted chemical losses.	EPRI (Electric Power Research Institute), Palo Alto, CA.
MCPOSSM (Monte Carlo PCB onsite spill model)	Multimedia	MCPOSSM is a chemical spill exposure assessment methodology, providing a quantitative framework for estimating uncertainties of chemical levels associated with spills. The core of the methodology is the PCB onsite spill model (POSSM) and a Monte Carlo chemical transport and fate model capable of considering key processes controlling chemical losses from a spill site, including volatilization, leaching to ground-water, and chemical washoff from the land surface due to runoff/erosion.	MCPOSSM provides a distribution of concentrations over time and probabilities of exceeding specified levels (i.e., probability of a worst-case level).	EPRI (Electric Power Research Institute), Palo Alto, CA.
TOX-RISK (toxic risk)	Multimedia	TOX-RISK is a package that provides data entry and management, computation of maximum likelihood estimates of risk or dose with confidence bounds, and graphs of dose-response functions fitted to the	TOX-RISK is a menu-driven interactive software package for performing health-related risk assessments. These risk assessments produce quantitative estimates of risks from quantal dose-response data.	EPRI (Electric Power Research Institute), Palo Alto, CA.

		data provided by the user. TOX-RISK has the ability to fit 10 different models to data which is readily available in animal bioassays — including the Multistage, One-Stage, Two-Stage, Three-Stage, Four-Stage, Five-Stage, Six-Stage, Weibull, Mantal Bryan, and the LogNormal model.		
TOX-SCREEN	Multimedia	TOX-SCREEN is a multimedia screening-level model which assesses the potential fate of toxic chemicals released to the air, surface water, or soil. It incorporates equations developed for media-specific models and allows user to describe compartments directly.	TOX-SCREEN is a screening device used to identify chemicals that are unlikely to pose problems even under conservative assumptions.	Office of Pesticides and Toxic Substances, U.S. EPA. U.S. EPA (1984). User's Manual for TOX-SCREEN: A MultiMedia Screening-level Program for Assessing the Potential of Chemicals Released to the Environment, EPA-560/5-830024, Oak Ridge National Lab EPA Office of Toxic Substances.
MICROBE-SCREEN	Multimedia	MICROBE-SCREEN is a screening level multimedia model for assessing potential fate of microorganisms released to air, surface water, or soil. The model deals specifically with passively dispersed microorganisms such as bacteria.	MICROBE-SCREEN is a screening device to estimate transport and densities of microorganisms in various environments.	GSC (1985) User's Guide to MICROBE-SCREEN Execution in GEMS, GSC-TR8753 General Sciences Corp., MD.
UTM-TOX (unified transport model for toxic materials)	Multimedia	UTM-TOX is a multimedia model designed for predicting the dispersion of pollutants through air, soil, and water. It links together several media-specific models.	UTM-TOX can be used to track a pollutant through the ecosystem, budge for the partitioning of contaminants, calculate concentration of pollutant in many compartments of the ecosystem, and reliably assess the impact of the pollutant.	Oak Ridge National Laboratory, TN. ORNL (1984). A User's Manual for UTM-TOX, the Unified Transport Model, NTIS No. ORNL-6064/LT.

Table D.1 (continued)

Model Name	Environmental Compartment	Model Description	Model Use	Sources/Refs.
SMCM (spatial-multimedia-compartmental-model)	Multimedia (air, water, soil, biota)	SMCM is based on a modeling approach that estimates the multimedia partitioning of organic pollutants in local environments. It consists of coupled partial and ordinary differential equations that are solved simultaneously by finite-difference method and using operator splitting techniques. It is a hybrid transport and fate model that makes use of both uniform and nonuniform 1-D compartments. It has the capability of simulating a variety of pollutant transport phenomena.	SMCM is designed to predict the multimedia concentrations of organic chemicals in the environment. It is used for the analysis of the environmental multimedia distribution of organic pollutants.	The National Center for Intermedia Transport, UCLA, Los Angeles, CA.
RISK*ASSISTANT	Multimedia	RISK*ASSISTANT is designed to assist the user in rapidly evaluating exposures and human health risks from chemicals in the environment at a particular site. User need only provide measurements or estimates of the concentrations of chemicals in the air, surface water, groundwater, soil, sediment, and/or biota. It provides an array of analytical tools, databases, and information-handling capabilities for risk assessment.	RISK*ASSISTANT is used to evaluate exposures and human health risks from chemicals. Has the ability to tailor exposure and risk assessments to local conditions.	U.S. EPA, Research Triangle Park, NC. Hampshire Research Institute, Alexandria, VA. California Department of Health Services (CDHS), Toxic Substances Control Program. New Jersey Department of Environmental Protection.
MSRM (mixture and systemic toxicant risk model)	Multimedia	MSRM is an exposure assessment model that contains statistical methods and extrapolation models for using available toxicological and	MSRM is used to estimate human health risks from exposure by various routes.	Office of Health and Environmental Assessment, U.S. EPA, Washington, DC.

		epidemiological data. Primarily consists of noncancer risk assessment models and estimation categories, but also has cancer risk models included for completeness. It is applicable to single chemicals and mixtures.		
WET (wastes-environ-ments-technologies model)	Multimedia	WET is a RCRA risk/cost policy model that establishes a system to allow users to investigate how tradeoffs of costs and risks can be made among wastes, environments, and technologies (W-E-Ts) in order to arrive at feasible regulatory alternatives. The system assesses waste streams in terms of likelihood and severity of human exposure to their hazardous constituents and models their behavior in three media air, surface water, and groundwater. Groundwater exposure/risk score is tallied based on key flow and transport parameters.	WET is used to assist policymakers in identifying cost-effective options that minimize risks to health and the environment.	Office of Health and Environmental Assessment, U.S. EPA, Washington, DC.

Over the last two decades, the sophistication of air pollution models has grown rapidly. Many of the more recent models attempt to address various factors that influence atmospheric dispersion such as atmospheric stability, irregular or complex terrain, multiple emission sources, local building wake effects and short-term events. Infact, all but the simplest models take into account atmospheric stability, which represents the turbulence of the transporting wind and its ability to disperse emitted materials.

There are many levels of complexity implicit in the models of air pathways assessment. A brief description of some relatively simple air pathways models will be presented, without giving a detailed discussion of the mathematics of air pathways themselves.

Mathematical Model Structure

The transport and dilution of air contaminants is a complex process involving many variables. For instance, once gases are released into the atmosphere, their transport and dispersion are governed primarily by the strength and direction of the advecting wind and the level of turbulent motion. Due to the turbulence inherently present in both the atmosphere and within the emission source vector, and due to the turbulence created by the shearing action between the contaminant plume and the ambient air, air is entrained into the plume. In the case of a stack emission, air dilutes both the vertical momentum and the buoyancy of the plume and causes the plume to acquire the horizontal velocity of the wind. Thus, the plume from a stack rises because of initial vertical momentum and buoyancy, grows in diameter as ambient air is entrained, and has its path deflected horizontally by the wind. At some point downwind of the point of emission, the vertical velocity of the plume relative to the ambient air decays to zero or near zero and the plume moves with the local velocity of the wind. In turn, the local wind velocity is influenced by topographical features (if any). As the plume is transported by the wind, it is dispersed by atmospheric turbulence.

Simultaneously, pollutants may be removed from the plume by a number of natural processes including chemical reactions of the plume contaminants with the ambient air, complex photochemical reactions involving several pollutants, adsorption on the ground and/or on vegetation, washout by rainfall or snowfall, and settling out (particulates). The degree to which the various attenuating mechanisms are described within air pathways models is important in differentiating between the models. The sophistication of models has grown tremendously in recent years. The selection of which model to employ in a specific application is dependent on numerous features, including the physical situation being modeled, the availability of data, and the experiences of the individual doing the modeling. Some indications of the various available model types for the emissions component and the air pathways model component are described below.

Emissions Modeling

Hazardous waste site air emission sources may be classified as either point or area sources. Point sources include vents (e.g., landfill gas vents) and stacks (e.g., incinerator and air stripper releases); area sources are generally associated with fugitive emissions (e.g., from landfills, lagoons and contaminated surface areas).

Emission sources may be further classified as continuous/semi-continuous/instantaneous releases. A stack emission from a steel mill will act as a "continuous point source emission." A vapor leak from the puncture of a tank on a truck may act as an "area source but of a semicontinuous nature." Alternatively, an accident may involve an "instantaneous cloud or puff release of contaminants." The characteristics of the different source types are relevant to the task of selecting a mathematical model for a particular application.

Effects of Emission Variables

Stack emissions from industrial, chemical processing, and electric power plants are usually emitted to the atmosphere from stacks in order to spread the pollutant over a much wider downwind area and thereby reduce the GLC. The relevant emission variables associated with stack emissions then include:

1. The stack height
2. Emission temperature — an increase in the emission temperature (therefore, a decrease in the density and an increase in buoyancy) causes the emissions plume to rise higher, thus reducing the concentrations at the ground
3. Exit velocity — an increase in the exit velocity of nonbuoyant emissions increases plume rise, thereby reducing the GLCs

The incorporation of these many features into the dilution-entrainment of stack emissions is extensively covered in many technical sources. For example, the effective stack height can be calculated from the equation proposed by Briggs (1973) or using Hollands equation given in Turner's Workbook (Turner, 1970). Briggs (1973) suggested plume downwash when efflux vertical velocity of plume is less than 1.5 times the horizontal wind speed, and this can be approximately accounted for in effective stack height.

Alternatively, emissions from spills are generally modeled assuming the contaminant does not impact the flow field of the ambient air. Released liquids are assumed to form a pool from which chemicals are released to the atmosphere, with the concern being one of evaporation rate, Q, calculated as:

$$Q = \frac{KAMP}{RT}$$

where

Q = evaporation rate (vapor release rate, g/sec)
K = mass transfer coefficient (m/sec)
A = area of liquid pool (m^2)
P = vapor pressure (Pa)
M = molecular weight of contaminant (g/mol)
R = universal gas constant = 8.314 Pa m^3/mol K
T = temperature in (K)

This equation was developed for spills of particular chemicals (i.e., a single component system). The mass transfer coefficient K can be calculated from (MacKay and Matsugu, 1973)

$$K = 0.0048\, V^{0.78} d^{-0.11} S_c^{-0.67}$$

where

V = wind speed (m/sec)
d = pool diameter (m)
S_c = Schmidt number (dimensionless)

The Schmidt numbers, which represent the effect of fluid friction on mass transfer, are approximated based on the molecular weight of the compound. Compounds with molecular weight less than 100 were assigned an $S_c^{-0.67}$ of 0.7, those with molecular weight from 100 to 200 were assigned an $S_c^{-0.67}$ of 0.6, and those with molecular weight greater than 200 were assigned an $S_c^{-0.67}$ of 0.5.

Adjustments for evaporation rates at a different temperature, T_t may be accomplished as

$$Q = \frac{P_t}{P_{20°C}} \times \frac{293}{T_t}$$

since the partial pressure is inversely responsive to the vapor pressure. However, this necessitates determination of the vapor pressure response, which is temperature dependent.

The high degree of sensitivity of the vapor pressure to the temperature, and thus the chemical evaporation rate to release-scenario conditions, is noteworthy in that it demonstrates the uncertainties in air pathways modeling in particular and contaminant migration pathways modeling in general. Specifically, the vapor pressures for various temperatures have been characterized (e.g., as per Weast, 1973) as

$$\log_{10} P = -0.2185A / K + B$$

where K is temperature (K) and A and B are coefficients specific to individual chemicals. Values listed in Table D.2 demonstrate the considerable variation of vapor pressures for trichloroethylene, as an example. Given the scenario of a stockpile of trichloroethylene-contaminated soil awaiting incineration, on a hot summer day, the difference from a cold winter day is substantial. However, the degree of conservativeness in parameter assignments and modeling scenarios is a major point of interest.

Additional complexities enter the prediction of vapor emissions when more than a simple contaminant is present. Transport may be controlled by other materials present in the mixture. However, the impacts of these other chemicals are difficult to quantify and are likely to reduce the value of the mass transfer coefficient. In order to extend a single component system to a multiple component system, Raoult's Law

Table D.2 Vapor Pressure of Trichloroethylene in Response to Varying Temperatures

Temperature		Vapor Pressure
°C	K	(mm Hg)
0	273	20
20	293	57
40	313	142
60	333	317

for ideal vapor-liquid equilibria may be used. Raoult's Law for multicomponent mixtures is based on the assumptions that the gas behaves as an ideal gas, the liquid behaves as an ideal liquid, and the fugacity of the mixture is insensitive to variations in pressure. Raoult's Law is given by

$$y_i = \frac{X_i P_i}{P_o}$$

where

y_i = mole fraction of component i in the gas phase
x_i = mole fraction of component i in the liquid phase
P_i = vapor pressure of component i, mm Hg
P_o = ambient pressure, mm Hg

The modified emission rate equation becomes

$$Q_i = \frac{(K_{gi})(A)(P_i)(x_i)(M_i)}{RT}$$

where

Q_i = emission rate for species i (g/sec)
K_{gi} = the gas mass transfer coefficient for species i
P_i = vapor pressure for species i (P_a)
x_i = mole fraction for species i, (dimensionless)
M_i = molecular weight of species i, (g/mol)

Mass transfer of a chemical in a complex solution or mixture is a complicated process, involving mass transport across several phase boudaries. Explanations of these additional features are contained in Thomann and Mueller (1987). For complex situations, laboratory or field-scale experiments are probably necessary.

The preceding discussion related specifically to gas-phase emissions. However, of relevant concern for some risk assessment questions are particulate matter emissions.

Air Dispersion Models

An air dispersion model uses the source emission information (from an emission model such as described above), meteorological conditions, geographic boundaries, etc. as inputs and computes the dispersion of pollutants by the atmosphere, providing concentrations of air quality for specified time periods.

Box Models. One of the simplest forms of the air dispersion model is the so-called "Box" model. The Box model assumes that pollutant concentrations are uniform throughout a prescribed region. Let V = the box volume, C = the pollution concentration in the box and leaving the box, Q = the emission rate, F = the air flow rate through the box, and t = time. Then a simple mass balance equation states

rate of change of pollution in the box	=	*rate of pollution entering the box*	–	*rate of pollution leaving the box*

or

$$V\frac{dc}{dt} = Q - FC$$

If the initial concentration in the box is zero, then the equation can be integrated to

$$C = \frac{Q}{F}\left(1 - e\frac{-Ft}{V}\right)$$

Variations on this simple box model are frequently employed for determining contaminant exposure scenarios in the workplace.

Gaussian Point Source Diffusion Models. These models assume a normal or Gaussian distribution in the dispersion of pollutants. The Gaussian distribution is most commonly represented by the familiar bell-shaped curve which is symmetrical about the mean. All Gaussian point source diffusion models incorporate a feature which accounts for the rise of emissions after they leave the source. This characteristic, known as plume rise, is more pronounced in emissions from smokestacks and chimneys and is dealt with by the computational procedures indicated previously in the emissions sector discussions. The Gaussian models may be for emissions of continuous nature, semicontinuous nature, or as a puff or instantaneous release.

If a pollutant is released continuously from a point source at rate Q into a steady wind of uniform velocity V, the plume of pollutant will expand by diffusion downwind of the source as per

$$C(x,y,z,t) = \frac{Q}{2\pi V\sigma_y\sigma_z}\exp\left[\frac{-1}{2}\left(\frac{y}{\sigma_y}\right)^2\right]\left\{\exp\left[\frac{-1}{2}\left(\frac{Z-h}{\sigma_z}\right)^2\right] + \exp\left[\frac{-1}{2}\left(\frac{Z+h}{\sigma_z}\right)^2\right]\right\}$$

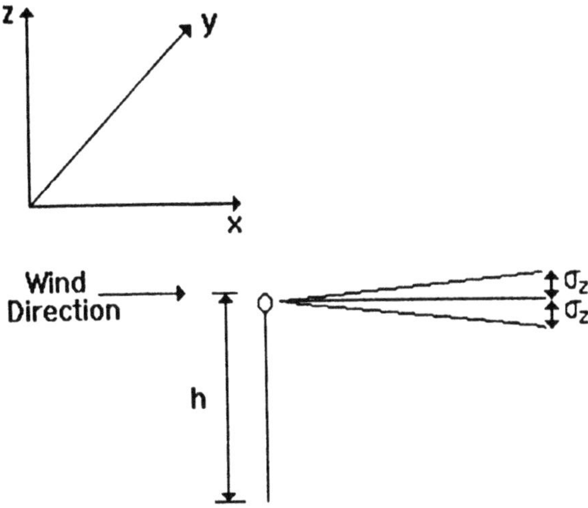

Figure D.1 Physical configuration and coordinate system utilized in Gaussian point source models.

where

x, y, and z specify location and t specifies time
C = concentration of vapor (g/m^3)
V = wind speed (m/sec)
x = downwind distance (m) defined as the direction of the wind
y = crosswind distance (m)
z = vertical distance (m)
h = height of source emission (m)
σ_y, σ_z = crosswind and vertical standard deviation of Gaussian concentration
t = time since release

This equation is valid for $x \leqslant Vt$ and otherwise $C(x, y, z, t) = 0$ for $x > Vt$. The physical configuration of the release height and the associated coordinate system are schematically depicted in Figure D.1. Assumptions implicit in the equation include

1. The vapor that is diffusing is neutrally buoyant. There is no gross movement of the vapor cloud caused by gravity or buoyancy.
2. Mixing with air is uniform throughout the vapor cloud.
3. The concentration obtained is time averaged. There is a near-field "patchiness" associated with Gaussian models that makes questionable their predictions of concentration points in close proximity to the source.
4. The wind is uniform throughout the vertical extent of the vapor cloud.
5. The terrain is flat.
6. The plume is not depleted (e.g., by deposition).
7. The height of the contaminant plume is not limited by a mixing layer in the atmosphere.

For concentration levels along the centerline, $y = 0$, and at ground levels, $z = 0$, and for a surface release, $h = 0$, the equation becomes

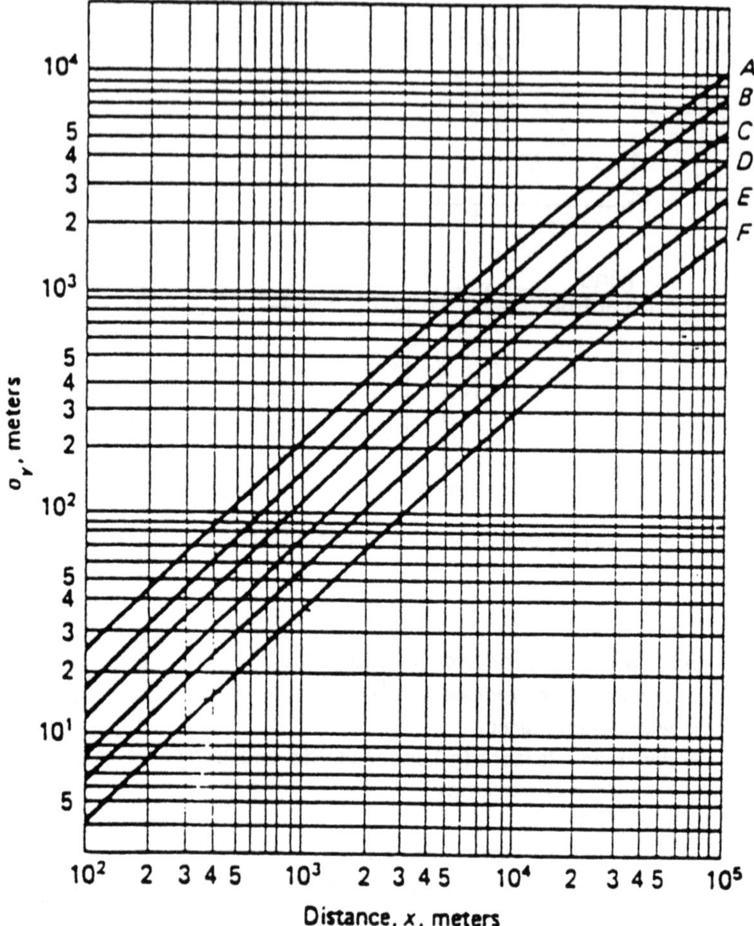

Figure D.2 Dispersion coefficient in the crosswind direction as a function of distance downwind (Turner, 1970).

$$C(x,o,o,t) = \sigma_y, \sigma_z \qquad x \leq Vt$$

$$= 0 \qquad x > Vt$$

Prior to use of these equations, it remains to specify the dispersion terms. The dispersion terms are directly related to the turbulence of the atmosphere. In stable atmospheric conditions, turbulence levels are low, resulting in little diffusion of pollutants. Materials emitted into a stable atmosphere will drift downwind as a thin, undisturbed ribbon for many kilometers. Consequently, a stable condition with light winds may be of great concern in terms of exposure risk. In unstable conditions there is a high level of turbulence which rapidly diffuses pollutants.

The values of the dispersion terms vary with the turbulent structure of the atmosphere. The turbulent structure of the atmosphere and wind speed are considered in the stability classes presented in Turner (1970). Class A is the most unstable

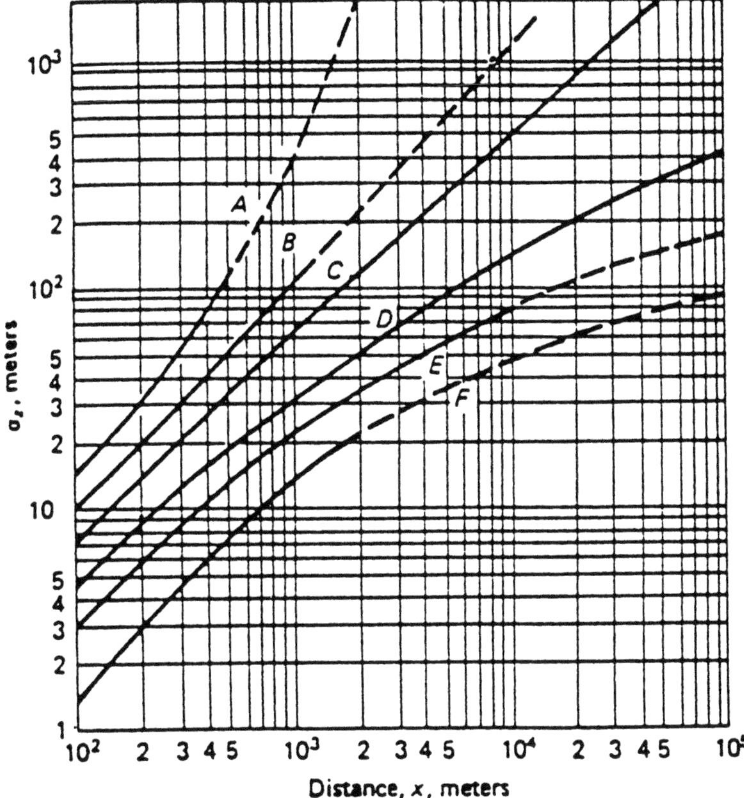

Figure D.3 Dispersion coefficient in the vertical direction as a function of distance downwind (Turner, 1970).

class, and Class F the most stable. The differences between open-country and urban conditions is due primarily to the influence of the city's larger surface roughness and heat island effects upon the stability regime. The greatest difference occurs on calm, clear nights. Quantification of the dispersion terms is typically accomplished using the Pasquill stability form, as depicted in Figures D.2 and D.3. Numerical equations for each of the stability types are provided in Tables D.3 and D.4.

The equations presented relate to a "point" source. To adjust these equations for an "area" source, it would be possible to utilize integration principles over many small sources to effectively represent an area source, but this would involve a considerable computational effort. Alternatively, for estimating concentrations at large distances (greater than two equivalent diameters of the source area), the following simplified approach may suffice. The area is replaced by a "virtual point source" of the same total strength but placed upwind by a distance x_y, as depicted in Figure D.4. The distance x_y is a function of the concentration itself and is estimated by setting the crosswind extent (or diameter (d)) of the area equal to the plume width (4.3 σ_y) which defines the 10% concentration edge of a plume) downwind of the virtual point source; computing σ_y then using Figure D.4, determining the distance x_y which corresponds to this rate of plume spread (Turner, 1970). Hence,

Table D.3 Key to Stability Categories

Surface Wind Speed at 10 M (m/sec) Class[a]	Incoming Solar Radiation (Day)			Cloud Cover (Night)	
	Strong (1)	Moderate (2)	Slight (3)	Mostly Overcast (4)	Mostly Clear (5)
<2	A	A–B	B	E	F
2–3	A–B	B	C	E	F
3–5	B	B–C	C	D	E
5–6	C	C–D	D	D	D
>6	C	D	D	D	D

Source: Turner (1970).

[a] The neutral class, D, should be assumed for overcast conditions during day or night. Class A is the most unstable and Class F is the most stable, with Class B moderately unstable and Class E slightly stable.

$$x^1 = x + x_y$$

If x_y is less than 100 m, then the following approximation has been suggested (after Raj and O'Farrel, 1977)

$$x^1 = x + 5d$$

For a "puff" release, the concentration at some location x, y, z and time t, downwind is given by

$$C(x,y,z,t) = \frac{2Q_T}{(2\pi)\frac{3}{2}\sigma_y\sigma_y\sigma_z} \cdot \exp\left[\frac{-1}{2}\left(\frac{s-Vt}{\sigma_x}\right)^2\right] \cdot \exp\left[\frac{-1}{2}\left(\frac{y}{\sigma_y}\right)\right]$$

$$\left\{\exp\left[\frac{-1}{2}\left(\frac{Z-h}{\sigma_z}\right)\right]\right\} + \exp\left[-\frac{1}{2}\left(\frac{Z+h}{\sigma_z}\right)^2\right]$$

where

Q_T = total mass of vapor liberated (kg)
C = concentration from an instantaneous (puff) release (kg/m^3)

and the other variables are as defined earlier. Relationships for the puff diffusion coefficient are assumed to be the same as the Pasquill plume coefficents (i.e., $\sigma_x = \sigma_y$).

Furthermore, for concentrations along the puff centerline, (y = 0) and at ground level (z = 0) from a surface release (h = 0), the equation reduces to

Table D.4 Equations of Dispersion for Alternative Stability Classes

Pasquill Stability Type	σ_y (m)	σ_z (m)
	Open-Country Conditions	
A	$0.22d\,(1 + 0.0001d)^{1/2}$[a]	$0.20d$
B	$0.16d\,(1 + 0.0001d)^{1/2}$	$0.12d$
C	$0.11d\,(1 + 0.0001d)^{1/2}$	$0.07d\,(1 + 0.0002d)^{1/2}$
D	$0.08d\,(1 + 0.0001d)^{1/2}$	$0.06d\,(1 + 0.0015d)^{1/2}$
E	$0.06d\,(1 + 0.0001d)^{1/2}$	$0.03d\,(1 + 0.0003d)^{-1}$
F	$0.04D\,(1 + 0.0001d)^{1/2}$	$0.016d\,(1 + 0.0003d)^{-1}$
	Urban Conditions	
A–B	$0.32d\,(1 + 0.0004d)^{1/2}$	$0.24\,(1 + 0.001d)^{1/2}$
C	$0.22d\,(1 + 0.0004d)^{1/2}$	$0.20d$
D	$0.16d\,(1 + 0.0004d)^{1/2}$	$0.14d\,(1 + 0.0003d)^{1/2}$
E–F	$0.11d\,(1 + 0.0004d)^{1/2}$	$0.08d\,(1 + 0.00015d)^{1/2}$

Source: Briggs (1973).

[a] d = Downwind distance in m and $10^2 < d < 10^4$ m.

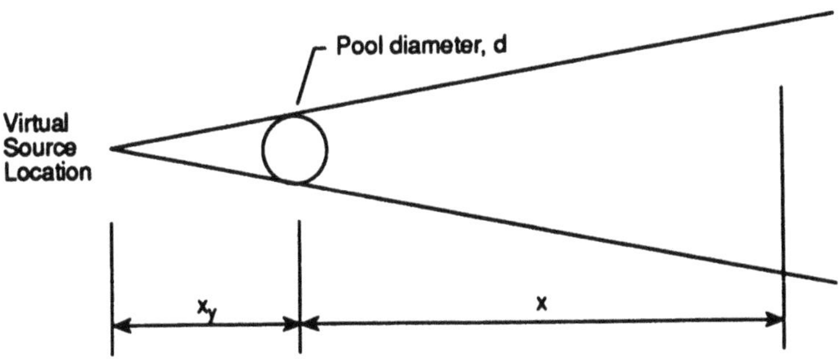

Figure D.4 Virtual source characterization for area source representation.

$$C(x,o,o,t) = \frac{4Q_t}{(2\pi)\frac{3}{2}\sigma_x\sigma_y\sigma_z}\exp\left[\frac{1}{2}\left(\frac{x - Vt}{\sigma_x}\right)^2\right]$$

Terrain can influence the transport and diffusion of plumes to a great degree, especially for heavier-than-air gases such as chlorine and bromine. Chaneling of plumes along valleys restrict lateral dispersion. Dense gas may be trapped in hollows — heavy gas may persist for a lengthy period. Sources of technical information regarding the modeling include Huerzeler and Fannelop (1991) and Griffiths and Fryer (1988). The "slumping" effect due to density changes the plume dispersion.

Numerical Models: Alternative models exist to the Box and Gaussian models that can address the density effects. These models utilize a direct solution of the three-dimensional conservation of mass equations using finite difference or finite element approaches. As a direct consequence of the mathematical rigor of this more complex approach to the modeling, there is a significant increase in numerical sophistication (see, for example, Hanna et al., 1991).

D.6 REFERENCES

Briggs, G., "Diffusion Estimation for Small Emissions," ATDL contribution File No. 79, Atmospheric Turbulence and Diffusion Laboratory (1973).

Griffiths, R. and L. Fryer. "A Comparison Between Dense-Gas and Passive Tracer Dispersion Estimates for Near-Source Toxic Effects of Chlorine Releases," *J. Hazard. Mater.* 19:169–181(1988).

Hanna, S., D. Strimatius, and J. Chang. "Evaluation of Fourteen Hazardous Gas Models with Ammonia and Hydrogen Fluoride Field Data," *J. Hazard. Mater.* 26:127–158(1991).

Huerzeler, B. and T. Fannelop. "Small Spills of Heary Gas from Continuous Sources," *J. Hazard. Mater.* 26:187–202(1991).

MacKay, D. and R. Matsugu. "Evaporation Rate of Hydrocarbon Spills on Water and Land," *Can. J. Chem. Eng.* 51:434–439(1973).

Raj, P. and P. O'Farrel. "Development of Additional Hazard Assessment Models," Arthur D. Little Inc., Cambridge, MA, report to U.S. Coast Guard, Washington, DC, AD-A042365, March (1977).

Thomann, R. V., and J. A. Mueller. *Principles of Surface Water Quality Modeling and Control* (New York: Harper & Row Publishers, 1987).

Turner, D. *Workbook of Atmospheric Dispersion Estimates*, Public Health Service Publication No. 99-AP-26, U.S. Department of Health, Education and Welfare, Washington, DC (1970).

Weast, R.C., Ed. *Handbook of Chemistry and Physics*, 53rd ed. (Boca Raton, FL: CRC Press, 1973).

APPENDIX E

Selected Units and Measures

E.1 SELECTED UNITS OF MEASUREMENT AND APPROXIMATE CONVERSIONS

Mass Units

g	gram(s)
ton (metric)	tonne = 1×10^6 g
kg	kilogram(s) = 10^3 g
mg	milligram(s) = 10^{-3} g
μg	microgram(s) = 10^{-6} g
ng	nanogram(s) = 10^{-9} g
pg	picogram(s) = 10^{-12} g

Volumetric Units

L	liter(s) = 10^3 cm^3
cc or cm^3	cubic centimeter(s) = 10^{-3} L
mL	milliliter(s) = 10^{-3} L
m^3	cubic meter(s) = 10^3 L

Environmental Concentration Units

ppm	parts per million
ppb	parts per billion
ppt	parts per trillion

Mass Conversions

1 g	= 0.035 oz
1 ton	= 2205 lb
1 kg	= 2.25 lb
1 mg	= 10^{-3} g
1 μg	= 10^{-6} g
1 ng	= 10^{-9} g
1 pg	= 10^{-12} g

Volume Conversions

1 cc	= 1 mL (approximately)
1 mL	= 10^{-3} L
1 L	= 0.95 liquid quart
m^3	= 35 cubic feet

Concentration Equivalents

1 ppm	= mg/kg or mg/L
1 ppb	= μg/kg or μg/L
1 ppt	= ng/kg or ng/L

Concentrations in Soils or Other Solid Media

mg/kg — mg of chemical per kg body weight, or mg chemical per kg of sampled media

Concentrations in Water or Other Liquid Media

mg/L — mg of chemical per liter volume, or mg chemical per liter of total fluid volume

Concentrations in Air Media

mg/m^3 — mg of chemical per cu. meter volume, or mg chemical per m^3 of total fluid volume

Units of Chemical Intake and Dose

mg/kg/day — milligrams of chemical exposure per unit body weight of exposed receptor per day

Commonly Used Expressions

"Order of Magnitude" — Used in reference to calculation of environmental quantities or risk and meaning a factor of 10.

Exponentials denoted by 10^2, 10^3, 10^6, etc. — Superscripts refer to the number of times "10" is multiplied by itself, e.g., $10^2 = 10 \times 10 = 100$; $10^3 = 10 \times 10 \times 10 = 1000$.

Exponentials denoted by 1.00E-01, 1.00E+01, X.YZEmn, etc. — Number after the "E" indicates the power to which 10 is raised, and then multiplied by preceding term; e.g., 1.00E-01 = $1.00 \times 10^{-1} = 0.1$; 4.44E+04 = $4.44 \times 10^4 = 44{,}400$.

E.2 CONVERSION UNITS FOR CONCENTRATION MEASUREMENTS

Frequently, units for concentrations are reported for air, water, and soil in parts per million and/or SI units. The following demonstrate the means of conversion.

Air

The conversion of parts per million (on a volume ratio) of chemical X to SI units is accomplished as follows. Assume all gaseous components behave as ideal gases and standard temperature and pressure (STP) conditions hold. Then

$$1ppm = \frac{1m^3 X}{10^6 m^3 air}\left(\frac{1000\Delta}{m^3}\right)\left(\frac{mol}{22.4L}\right)\left(\frac{1000m\ mol}{mol}\right)$$

$$= \frac{1}{22.4}\frac{m\ mol\ X}{m^3}\frac{M_x mg}{m\ mol} = \frac{M_x}{22.4}\frac{mg\ X}{m^3}$$

where M_x is the molecular weight of X, in grams per mole.

Example — 1 ppm (on a volume ratio) of trichloroethylene is equivalent to

$$1\ ppm(trichloroethylene) = \frac{131}{22.4} = 5.85\frac{mg}{m^3}$$

or

$$1\ ppb(trichloroethylene) = 5.85\frac{\mu g}{m^3} \qquad (at\ STP)$$

Water

The conversion of parts per million (on a mass ratio) of chemical X to SI units is accomplished as follows:

$$1ppm = \frac{1g\ X}{10^6 gH_2O}\left(1\frac{g}{cm^3}\right)\left(100\frac{cm}{m}\right)^3 = 1\frac{gX}{m^3} = 1\frac{mg}{L}$$

Example —

$$1\,ppm(trichloroethylene)in\ water = 1\frac{mg}{L}$$

Soil

The conversion of parts per million (on a mass ratio) of chemical X to SI units is accomplished as follows:

$$1\,ppm = \frac{1g\ X}{10^6 g\ soil}\left(10^6\frac{mg}{kg}\right) = \frac{1mgX}{kg\ \ soil} = 1\ \text{mg/kg}$$

Example:

$$1\,ppm(trichloroethylene)within\ soil = 1\frac{mg}{kg\ soil}$$

INDEX

D

E

I

Milton Keynes UK
Ingram Content Group UK Ltd.
UKHW021823071024
449327UK00021B/1410

9 780367 449971